Faecal Sludge Management

Systems Approach for Implementation and Operation

Faecal Sludge Management

Systems Approach for Implementation and Operation

Editors

Linda Strande

Mariska Ronteltap

Damir Brdjanovic

Published by **IWA Publishing**
Alliance House
12 Caxton Street
London SW1H 0QS, UK
Telephone: +44 (0)20 7654 5500
Fax: +44 (0)20 654 5555
Email: publications@iwap.co.uk
Web: www.iwapublishing.com

First published 2014

British Library Cataloguing in Publication Data
A CIP catalogue record for this book is available from the British Library

Library of Congress Cataloging- in-Publication Data
A catalog record for this book is available from the Library of Congress

Cover design: Peter Stroo
Graphic design: Hans Emeis

ISBN: 9781780404721 (Hardback)
ISBN: 9781780404738 (eBook)

About the editors

Linda Strande

Dr. Linda Strande leads the Excreta and Wastewater Management group at EAWAG (the Swiss Federal Institute of Aquatic Science and Technology) in SANDEC (the Department of Water and Sanitation in Developing Countries). The overarching goal of Dr. Strande's research is to increase scientific knowledge that will advance and increase sustainable urban faecal sludge management technologies. In engineering and development research, she believes it is always important to consider how fundamental research can translate into real-life implementations. To achieve this, she has pursued a systems based approach to faecal sludge management, including technology, management and planning, so all aspects can integrate into complete and functional systems. Currently, the research focus of Dr. Strande's group includes optimisation of treatment technologies, innovation in resource recovery, and methods for sustainable systems level implementations. Dr. Strande has been working in the environmental sector for over 15 years and holds interdisciplinary degrees in engineering, soils science and mathematics. Her academic background, together with wide-ranging international experiences, has provided her with a global perspective, and an ability to research and apply environmental engineering fundamentals in complex, interdisciplinary situations.

Mariska Ronteltap

Dr. Mariska Ronteltap is a Senior Lecturer in Sanitary Engineering at UNESCO-IHE (Institute for Water Education), with 12 years of experience working in the field. She holds a Master's degree in Environmental Engineering from the University of Wageningen, and a PhD jointly from ETH (the Swiss Federal Institute of Technology Zurich) and EAWAG (the Swiss Federal Institute of Aquatic Science and Technology). Her PhD research involved urine separation as a novel approach in the field of wastewater technology, with a strong chemical focus including thermodynamic modelling. Her practical knowledge in the field of struvite precipitation from urine has been employed in several research pilot projects in low-income countries as well as the Netherlands. Dr. Ronteltap's main research topics include nutrient and energy recovery, water conservation and reclamation, and sustainable and ecological sanitation. Dr. Ronteltap is supervising several PhD and master's research projects in these topics. Through connecting with international organisations and platforms, she aims to contribute to global knowledge in these fields. Dr. Ronteltap also coordinates several online and short courses at UNESCO-IHE, including the online course and the short course on Faecal Sludge Management.

Damir Brdjanovic

Prof. Damir Brdjanovic is the Head of the Environmental Engineering and Water Technology Department of UNESCO-IHE (Institute for Water Education). The professional mission of Prof. Brdjanovic is to contribute to a balance of knowledge development, research and capacity building in the urban sanitation field, with a clear view of the needs of low- and middle-income countries. The unifying vision of his research activities is integrated management of the urban water cycle, which includes provision of sanitation to the urban poor, onsite decentralised sanitation, urban drainage, wastewater collection, treatment and reclamation/reuse, and residuals management. His approach includes centralised to decentralised approaches, advanced versus low-cost technologies, and engineered versus natural systems. Prof. Brdjanovic's research group is also conducting research in emergency sanitation, resource oriented sanitation, faecal sludge management, anaerobic treatment, membrane bio-reactors and infrastructure asset management. His research is conducted through experimental work at laboratory, pilot, and field scale as well as mathematical modelling, decision support and process optimisation in municipal and industrial applications. Prof. Brdjanovic is currently leading a large research and education project for pro-poor sanitation funded by the Bill & Melinda Gates Foundation.

Authors in alphabetical order

Magalie Bassan, EAWAG – Swiss Federal Institute of Aquatic Science and Technology, Switzerland

Damir Brdjanovic, UNESCO-IHE Institute for Water Education, the Netherlands

Bipin Dangol, ENPHO – Environment and Public Health Organization, Nepal

Pierre-Henri Dodane, Independent Consultant, France

Christine Maria Hooijmans, UNESCO-IHE Institute for Water Education, the Netherlands

Carlos Manuel Lopez-Vazquez, UNESCO-IHE Institute for Water Education, the Netherlands

Mbaye Mbéguéré, ONAS – Office National de l'Assainissement du Senegal, Senegal

Georges Mikhael, WSUP – Water and Sanitation for the Urban Poor, United Kingdom

Berta Moya Diaz-Aguado, Independent Consultant, Spain

Charles Buregeya Niwagaba, Makerere University, Uganda

Ives Magloire Kengne, University of Yaoundé I, Cameroon

James Edward Ramsay, Independent Consultant, United Kingdom

Philippe Reymond, EAWAG – Swiss Federal Institute of Aquatic Science and Technology, Switzerland

David M. Robbins, Independent Consultant, USA

Mariska Ronteltap, UNESCO-IHE Institute for Water Education, the Netherlands

Linda Strande, EAWAG – Swiss Federal Institute of Aquatic Science and Technology, Switzerland

Elizabeth Tilley, ETH – Swiss Federal Institute of Technology Zurich, Switzerland

Foreword

Doulaye Kone

After decades promoting sanitation in low- and middle-income countries, several countries and the global sanitation community have come to realise that it is time to rethink the approach to accelerating access to quality services. Since 2000, the Joint Monitoring Program (WHO/UNICEF) of the Millennium Development Goals (MDG) has consistently reported that the share of the population in low- and middle-income countries that use pit latrines, septic tanks, and systems termed as 'unimproved' sanitation facilities is growing. It is now estimated that between 2.1 – 2.6 billion people in low- and middle-income countries rely on onsite technologies that produce tons of untreated faecal sludge (FS) every day. When septic tanks and pit latrines become full, the sludge that is collected from them is largely discharged untreated into open drains, irrigation fields, open lands, or surface waters. The amount of untreated FS discharged into the open environment poses a serious public health risk. A 5 m^3 truck load of FS dumped into the environment is the equivalent of 5,000 people practicing open defecation. Adding to this is the heavy load from open defecation of raw faeces excreted in the open by an additional 1.1 billion people who still do not have access to any toilet. The consequences of this waste entering the environment are staggering. The World Bank estimates that poor sanitation costs the world 260 billion USD annually. Poor sanitation contributes to 1.5 million child deaths from diarrhoea each year. Chronic diarrhoea can also hinder child development by impeding the absorption of essential nutrients that are critical to the development of the mind, body, and immune system. It can also impede the absorption of life-saving vaccines.

In the 1980s, under the leadership of Roland Scherteinleib and Martin Strauss, the Swiss Federal Institute of Aquatic Science and Technology (EAWAG) established the Department of Water and Sanitation for Developing Countries (SANDEC) with a strong research and development focus on FS management (FSM). Since then, SANDEC has been a research pioneer in developing, evaluating and testing sanitation solutions, complemented by a strong policy and advocacy program. It has both informed and driven a global call to action on the issue.

This book is an impressive resource that capitalises on recent scientific evidence and practical solutions tested at scale by sector professionals. It compiles lessons drawn from rigorous scientific and case study investigations to formulate operational approaches and solutions for planners, engineers, scientists, students, and researchers. I personally coordinated an intensive and very exciting part of this work while working at SANDEC as a program officer and as team leader of the FSM team, which later became the Excreta and Wastewater Management (EWM) Group. This book builds on lessons gathered from Latin America (Argentina), Africa (Benin, Burkina Faso, Cameroun, Cote d'Ivoire, Ghana, Kenya, Mali, Nigeria, Senegal, Togo, Uganda, South Africa) and Asia (Thailand, Cambodia, China, India, Indonesia, Malaysia, Philippines, Thailand, Vietnam). It fills important FSM knowledge gaps, while at the same time acknowledging persistent gaps and identifying new areas of innovation for future research. It is a valuable handbook for any sanitation professional or academic. It is solution-oriented and addresses the issues that real practitioners face (e.g. city managers, engineering companies, development organisations).

From its inception, the Bill & Melinda Gates Foundation's Water Sanitation and Hygiene (WSH) programme has emphasised the strategic importance of improving FSM globally. We have engaged new partners and supported established organisations such as EAWAG/SANDEC and UNESCO-IHE to propose and promote catalytic solutions that can positively impact the lives of billions of people in low- and middle-income countries who do not have access to FSM services. The technologies, project planning tools, and FSM business operation and management practices shared in this book will help stakeholders globally begin to build functional and viable sanitation service chains that benefit poor communities. Key insights on the potential and limitations of technologies, FSM operations, businesses, and the financial and economic value that can be recovered from FS processing will all help to transform sanitation service provision into a more sustainable and profitable business service chain. As the global community is currently looking forward to the 2015 post-MDG solutions, this paradigm will inform new public-private partnership models that promote quality and affordable sanitation services, especially in poor communities where the large majority still live with toilets that are not connected to any infrastructure or public utility services.

Doulaye Kone, PhD
Bill & Melinda Gates Foundation
Seattle, March 2014

Acknowledgments

Funders

Swiss Development Cooperation - SDC
Bill & Melinda Gates Foundation - BMGF

Contributors (in alphabetical order)

Benedict Borer
Sally Brown
Chris Buckley
Grover Mamani-Casilla
Kartik Chandran
Manus Coffey
Stefan Diener
Moritz Gold
John Harrison
Halidou Koanda
Doulaye Kone
Neil Macleod
Kate Medlicot
Susan Mercer
Martin Mulenga
Josiane Nikiema
Peter Penicka
Selvi Pransiska
Apurva Sahu
Lars Schoebitz
Alyse Schrecongost
Dave Still
Claire Taylor
Lukas Ulrich
Melanie Valencia
Konstantina Velkushanova
Chris Zurbrügg

Reviewers (in alphabetical order)

Isabel Blackett
Olufunke Cofie
George Ekama
Guy Hutton
Florian Klingel
Thammarat Koottatep
Christoph Lüthi
Jennifer McConville
Ashley Murray Muspratt
Kara Nelson
Guy Norman
Jonathan Parkinson
David Robbins
Pippa Scott
Martin Strauss
Steve Sugden
Kevin Taylor
Björn Vinnerås

Contents

Chapter 1

The Global Situation

Linda Strande

1.1 INTRODUCTION

Solutions for effective and sustainable faecal sludge management (FSM) presents a significant global need. FSM is a relatively new field, however, it is currently rapidly developing and gaining acknowledgement. This chapter provides an introduction to what FSM is, some of the unique challenges of FSM, an overview of the systems level approach for implementation and operation presented in this book, and additional resources that are available on the internet.

1.2 WHAT IS FAECAL SLUDGE?

Faecal sludge (FS) comes from onsite sanitation technologies, and has not been transported through a sewer. It is raw or partially digested, a slurry or semisolid, and results from the collection, storage or treatment of combinations of excreta and blackwater, with or without greywater. Examples of onsite technologies include pit latrines, unsewered public ablution blocks, septic tanks, aqua privies, and dry toilets. FSM includes the storage, collection, transport, treatment and safe enduse or disposal of FS. FS is highly variable in consistency, quantity, and concentration.

1.3 GLOBAL RELEVANCE

The sanitation needs of 2.7 billion people worldwide are served by onsite sanitation technologies, and that number is expected to grow to 5 billion by 2030 (Figure 1.1). It is a common perception that onsite technologies fulfil sanitation needs for rural areas, but in reality, around one billion onsite facilities worldwide are in urban areas. In many cities, onsite technologies have much wider coverage than sewer systems. For example, in Sub-Saharan Africa, 65-100% of sanitation access in urban areas is provided through onsite technologies (Strauss *et al.*, 2000). However, despite the fact that sanitation needs are met through onsite technologies for a vast number of people in urban areas of low- and middle-income countries, there is typically no management system in place for the resulting accumulation of FS. It is evident that the management of FS is a critical need that must be addressed, and that it will continue to play an essential role in the management of global sanitation into the future.

In the past, sludge management from onsite facilities has not been a priority of engineers or municipalities, and has traditionally received little to no attention. Several generations of engineers have considered waterborne, sewer-based systems as the most viable, long-term solution to fulfil sanitation needs. Onsite technologies have traditionally been viewed as only temporary solutions until sewers could be built. This practice is a result of the effectiveness of sewer-based approaches throughout Europe and North America in cities where water is for the most part readily available, as

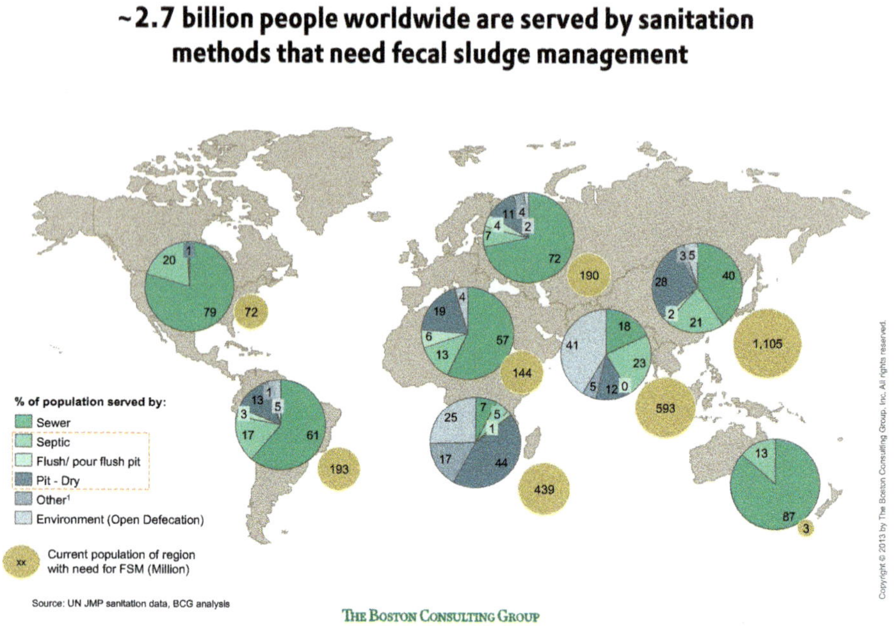

Figure 1.1 Percent of population served by onsite sanitation technologies (Reproduced with permission from the Boston Consulting Group; 2013).

well as out-of-date engineering curricula, and the preference for large-scale infrastructure investments by development banks and governments. However, the expansion and development of functioning, conventional sewer networks is not likely to keep pace with the rapid urban expansion typical of low- and middle-income countries. In addition, where sewers and wastewater treatment plants (WWTPs) have been constructed in low-income countries they have most frequently resulted in failures. Over the last 15 years, the thinking of engineers worldwide has started to shift, and people are starting to consider onsite or decentralised technologies as not only long-term viable options, but possibly the more sustainable alternative in many ways compared to sewer-based systems which are prohibitively expensive and resource intensive. In urban areas, it has been demonstrated that, depending on local conditions, the cost of FSM technologies are five times less expensive than conventional sewer-based solutions (Dodane *et al.*, 2012).

Increasing access to sanitation is a global priority. Currently one in five children die from diarrheal-related diseases, which is more than that of aids, malaria, and measles combined (UNICEF and WHO 2009). In addition to health benefits, improved sanitation has significant economic benefits, for example the return on one USD spent on water and sanitation improvements in low-income countries is 5-46 USD depending on the intervention (Hutton *et al.*, 2007). Progress towards the Millennium Development Goals (MDGs) has been successful in increasing access to *improved*[1] sanitation facilities. However, providing adequate access to sanitation facilities does not end when onsite technologies are built. The promotion of onsite technologies has greatly reduced open defecation, but without solutions

1 Target 7C - reducing by half the number of people without access to 'improved' sanitation. Improved is defined as systems that hygienically separate human excreta from human contact, and includes: flush toilets, connection to a piped sewer system, connection to a septic system, flush/pour-flush to a pit latrine, ventilated improved pit (VIP) latrines, and composting toilets.

or funding to maintain their functionality through appropriate FSM, it has also resulted in numerous cases in a sludge management crisis, having significant impacts on human and environmental health. Onsite technologies can represent viable and more affordable options, but only if the entire service chain, including collection, transport, treatment and safe enduse or disposal, is managed adequately. Without an FSM structure in place, when the containment structure fills up, the untreated FS most likely ends up directly in the local environment (Figure 1.2). This results in the pervasive contamination of the environment by pathogens and is not providing a protective barrier to human contact and hence protection of public health. For example, in Dakar only 25% of FS that accumulates in onsite facilities is being collected and transported to legitimate FSTPs (BMGF, 2011). When developing sanitation goals and implementing sanitation projects, it is imperative to consider *downstream sanitation*, beyond only a focus at the household level and only providing toilets.

Effective management of FS systems entails transactions and interactions among a variety of people and organisations from the public, private and civil society at every step in the service chain, from the household level user, to collection and transport companies, operators of treatment plants, and the final enduser of treated sludge. Sewer systems and FSM can be complementary, and frequently do exist side-by-side in low-income countries. A very successful example of this management model is in Japan where the systems successfully co-exist in urban areas (Gaulke, 2006).

Figure 1.2 Faecal sludge from a pit latrine being directly emptied into a drainage channel in Kampala, Uganda (left), and emptying of faecal sludge from a septic tank next to the house in Dakar, Senegal (photo: Linda Strande).

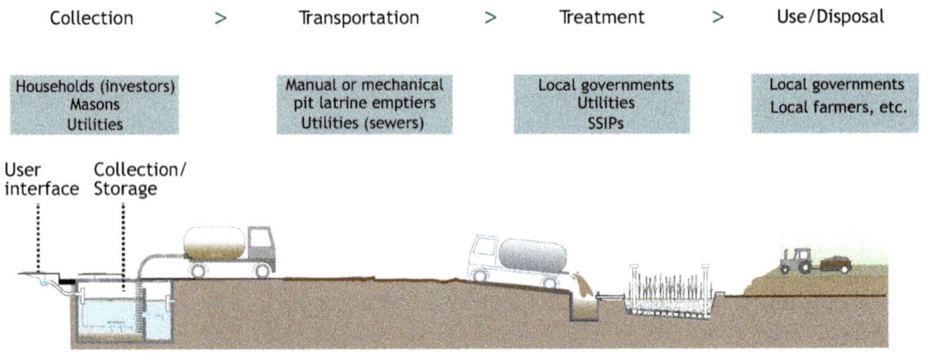

Figure 1.3 Sanitation and faecal sludge management service chain (Parkinson *et al.*, 2013).

The complete sanitation service chain is shown in Figure 1.3, the FSM component is specifically the emptying, collection, transport, treatment and enduse or disposal of FS. Factors such as technology designs and options for user interfaces, or onsite collection and storage methods to reduce sludge volumes are covered in more detail in The Compendium of Sanitation Systems and Technologies, which is also available free of charge from the SANDEC website (Tilley *et al.*, 2014). Weak links in the FSM service chain include many factors, such as household level users not being able to afford professional emptying services; collection and transport trucks not being able to access narrow lanes and paths leading to houses; operators not able to afford the transport of FS over large distances to treatment facilities; and the lack of legitimate FS discharge locations or treatment facilities. The solution to overcoming these problems and designing functioning and sustainable FSM requires a systems-level approach that addresses every step in the service chain. To move towards complete and functioning FSM service chains, this book develops an integrated systems-level approach that incorporates technology, management and planning.

1.4 BOOK OBJECTIVE

Developing solutions for FSM is a serious global problem that has received limited attention over the past twenty years (Strauss and Heinss, 1996). Compared to wastewater management practice, there is a hundred year gap in knowledge of FSM in urban areas. However, the FSM field is now rapidly developing and gaining acknowledgement, as shown by many recent examples where municipalities are adopting FSM into their urban planning (e.g. Dakar, Senegal and Ouagadougou, Burkina Faso), and the commitment of organisations like the Bill & Melinda Gates Foundation placing significant resources into research of FSM. Recently, experience through pilot and full-scale systems has started becoming available (Figure 1.4), but practice is still not up to desired speeds. As awareness of the need for FSM has increased, so has the need for solutions. However, information on FSM is generally not readily available and therefore the objective of this book is to present an approach for the comprehensive and integrated management of FS in urban and peri-urban areas of low- and middle-income countries. This book aims to contribute to filling the knowledge gap by bringing together and presenting the current state of knowledge in the field.

The target audience of this book includes students and practitioners in the field who are or will be designing, planning, promoting, or managing FSM systems. The book provides a comprehensive approach that includes an overview, design guidelines of treatment technologies, important considerations of operations and maintenance for successful operation of implemented technologies, and a planning approach so that all necessary requirements are met to ensure a long-term, sustainable system. The book assumes the reader has basic knowledge of sanitation and wastewater treatment.

The book is expected to contribute to readers' better understanding of treatment, management and planning aspects of FSM; enable them to identify suitable treatment options; understand the mechanisms and designs of specific treatment technologies; and enable them to communicate important aspects of FSM to stakeholders involved in the process including managers and decision makers. The book is also relevant for employees of municipalities, national sanitation utilities, consultants, donor agencies, decision makers, and waste-related businesses in order to expand their knowledge, understanding and overview of integrated FSM systems.

The book was designed as a learning tool with many elements of a textbook. Each chapter includes learning objectives so it is clear what readers can gain from the chapter. The end of chapter study questions help to evaluate whether the learning objectives are achieved. Where relevant, example problems are also included to illustrate how any calculations were made, and case studies are included to describe the importance of real-life lessons learned in each of the covered areas. As such, the book can also be used in any classroom setting, and is currently used in the context of a newly developed three week course on FSM as well as in a new online course on FSM offered within the program of UNESCO-IHE Institute for Water Education.

Figure 1.4 Class 2014-2016 of the Master Specialisation in Sanitary Engineering at UNESCO-IHE (photo: UNESCO-IHE).

1.5 DESIGNING FOR FAECAL SLUDGE MANAGEMENT TREATMENT ENDUSE

When designing treatment technologies, the final enduse or disposal option of sludge and liquid streams should first be determined, so that obtaining the appropriate level of treatment for the desired enduse can be incorporated into the design. Once the final enduse or disposal options have been selected, it becomes possible to work backwards starting from the final treatment requirements to design a system that achieves the treatment objectives. For example, pathogen reduction and level of sludge dryness requirements will be very different if the intended endproduct is compost for use on food crops or if it is fuel for use as combustion in industrial processes. These decisions are context specific, and need to be made based on local regulations and the market demand for endproducts. Similar to designations for Class A and Class B biosolids in the United States, FS is treated for levels of pathogen reduction that make it appropriate for different enduses. This approach is important to ensure that effluents and endproducts achieve adequate and appropriate levels of treatment; systems are not over-designed, wasting financial resources; and that systems are not under-designed risking public and environmental health.

Resource recovery from treatment products should be considered as a treatment goal whenever possible, but the number one goal is obviously the protection of public health. In many low- and middle-income countries, regulations for the enduse of sludge do not exist and/or are not enforced. In the apparent lack of a regulatory environment, the required levels of treatment becomes a societal decision. On the other hand, standards that are too strict may also have a negative impact if they prevent action from being taken because they cannot be met. To ensure adequate protection of human health, a multi-barrier approach is recommended, as described in Chapter 10, Enduse of Treatment Products. Financial flows from the sale of endproducts can also help to achieve the sustainability of treatment options, as they offset sludge disposal costs, potentially provide a revenue stream, help to ensure treatment plants are operated well to provide quality products, and provide a benefit to society through resource recovery. This type of context-specific solution needs to take into account the local market demand, and ways to increase the value of treatment products as markets vary significantly among locations (Diener *et al.*, 2014).

1.5.1 Systems approach

For sustainable implementation and ongoing operation, FSM requires an integrated systems approach incorporating technology, management and planning, as depicted in Figure 1.5. In this book, chapters fall under each of the technology, management or planning sections, as is clearly presented throughout the book by the colour scheme, but what is of utmost importance is how all three of these fields come together to provide a framework that will guide practitioners from the initial project planning phase to implementation and ongoing operations and maintenance phases. A multi-disciplinary, systems-level approach to FSM like that developed here is required to ensure that untreated FS is removed from the community, not remaining at the household level, and that it is treated in a safe and effective manner. For example, removing sludge from the household is a private interest, but the FSM service chain is a public interest, requiring regulation and enforcement by an authority that is responsible for the public good. If only a few people in a community properly manage FS, it would not have a net impact on the community as a whole; there needs to be collective participation at the community scale to ensure that public health benefits are realised. This requires sustained public sector commitment, effective policies, appropriate implementation and enforcement to promote understanding and adherence (Klingel *et al.*, 2002), topics that are covered in the Planning and Management sections.

Although technologies are an integral and essential component of FSM, they should not be considered in isolation. Planning and management methodologies presented in this book will help form the fundamental foundation that long-term successful FSM systems are built on. They not only represent the first phase of designing a system, but are necessary to ensure a continuum of success throughout the life of a project. As presented in the planning section, ideally all key stakeholders will realise the need for and have desire to participate in the planning stages, including public authorities, entrepreneurial collection, transport, and treatment service providers, and the serviced and impacted communities. Methods for increasing stakeholder engagement will help to ensure that stakeholders have a long-term investment in the success of the system, and will continue to provide feedback that results in further improved solutions. This can be aided by clearly defining responsibilities, communication, and coordination mechanisms during the planning phases. Including an integrated planning approach helps to ensure vested participation and management, without which technologies implemented in low-income countries will fail over the long term.

In this book, the planning process includes exploring the situation (identify stakeholders and their interactions; understand the existing situation; develop goals and objectives); developing solutions (including institutional, financial, and technical aspects); and defining measures for implementation (Klingel *et al.*, 2002). This covers organisational, institutional, financial, legal and technical aspects of the entire FSM service chain, from the collection and transport, to the final disposal or enduse of treatment products, and is necessary to coordinate and ensure varied and complex levels of service, among stakeholders that have diverse interests. This FSM planning approach includes understanding and matching stakeholders' interests, needs and constraints with an appropriate institutional framework, financial mechanisms, capacity and the initial situation. This type of integrated planning can prevent previous failures, for example, locating a FS treatment plant on the outskirts of a city where land is available and relatively inexpensive, meaning that the costs associated with haulage time and distance for collection and transport companies is prohibitive, resulting in direct dumping of FS in the environment, and the treatment plant not being used.

Management factors presented in this book, such as institutionalisation, technical capacity, legal frameworks and cost recovery mechanisms, will help to ensure long-term success of FSM technologies (Bassan *et al.*, 2014). Management concerns need to be incorporated into technology decisions, for

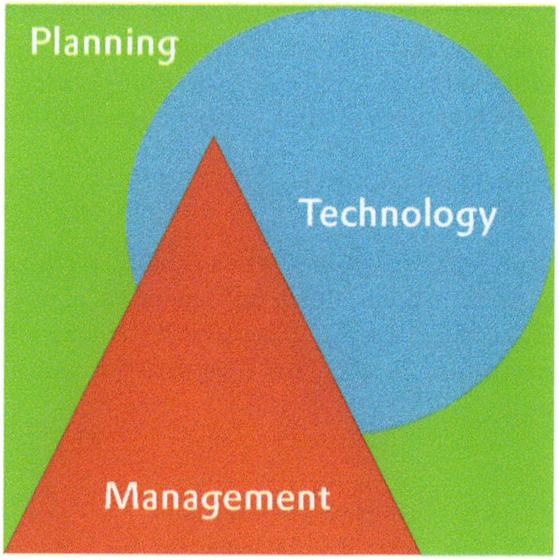

Figure 1.5 Faecal sludge management requires an integrated systems level approach, incorporating technology, management and planning.

example locally available or reparable pumps to ensure ongoing operation of technologies when pumps break down. Environmental regulations can be in place, but will require adequate enforcement for adherence. Evaluating and implementing financial structures that can sustain the system ensures financial viability and long-term operation, including appropriate financial incentives and sanctions (Wright, 1997). Methods to ensure running costs are covered so the entire system can operate in an affordable fashion need to be determined, as well as ways that financial transfers throughout the service chain can provide adequate funding for each step in the chain.

This systems-level approach includes evaluating in existing systems what can be done for improvement at each step in the chain, and then most importantly, how all the steps integrate and influence each other. For example, could resource recovery from treatment products be a financial driver for the service chain, ultimately reducing collection fees at the household level, and thereby increasing access to sanitation? Could a significant market demand for treatment endproducts such as use as an industrial fuel provide a financial incentive for collection and transport companies to deliver FS to treatment facilities instead of discharging it untreated directly into the environment?

This book contains 18 chapters, subdivided into the technology, management and planning sections, plus The Global Sitation and The Way Forward chapters. This approach covers individual topics in a focused matter, yet embeds and links them to each other throughout the book. Chapters 2 – 10 are focused on technical aspects of collection, transport and treatment, Chapters 11 – 13 are focused on examples of on-going management of FSM systems, and Chapters 14 – 17 are focused on planning integrated FSM systems. Each chapter presents different aspects of that field, and then how they can all be combined into one integrated approach is brought together in Chapter 17 Planning Integrated FSM Systems where a logical framework is presented that highlights the tasks and activities that need to be included in designing a comprehensive system.

Chapter 2 Quantification, Characterisation and Treatment Objectives

This chapter presents an overview of the challenges and objectives of FSM from the technology perspective. It covers the difficulties in obtaining reliable data for estimating the quality and quantity of FS that is produced in a city, introduces parameters that are important in FS characterisation and how they are analysed. Examples are provided to illustrate the wide range of high, medium and low strength FS that has been observed in the field, and explains some of the operational factors responsible for this variability. The chapter then explains what treatment targets and objectives are in an FSM system.

Chapter 3 Treatment Mechanisms

This chapter presents the basic scientific mechanisms that existing technologies rely on for the treatment of FS, to provide the reader with a more in-depth understanding of how technologies function, and their operation and maintenance requirements. It explains key parameters that need to be monitored and optimised to ensure treatment efficiency, and how to assess which treatment mechanisms are appropriate for a given context.

Chapter 4 Methods and Means for Collection and Transport

This chapter presents the current state of knowledge for how FS can be collected from onsite technologies, and transported to treatment facilities, including the role of transfer stations. Technologies are explained including social, procedural and technical aspects. Manual (Figure 1.6), manually operated mechanical and fully mechanised technologies are presented. The importance of health and safety issues regarding FS collection and transport are also presented.

Figure 1.6 Transport of faecal sludge in an informal settlement in Nairobi, Kenya (photo: Linda Strande).

Chapter 5 Overview of Treatment Technologies

This chapter presents an overview of potential treatment technologies. It presents well established technologies, which are then covered in more detail in respective chapters, technologies that appear to be very promising but have had limited implementations, and promising technologies that are still in the research phase. The advantages, constraints and field of application of each treatment option are presented, and information provided so the reader can compare and contrast the potential performance and scope of appropriate application. It also emphasises the importance of finding a context-adapted combination of technologies, and the parameters that are important to consider when designing a system.

Chapter 6 Settling-Thickening Tanks

This chapter provides information on the design and ongoing operations and maintenance of settling-thickening tanks. It presents information on when settling-thickening tanks are adequate treatment technologies, the fundamental mechanisms of how they function, and their potential advantages and disadvantages. Information is provided on how to design a settling-thickening tank according to the desired treatment goal.

Chapter 7 Unplanted Drying Beds

This chapter presents an overview of unplanted drying beds for sludge dewatering. It explains their main components and how they affect the performance of drying beds. It provides an understanding of the appropriate level of operations, maintenance and monitoring for their performance. Information is provided on how to design drying beds according to the desired treatment goal.

Chapter 8 Planted Drying Beds

This chapter presents an understanding of planted drying beds for sludge dewatering and stabilisation. It presents an overview of the vegetation types that can be used, and the role they play in sludge dewatering. Information is provided on the appropriate level of operations, maintenance and monitoring for their performance. Information is provided on how to design planted drying beds according to the desired treatment goal and context-specific parameters.

Chapter 9 Co-treatment with Wastewater

This chapter presents information on the co-treatment of FS with municipal wastewater. The possibilities discussed include activated sludge, anaerobic digestion, and anaerobic ponds. This chapter presents information on the extreme care that must be exercised when considering combined FS and wastewater treatment, as the system can be overloaded and fail even when relatively small volumes of FS are added to the wastewater treatment plant. Information is presented on the fractionation of organic matter and nitrogen compounds in FS. Key considerations and potential impacts of FS co-treatment in wastewater treatment systems are explained, and results of calculations of the volume/load of FS that can effectively be co-treated with wastewater are presented.

Chapter 10 Enduse of Treatment Products

This chapter presents information on the safe enduse or disposal of FS treatment products. The importance of resource recovery, combined with adequate protection of human and environmental health is stressed. Existing options for resource recovery are presented, along with innovative options that are still in the research stage of development. Information is presented on how to determine rates for the land application of sludge, and possibilities for the enduse or disposal of liquid streams.

Chapter 11 Operation, Maintenance and Monitoring

This chapter presents information on critical operations and maintenance factors that should be considered when building and operating a FSTP. It introduces operations and maintenance manuals, the importance of monitoring activities, and why and how administrative management is crucial to the long-term successful operation of an FSTP.

Chapter 12 Institutional Frameworks

This chapter presents information on the institutional framework that needs to be in place for an effective FSM-enabling environment. It presents regulations and contracts that can be used for enforcing adequate service. The main strength and weaknesses of stakeholders related to the institutional framework are explained, and an overview of the potential institutional arrangements for the distribution of responsibilities in the service chain is provided. Advantages and drawbacks of different institutional arrangements are also discussed.

Chapter 13 Financial Transfers and Responsibilities

This chapter presents information on possibilities for different models of financial flows among stakeholders in the FSM service chain. It explains what types of financial transfers play a role, necessary incentives, sustainable tariffs, and what legal and institutional frameworks have to be in place. It also explains the complexity and difficulty of designing, implementing, monitoring and optimising the financial flows for an entire FSM system.

Chapter 14 Assessment of the Initial Situation

This chapter presents the first step in the planning process, how to understand what is important to know at the beginning of the FSM planning process, and what information needs to be collected. It explains different methods and tools for collecting the relevant data, and how to identify shortcomings and challenges of the existing FSM systems and enabling environment.

Chapter 15 Stakeholder Analysis

This chapter presents why stakeholder analysis is important in FSM project design, and how to perform a stakeholder analysis including identifying and characterising the key stakeholders and relationships. It also explains how the stakeholder selection evolves throughout the planning process, and how to determine stakeholders that need empowerment, incentives, capacity-building or other forms of information.

Chapter 16 Stakeholder Engagement

This chapter presents why it is important to engage stakeholders from the very beginning of project implementation, and how this is effective in easing project implementation and to enhance long-term sustainability. It explains how to use information gathered during the stakeholder analysis to plan stakeholder involvement and how to distribute and formalise roles and responsibilities, and it provides tools to inform, consult and collaborate with stakeholders.

Chapter 17 Planning Integrated Systems

This chapter presents the importance of tying together all of the information presented in the book into one integrated planning approach. It links all the technology, management, and planning aspects developed in the book and explains how they are all connected and influence each other. A logical framework is presented that highlights the tasks and activities that need to be included in designing a comprehensive system. The chapter illustrates how to plan an integrated FSM system at the city level and how to select context-appropriate options.

Chapter 18 The Way Forward

This chapter discusses lessons that have been learned, what is still lacking in the field of FSM, and how we can move forward in developing and obtaining the necessary knowledge.

Available Resources

In addition to this book, there are many resources available free of charge via the internet for designing and improving complete access to environmental sanitation. All of these tools should be used in conjunction with each other to ensure the most sustainable and comprehensive approach possible. Resources include:

Community-Led Urban Environmental Sanitation (CLUES), EAWAG/WSSCC/UN-Habitat

CLUES presents a complete set of guidelines for sanitation planning in low-income urban areas. It is the most up-to-date planning framework for facilitating the delivery of environmental sanitation services for urban and peri-urban communities. CLUES features seven easy-to-follow steps, which are intended to be undertaken in sequential order. Step 5 of the planning approach relies on the Compendium, applying the systems approach to select the most appropriate technological option(s) for a given urban context. The document also provides guidance on how to foster an enabling environment for sanitation planning in urban settings.

Published in 2011, 100 pages, with a memory key. It is available for download at: www.sandec.ch/clues.

Compendium of Sanitation Systems and Technologies

The Compendium is a guidance document for engineers and planners in low- and middle-income countries, primarily intended to be used for communicative planning processes involving local communities. It is also intended for persons/experts who have detailed knowledge about conventional high-end technologies and require information, for instance, on infrastructure and different system configurations. It is not intended as a standalone document for engineers, making decisions for the community, e.g., expert-driven decision-making.

The Compendium of Sanitation Systems and Technologies was first published in 2008 during the International Year of Sanitation. This new version contains more technologies, a simplified user guide as well as a section on emerging technologies while keeping the same structure: brief, concise and connected. It will be available for download at: www.eawag.ch/sandec.

How to Design Wastewater Systems for Local Conditions in Developing Countries (RTI/IWA)

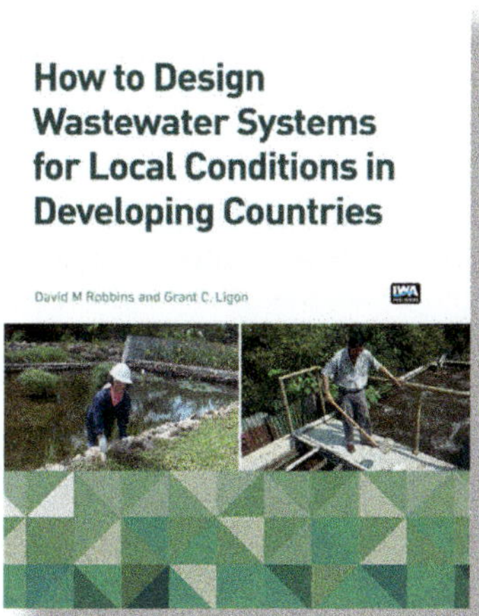

This manual provides guidance in the design of wastewater systems in low-income country settings. It promotes a context-specific approach to technology selection by guiding the user to select the most suitable technologies for their area. It provides tools and field guides for source characterisation and site evaluation, as well as technology identification and selection. This manual is primarily addressed to private and public sector service providers, regulators and engineers/development specialists in charge of implementing wastewater systems. RTI edited the manual, and IWA published it in 2014. It is available for download at: http://www.iwapublishing.com/template.cfm?name=isbn9781780404769_&type=new.

Expanding your knowledge in a course

Over the last few years, the knowledge and understanding of FSM has advanced extensively. For the new generation of scientists and engineers entering the sanitation profession, the quantity, complexity and diversity of these new developments can be overwhelming, particularly in low-income countries where access is not readily available to advanced level courses in FSM. This book seeks to address that deficiency. It assembles and integrates materials of experts around the world that have made significant contributions to the advances in FSM. The book also forms part of a three-week course at UNESCO-IHE Institute for Water Education as well as of an internet-based curriculum (online course) in FSM and, as such, may also be used together with lecture handouts, filmed lectures by the authors and tutorial exercises for readers' self-study. Upon completion of this curriculum, modern approaches to FSM can be embraced with deeper insight, advanced knowledge and greater confidence.

Figure 1.7 Graduating class of Masters of Science from UNESCO-IHE. In addition to being used in the Masters curricula, this book makes part of the distance-learning course on faecal sludge management and of the newly established Postgraduate Diploma Program in Sanitation and Sanitary Engineering at UNESCO-IHE (photo: UNESCO-IHE).

1.6 BIBLIOGRAPHY

Bassan, M., Mbéguéré, M., Koné, D., Holliger, C., Strande, L. (2014). Success and failure assessment methodology for wastewater and faecal sludge treatment projects in low-income countries. Journal of Environmental Planning and Management (in press).

Bill & Melinda Gates Foundation (BMGF) (2011). Landscape Analysis & Business Model Assessment in Fecal Sludge Management: Extraction & Transportation Models in Africa - Senegal.

Diener, S., Semiyaga, S., Niwagaba, C., Muspratt, A., Gning, J.B., Mbéguéré, M., Ennin, J.E., Zurbrügg, C., Strande, L. (2014). A value proposition: resource recovery from faecal sludge – can it be the driver for improved sanitation? Resources Conservation & Recycling (in press).

Dodane, P.H., Mbéguéré, M., Ousmane, S., Strande, L. (2012). Capital and Operating Costs of Full-Scale Faecal Sludge Management and Wastewater Treatment Systems in Dakar, Senegal. Environmental Science & Technology 46(7), p.3705-3711.

Gaulke, L., Johkasou S. (2006). On-site Wastewater Treatment and Reuses in Japan. Proceedings of the Institute of Civil Engineers - Water Management 159(2), p.103-109.

Hutton, G., Haller, L., Bartram, J. (2007). Global Cost-benefit Analysis of Water Supply and Sanitation Interventions. Journal of Water and Health 5(4), p.481-502.

Klingel, F., Montangero, A., Koné, D., Strauss, M. (2002). Fecal Sludge Management in Developing Countries. A planning manual. EAWAG: Swiss Federal Institute for Environmental Science and Technology SANDEC: Department for Water and Sanitation in Developing Countries.

Parkinson, J., Lüthi, C., Walther, D. (2013). Sanitation 21: A Planning Framework for Improving City-wide Sanitation Services. Published by IWA.

Strauss, M., Heinss, U. (1996). Faecal Sludge Treatment, SANDEC News no. 2.

Strauss, M., Larmie, S.S., Heinss, U. and Montangero, A.(2000). Treating Faecal Sludges in Ponds. Water Science & Technology 42(10), p.283–290.

Tilley, E., Ulrich, L., Lüthi, C., Reymond, P., Zurbrügg, C. (2014). Compendium of Sanitation Systems and Technologies. Dübendorf: Swiss Federal Institute of Aquatic Science & Technology (EAWAG) 2nd revised edition.

UNICEF and WHO (2009). Diarrhoea: Why children are still dying and what can be done.

Wright, A.M. (1997). Toward a Strategic Sanitation Approach: Improving the Sustainability of Urban Sanitation in Developing Countries. UNDP-World Bank Water and Sanitation Program.

Technology

Chapter 2

Faecal Sludge Quantification, Characterisation and Treatment Objectives

Charles B. Niwagaba, Mbaye Mbéguéré and Linda Strande

Technology

Learning Objectives

- Understand the difficulties in obtaining reliable data on the quality and quantity of faecal sludge production on a citywide scale.
- Know which parameters are important for faecal sludge characterisation, how they are analysed, and which ranges determine high, medium and low strength faecal sludge.
- Be able to describe how operational factors impact the variability of faecal sludge.
- Have an understanding of faecal sludge management and treatment targets and objectives.

2.1 INTRODUCTION

The first step in designing faecal sludge (FS) treatment technologies that will meet defined treatment objectives is to quantify and characterise the FS to be treated. Ideally, this should be carried out as part of the Feasibility Study as described in Chapter 17, but is however difficult due to the lack of standardised methodologies for the quantification or characterisation of FS. This complicates the design of adequate and appropriate systems.

The quantities of FS generated and the typical FS characteristics are difficult to determine due the variety of onsite sanitation technologies in use, such as pit latrines, public ablution blocks, septic tanks, aqua privies, and dry toilets. In many cities, a mixture of these technologies often exist side-by-side, and there is generally a prevalence of different technologies in different geographical regions. For example, in Bangkok, Thailand; Dakar, Senegal; Hanoi, Vietnam, and Buenos Aires, Argentina septic tanks are the predominant form of onsite FS containment technology; whereas in Kampala, Uganda; Nairobi, Kenya; and Dar es Salaam, Tanzania, various types of pit latrines are the predominant form of FS containment technology (e.g. improved and unimproved private latrines, shared and public latrines).

The quantity and characteristics of FS also depends on the design and construction of the sanitation technology, how the technology is used, how the FS is collected, and the frequency of collection. All of these variables results in a significant difference in FS characteristics within cities, and within the same type of containment technology in different locations.

This chapter therefore aims to provide an overview of the current state of knowledge on the quantification and characterisation of FS, to identify gaps in the existing body of knowledge, and to put these into perspective with regards to FS treatment objectives.

2.2 QUANTIFICATION OF FAECAL SLUDGE

Deriving accurate estimates for the volume of FS produced is essential for the proper sizing of infrastructure required for collection and transport networks, discharge sites, treatment plants, and enduse or disposal options. Due to the variability of FS volumes generated it is important to make estimates based on the requirements specifically for each location and not to estimate values based on literature. However, no proven methods exist for quantifying the production of FS in urban areas, and the data collection required in order to accurately quantify FS volumes would be too labour intensive, especially in areas where there is no existing information. There is therefore a need to develop methodologies for providing reasonable estimates.

Two theoretical approaches that have been developed are the Sludge Production Method, and the Sludge Collection Method, depending on whether the goal is to determine total sludge production, or the expected sludge loading at a treatment plant. The Sludge Production Method for estimating FS quantities starts at the household level with an estimate of excreta production (i.e. faeces and urine), the volume of water used for cleansing and flushing and in the kitchen, and accumulation rates based on the type of onsite containment technology. The Sludge Collection Method starts with FS collection and transport companies (both legal and informal), and uses the current demand for services to make an estimate of the volume of FS. Unfortunately, many assumptions have to be made in both methods due to a lack of available information. The following sections provide an example of how these methods are used to estimate the quantity of FS.

Table 2.1 Reported faecal production rates

Location	Wet weight (g/person/day)
high income countries[1]	100-200
low income countries, rural[2]	350
low income countries ,urban[2]	250
China[3]	315
Kenya[4]	520
Thailand[5]	120-400

[1] Lentner *et al.* (1981); Feachem *et al.* (1983); Jönsson *et al.* (2005); Vinnerås *et al.* (2006)
[2] Feachem *et al.* (1983)
[3] Gao *et al.* (2002)
[4] Pieper (1987)
[5] Schouw *et al.* (2002)

2.2.1 Sludge production method

The quantity of faeces produced on a daily basis can vary significantly based on dietary habits. People with a diet consisting of unprocessed food with a high fibre content will produce a higher quantity of faeces (mass and volume) compared to people who have a proportionally higher meat based and highly processed food diet (Guyton, 1992). The frequency of faecal excretion is on average one stool per person per day, but can vary from one stool per week up to five stools per day (Lentner *et al.*, 1981; Feachem *et al.*, 1983). Reported values for faeces production are presented in Table 2.1.

The volume of urine excreted daily also varies significantly, based on factors such as liquid consumption, diet, physical activity and climate (Lentner *et al.*, 1981; Feachem *et al.*, 1983). Reported values for urine production are presented in Table 2.2.

In addition to the volume of excreta generated daily, FS accumulation depends on time and spatial habits that influence where people use the toilet, such as work schedule, eating and drinking habits, patterns of societal cohesiveness, and frequency of toilet usage. The volume of solid waste and other debris that is disposed of in the system also needs to be taken into account.

In order to obtain a good estimate of FS production, the following data is required:

- number of users;
- location;
- types and number of various onsite systems;
- FS accumulation rates; and
- population of socio-economic levels.

The collection of data can pose some challenges depending on the available information, as frequently, onsite systems are built informally, so there is no official record of how many, or what type, of systems exist on a city-level scale. An accurate estimate of this would require intensive data collection at the level of household questionnaires. In some cases detailed demographic information is available, while in others it does not exist. A further complication is the rapid population growth in urban areas of low-income countries. Estimating the volume of FS to be delivered to treatment plants also needs to take into account that vacuum trucks do not always empty the contents of the entire sanitation containment system (Koanda, 2006).

Table 2.2 Reported urine production rates

Location	Volume (g/person/day)
General value for adults[1]	1,000 - 1,300
Sweden[2]	1,500
Thailand[3]	600-1,200
Switzerland (home, weekdays)[4]	637
Switzerland (home, weekends)[4]	922
Sweden[5]	610-1,090

[1] Feachem *et al.* (1983)
[2] Vinnerås *et al.* (2006)
[3] Schouw *et al.* (2002)
[4] Rossi *et al.* (2009)
[5] Jönsson *et al.* (1999)

This method for estimating total FS production will result in an overestimation of the potential volumes to be delivered to a FSTP. Although the ultimate goal is for all FS to be delivered to a treatment plant, it is not realistic to assume that all of the FS produced will initially be collected and transported for discharge at a FSTP.

2.2.2 Sludge collection method

The quantity of FS that is currently being collected from onsite systems in an area will vary depending on the FSM infrastructure, based on factors such as acceptance and promotion of FSM, demand for emptying and collection services, and availability of legal discharge or treatment sites. The volume that is currently being collected can be estimated based on interviews, site visits, and a review of internal records of FS collection and transport companies. Estimates can be based on the number of collections made each day, the volume of FS per collection, the average emptying frequency at the household level, and the estimated proportion of the population that employ the services of collection and transport companies (Koanda, 2006). The activity of informal or illegal collection should also be taken into account, as the volumes collected can be quite significant.

Estimating generation of FS based on this method is complicated by many factors such as the presence of a legal discharge location or treatment plant (see Figure 2.1), if the discharge fees are affordable, and whether there are enforcement measures to control illegal dumping. If all of these factors are in place, then it is possible that the majority of the FS collected will be transported and delivered to a treatment site. If a legal discharge location exists, a flow meter can be installed in order to provide an indication of the volume of FS that is being discharged. However, there is currently a lack of legal discharge locations, and, collection and transport companies are hesitant to cooperate in an official study that effectively documents their illegal activities. It is difficult to quantify the volume of FS being dumped illegally directly into the environment, either by collection and transport companies, or by households that hire manual laborers to remove FS. In addition, if volumes are being estimated for a treatment plant in an area where no legitimate discharge option currently exists, once it is built, it is expected to rapidly increase the market for these services, and hence the volume that will be delivered will also increase. This could result in an underestimation of the required capacity for the FSTP.

Figure 2.1 Discharge of faecal sludge at Duombasie landfill and faecal sludge treatment site in Kumasi, Ghana (photo: Linda Strande).

The accuracy of any method to estimate the volume of FS generated will depend on the quality of the available data, and the reasonableness of assumptions that are made. Methods to estimate volumes of FS will hopefully improve rapidly as more FSTPs are built, and as Faecal Sludge Management (FSM) gains acceptance and legitimacy.

2.3 CHARACTERISATION OF FAECAL SLUDGE

Parameters that should be considered for the characterisation of FS include solids concentration, chemical oxygen demand (COD), biochemical oxygen demand (BOD), nutrients, pathogens, and metals. These parameters are the same as those considered for domestic wastewater analysis, however, it needs to be emphasised that the characteristics of domestic wastewater and FS are very different. Table 2.3 presents examples from the literature illustrating the high variability of FS characteristics and provides a comparison with sludge from a wastewater treatment plant. A more detailed comparison of wastewater sludge and FS COD fractionation is presented in Chapter 9. The organic matter, total solids, ammonium, and helminth egg concentrations in FS are typically higher by a factor of ten or a hundred compared to wastewater sludge (Montangero and Strauss, 2002).

There is currently a lack of detailed information on the characteristics of FS. However, research is actively being conducted in this field. Research results, together with empirical observations, will continue to increase the knowledge of FS characteristics, and allow more accurate predictions of FS characteristics using less labour intensive methods. Section 2.4 discusses the operational factors that affect the variability of FS. In addition to these factors, the high variability of the observed results is also due to the lack of standardised methods for the characterisation of FS.

Case Study 2.1 : Variability of feacal sludge characteristics in Ouagadougou, Burkina Faso

The variability of FS characteristics is illustrated by Bassan *et al.* (2013a). A sampling campaign was set up to sample in the dry and the rainy season in Ouagadougou, Burkina Faso (see Figure 2.4). The TS concentration in the dry season was found to be 10,658 mg/L with a standard deviation of 8,264. Due to the high variability between the samples, a significant difference in strength of FS collected in the wet or dry season could not be detected. Yet, the campaign revealed that during the rainy season a much higher number of trucks arrived at the dumping locations, up to three times as many – indicating that pit latrines and septic tanks were filling up much faster due to leakages and run-off.

Given the significant variability of FS characteristics, it is important to collect data for specific locations when designing a FS treatment system. For example, in 2010, due to a lack of locally available data the design of a FSTP in Ouagadougou, Burkina Faso was based on general characteristics from the literature. The FSTP was designed to treat 125 m^3/day with a TS load of 21,000 mg/L, resulting in 96 drying beds with a surface area of 128 m^2. Follow-up studies on the characterisation of FS in Ouagadougou revealed that the plant was over-designed by a factor of two, and was hence actually able to treat 250 m^3/day (Bassan *et al.*, 2013b). Understanding the local FS characteristics prior to design would have significantly lowered the investment costs of the FSTP. This illustrates how important it is to understand local FS characteristics prior to designing treatment facilities.

Table 2.3 Reported characteristics of faecal sludge from onsite sanitation facilities and wastewater sludge

Parameter	FS source		WWTP sludge	Reference
	Public toilet	Septic tank		
pH	1.5-12.6			USEPA (1994)
	6.55-9.34			Kengne *et al.* (2011)
Total Solids, TS (mg/L)	52,500	12,000-35,000	–	Koné and Strauss (2004)
	30,000	22,000	–	NWSC (2008)
		34,106		USEPA (1994)
	≥3.5%	<3%	<1%	Heinss *et al.* (1998)
Total Volatile Solids, TVS (as % of TS)	68	50-73	-	Koné and Strauss (2004)
	65	45	-	NWSC (2008)
COD (mg/L)	49,000	1,200-7,800	-	Koné and Strauss (2004)
	30,000	10,000	7-608	NWSC (2008)
	20,000-50,000	<10,000	500-2,500	Heinss *et al.* (1998)
BOD (mg/L)	7,600	840-2,600	-	Koné and Strauss (2004)
	-	-	20-229	NWSC (2008)
Total Nitrogen, TN (mg/L)	-	190-300	-	Koné and Strauss (2004)
			32-250	NWSC (2008)
Total Kjeldahl Nitrogen, TKN (mg/L)	3,400	1,000	-	Katukiza *et al.* (2012)
NH_4-N (mg/L)	3,300	150-1,200	-	Koné and Strauss (2004)
	2,000	400	2-168	NWSC (2008)
	2,000-5,000	<1,000	30-70	Heinss *et al.* (1998)
Nitrates, NO_3^- (mg N/L)	-	0.2-21	-	Koottatep *et al.* (2005)
Total Phosphorus, TP (mg P/L)	450	150	9-63	NWSC (2008)
Faecal coliforms (cfu/100 mL)	$1x10^5$	$1x10^5$	$6.3x10^4$ - $6.6x10^5$	NWSC (2008)
Helminth eggs (Numbers/L)	2,500	4,000-5,700	-	Heinss *et al.* (1994)
	20,000-60,000	4,000	300-2,000	Heinss *et al.* (1998)
		600-6,000		Ingallinella *et al.* (2002)
		16,000		Yen-Phi *et al.* (2010)

2.4 OPERATIONAL FACTORS THAT IMPACT THE VARIABILITY OF FAECAL SLUDGE

The wide variability of observed FS characteristics is due not only to the range of different onsite technologies used, but also the way in which the system is used, the storage duration (filling rates and collection frequencies), inflow and infiltration, and the local climate. All of these factors should be taken into account when determining FS characteristics.

2.4.1 Toilet usage

Household habits associated with toilet usage influence the variability of FS in the onsite containment technology. The TS concentration is dependent on factors such as dry versus flush toilet, volume of flush water used, cleansing method ('washers' versus 'wipers'), and inclusion or exclusion of grey water from bathing or cooking. The fat, oil and grease concentration will increase with inclusion of kitchen wastewater without properly maintained oil and grease traps, and odors will also increase with additional organic waste streams. The filling rate will increase as more waste streams enter the toilet (e.g. solid waste from kitchen, rubbish), and with the number of people using the toilet. People also use additives to attempt to reduce filling rates such as supplemental microorganisms, salt, sugar, ash, fertiliser, and kerosene. Some additives can be quite harmful, and in general have not been found to be effective (Foxon *et al.*, 2012).

2.4.2 Storage duration

The filling rate and storage duration depends on the type of technology, quality of construction, toilet usage, and inflow and infiltration. The length of time that FS is stored in onsite containment systems before being collected and transported will greatly affect the characteristics due to the digestion of organic matter that occurs during storage. For example, in informal settlements in general, a large proportion of the population relies on public latrines that are highly frequented, and so require frequent emptying. In Kampala, 30 individuals (or 7 households) are on average sharing one single latrine (Günther *et al.*, 2011). In Kumasi, Ghana, 40% of the population relies on unsewered public toilets, which are emptied every few weeks. Hence, FS collected from public latrines tends to not be stabilised, and have high concentrations of BOD and NH_4^+-N (see Table 2.3). The emptying frequency of septic tanks varies greatly based on the volume and number of users, and can be anywhere from weeks to years. FS that has been stored in a septic tank for a period of years will have undergone more stabilisation than FS from public toilets. During the filling of onsite containment systems, the FS gets denser at the bottom due to compaction. This FS is more difficult to remove by pumping, and is therefore frequently not emptied and left at the bottom of the containment system.

2.4.3 Inflow and infiltration

The concentration and volume of FS is also greatly influenced by inflow and infiltration of leachate into the environment from the system and / or ground water into the system. The filling rate of systems will be slower if there is more leaching, resulting in a thicker FS. The permeability of containment systems is influenced by whether they are unlined, partially lined, completely lined, connected to drainfields or soakpits, and the quality of construction. If systems are permeable, the amount of inflow and infiltration will be influenced by the type of soil and the groundwater level. The exchange of ground water with FS can result in groundwater contamination which is made worse during periods of heavy and extensive rain due to an increase in flooding and the rising of the groundwater table. This is of particular concern in low-income countries where pit latrine and septic tank builders frequently come from the informal sector, and are not aware of the consequences of FS leaching into ground water, or may not have the means to determine the groundwater level.

2.4.4 Collection method

The FS collection method also influences its characteristics. FS at the bottom of containment systems that is too thick to pump will only be collected if it is manually emptied with shovels, or if water is added to decrease the viscosity and enable pumping (see Figure 2.2.). Pit latrines which are unlined or partially lined also frequently require flushing with large amounts of water in order to pump the FS, as liquid leaching from the pit increases the thickness of the FS. FS that has been removed by pumping is generally more dilute and less viscous than FS that is manually collected.

FS emptied from septic tanks will be more dilute if more supernatant than sludge is collected, or if the pump is not strong enough to remove all of the accumulated sludge. For example, in Dakar, Senegal, 83% of collection and transport vehicles are equipped with pumps and not strong vacuums, and are therefore unable to remove solids settled at the bottom of septic tanks (Diongue, 2006; Sonko, 2008). Where soak pits are used for the infiltration of septic tank effluent, these may also require emptying of sludge due to clogging. Collection methods are covered in more detail in Chapter 4, Methods and Means for Collection and Transport.

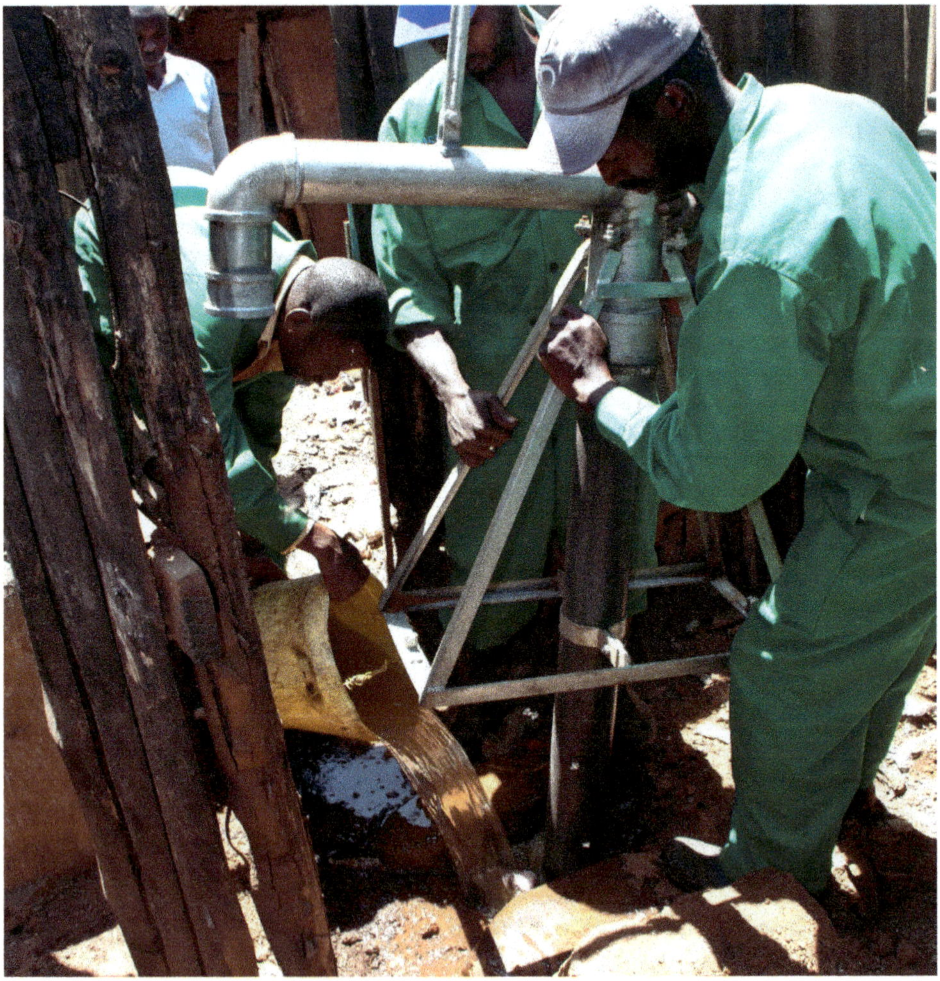

Figure 2.2 Addition of water to aid in collection of feacal sludge from a pit latrine utilising a "Gulper" in the Kibera informal settlement, Nairobi, Kenya (photo: Linda Strande).

2.4.5 Climate

Climate has a direct influence on FS characteristics, mainly due to temperature and moisture. Tropical countries may have one season of heavy rainfall, referred to as the wet season, while others have a bi-modal rainfall and/or dry season. Temperatures may be at their lowest during the wet season and at their highest during the dry season. Frequently the highest demand on collection and transport services occurs during the rainy season, as heavy rainfalls result in overflowing and flooding of onsite systems. Rates of biological degradation are also temperature dependent, and rates increase with warmer temperatures.

2.5 TREATMENT TARGETS

The main objective of FS treatment is to ensure the protection of human and environmental health. Legislation that establishes regulations specifically for the treatment and discharge, enduse, or disposal of FS is therefore essential. Institutional frameworks are described in more detail in Chapter 12. However, in many cases legislation specific to the treatment of FS is borrowed from wastewater treatment legislation such as National Wastewater Discharge Standards or Environmental Protection Agency Guidelines, which do not take the very different nature of FS into account. Treatment targets for FS should be set based on the intended enduse or disposal goal of the sludge, and enduse or discharge of liquid effluents. A multi-barrier approach is preferred over setting prescriptive, target-based requirements, and is covered in more detail in Chapter 10, Enduse of Treatment Products.

2.6 TREATMENT OBJECTIVES

Dewatering (or "thickening") of FS is an important treatment objective, as FS contains a high proportion of liquid, and the reduction in this volume will greatly reduce the cost of transporting water weight and simplify subsequent treatment steps. Environmental and public health treatment objectives are achieved through pathogen reduction, stabilisation of organic matter and nutrients, and the safe enduse or disposal of treatment endproducts.

2.6.1 Dewatering

Common methods for dewatering of FS include gravity settling, filter drying beds, and evaporation / evapotranspiration. FS has different dewatering characteristics compared to wastewater sludge, in that it tends to foam upon agitation, and resist settling and dewatering (USEPA, 1999). The duration of onsite storage, and the age of FS also affects the ability to dewater the sludge. Empirical evidence shows that 'fresh' or 'raw' FS is more difficult to dewater than older, more stabilised FS. The dewatering, or thickening process can also include adding dry materials such as sawdust to increase the solids content, This is a common practice in processes such as composting where the sawdust also increases the carbon to nitrogen (C:N) ratio. The liquid stream that is produced during dewatering also requires further treatment, as it can be high in ammonia, salts, and pathogens. Dewatering mechanisms are presented in further detail in Chapter 3, and treatment technologies in Chapters 5-8.

2.6.2 Pathogens

FS contains large amounts of microorganisms, mainly originating from the faeces. These microorganisms can be pathogenic, and exposure to untreated FS constitutes a significant health risk to humans, either through direct contact, or through indirect exposure. FS needs to be treated to an adequate hygienic level based on the enduse or disposal option. For example, exposure pathways are very different for treated sludge discharged to the environment, used in agriculture, or combusted as a fuel. Pathogens are covered in more detail in Section 2.10. Mechanisms for pathogen reduction and / or inactivation include starvation, predation, exclusion, desiccation, partitioning, and temperature, and are covered in more detail in Chapter 3.

2.6.3 Nutrients

FS contains significant concentrations of nutrients, which can be harnessed for beneficial resource recovery, but if not properly managed can result in environmental contamination. The nutrients in FS can supplement synthetic nitrogen based fertilisers that are heavily dependent on fossil fuels and phosphorus, which is a mined resource of which finite supplies are estimated to reach their peak availability within 100 years (the point at which demand outstrips supply) (Bentley, 2002; Steen, 1998). Environmental impacts from nutrients include eutrophication and algal blooms in surface waters (see Figure 2.3), and contamination of drinking water (e.g. nitrates leading to methemoglobinemia). Nutrients are covered in more detail below in Section 2.9.1. Further information on concerns and benefits of resource recovery of FS treatment endproducts are presented in Chapter 10.

2.6.4 Stabilisation

Untreated FS has a high oxygen demand due to the presence of readily degradable organic matter that consumes significant amounts of oxygen during aerobic respiration. If FS is discharged to the environment, it can result in depletion of oxygen in surface waters. The process of stabilisation results in a FS containing organic, carbon-based molecules that are not readily degradable, and which consists of more stable, complex molecules (e.g. cellulose and lignin). Stabilisation is achieved through the biodegradation of the more readily degradable molecules, resulting in a FS with a lower oxygen demand. Common indicators of stabilisation include measurement of Volatile Suspended Solids (VSS), BOD, and COD. In addition, stabilisation ensures that organic forms of nutrients present in treatment endproducts are stable, and can be more predictably and reliably used. Stabilisation also reduces foaming of FS, leading to better dewatering. Stabilisation is explained further in Chapter 3, Treatment Mechanisms.

Figure 2.3 Eutrophic river resulting from direct discharge of untreated faecal sludge and wastewater, Yaoundé Cameroon (photo: Linda Strande).

2.7 TREATMENT CONCERNS

The source of FS coming into a FSP should be monitored to ensure that toxic constituents are not introduced from industrial sources. Heavy metals are not removed during the treatment process, and it is therefore important to avoid contamination of the FS in the first place. Heavy metals are not usually a concern when dealing with domestic FS as these compounds typically come from industrial sources, although some contamination can occur from domestic sources, if for example batteries are disposed of into the toilet. Leachate from sludge drying beds and stabilisation ponds can be high in salinity. This is of concern if the effluent is to be used for irrigation due to impacts on plant growth, reduced soil permeability, and surface crusting. Use of a manifest system for tracking the source of FS is covered in Chapter 11, Operations and Maintenance, and the impact of metals and salinity is discussed in Chapter 10, Enduse of Treatment Products.

2.8 SAMPLING PROCEDURES AND PROGRAMMES

When characterising FS, the quality of the results are strongly influenced by the way in which samples are collected, as well as the laboratory methods and practices followed. This is exacerbated by the difficulty of trying to sample from within a closed onsite system. FS in containment structures is normally not mixed, so the sludge tends to form a scum layer on the top and is often more dense at the bottom. Where and how the sample is taken depends on the question that is being asked, and the sampling method should be reported alongside the results. For example, if the goal is to understand FS characteristics within the containment structure, versus understanding characteristics of FS that will be transported by trucks to treatment plants. For the former sampling would be done directly from the containment structure, for the later from the collection and transport truck. If the goal is to characterise FS within the containment structure, then a representative sample should be collected by taking multiple samples in the different zones (i.e. top, middle, bottom), and making a volumetrically representative composite sample.

If the goal is to characterise FS that will be delivered to a treatment plant, grab samples can be taken directly from the collection and transport truck at various intervals as the truck is discharging, and then mixed to obtain a composite sample (see Figure 2.4). The most accurate method is to take grab samples over set time periods (e.g. every 2 minutes), but samples can also be taken at the beginning, middle and end of discharge since the exact volume in the truck is not known (this is a more qualitative measurement, but has been shown to be reasonably accurate). The volume of the FS sample should be proportional to the volume of the truck from which the sample is taken as is carried out for flow proportional wastewater sampling (Vonwiller, 2007). If feasible, samples can also be taken following discharge to a storage tank fitted with a mixer in order to obtain a representative sample. Once FS is collected, settling occurs rapidly and this needs to be taken into consideration when selecting the timing and interval of sludge collection.

Other aspects that need to be taken into account include the need for sampling to take place over a short period of time, for the sample container to be closed to prevent volatilisation or contamination, and for the sample to be kept cool to prevent microbial activity. Samples should be analysed within eight hours from the time of collection, or if this is not possible, the sample should be preserved by refrigeration or freezing, or by the addition of a chemical fixative, depending on the standard method for the parameters to be measured.

Figure 2.4 Sampling campaign to quantify and characterise faecal sludge in Ouagadougou, Burkina Faso for the design of a new faecal sludge treatment plant. Currently, none exist in the city and the only option for collection and transport companies is direct discharge (photo: Hanspeter Zoellig).

Case Study 2.2: Difficulties associated with sampling for FS characterisation

(Kartik Chandran and Melanie Valencia, Columbia University, New York)

Starting in 2011, the Bill & Melinda Gates foundation funded a project to assess the possibility to produce biodiesel from FS. A partnership between KNUST in Kumasi-Ghana, Columbia University in New York and Waste Enterprisers was formed. The first step was the characterisation of f FS in Kumasi. A group of graduate students from KNUST conducted the "100 sample study" to assess solid content, ammonia, lipid content, pH and COD. The comparison was done among pit latrines, private and pubic toilets and was collected at site of discharge by the trucks. The analysis determined that toilets were the best feedstock for the purpose of maintaining a healthy fermenter system and getting a better yield of long chain fatty acids that could be precursors for biodiesel.

As the project evolved, volatile fatty acids were also considered as precursors for lipid production. This raised some concerns in terms of transporting the samples from the site to the laboratory due to the volatilisation of the material in question as well as the length of time between obtaining the sample and testing. The sample vials would then be filled up to the rim and taken to the laboratory immediately for same day VFA measurement. To obtain a homogeneous sample, especially for VFA production in the anaerobic digesters, four out of the six fermenters were mixed through pumps. Samples for gas production were collected before mixing while FS samples were collected afterwards.

Several managerial challenges were also encountered during the collection of the loads as well as during sampling. The facility is located within a pond system that serves as the wastewater treatment facility for the municipality. The first challenge was to get the trucks to discharge their FS in the experimental system rather than in the ponds. At the beginning, 5 USD (GHC10) were given to the truck drivers in compensation for their time, as it took longer to discharge the truck into the experimental system compared to the ponds. Nonetheless, during the rainy season many truck drivers refused to deliver FS because the road into the facility was damaged due to the rain. The incentive of 5 USD was no longer enough. Only a few truck drivers that became acquainted with the project continued to provide FS. Once the rainy season was over, trucks continued to not come to the facility even when the road had been fixed. Many alternatives to address this situation were considered, including making the sludge discharge opening larger for faster emptying. In the end, the fermentation team opted for doubling the pay. This attracted many more drivers and the word spread fast. More information about this and other FSM projects can be found at www.susana.org.

Case Study 2.3: Faecal sludge sample selection and characterisation from onsite sanitation facilities in eThekwini Municipality, Durban

Characteristics of faecal sludge may vary greatly between different locations and types of facilities. In order to verify how greatly, the Pollution Research Group (PRG) at the University of KwaZulu-Natal in Durban, South Africa carried out a study of FS properties from different types of onsite sanitation facilities in the Durban Metro area, including urine diverting toilets (UDDTs), household (HH) and community ablution block (CAB) VIP latrines. The first phase involved a sampling programme (Table 2.4) to empty pits and to obtain sludge samples from selected onsite sanitation facilities followed by chemical, physical, mechanical and biological properties analyses.

A sampling method was developed for different depths at the 'front' and 'back' sections of the pit for all dry VIPs and UDDTs. The wet VIPs had a high liquid level and the sludge was concentrated as a crusty layer at the surface of the liquid. Samples were selected from the 'crust' and from the liquid beneath the sludge layer; no distinction was made between the 'front' and the 'back' of the pit. The CAB pits were full of liquid similarly to the wet VIPs. On average, eight samples were selected from each dry VIP, between four to six samples from each wet VIP, two to six samples from each UD toilet (active and inactive vault) and about 12 samples from each CAB. The samples of about one litre were stored in plastic containers at 4 °C for further analytical tests, including: Chemical Oxygen Demand (COD), ammonia, Total Kjeldahl Nitrogen (TKN), (ortho)phosphate, potassium, water content, total, volatile, fixed and suspended solids, sludge volume index (SVI), pH, thermal conductivity, specific heat, rheology (viscosity), density, particle size distribution, calorific value, heavy metals and parasites.

Some of the analytical results are summarised in Table 2.5 as an average of all samples. They indicated a relatively high moisture content of all VIP samples – about 80%. Expectedly, the UDDT samples had a lower moisture content (60%). The combination of moisture content (respectively total solids) and viscosity of the materials provides a useful starting point for assessment and design of mechanical pit emptying devices. The maximum measured sludge viscosity from a dry VIP was about $6x10^6$ Pa.s, with an average value of about $3x10^5$ Pa.s. However, these values are based on the untreated FS: for pit emptying purposes the content of non-faecal wastes should also be taken into account. The COD values were the lowest in the UD toilets.

Table 2.5 Average values of faecal sludge properties based on lab analyses

Type	Moisture %	SS mg/L	VSS g/g DW	Ash g/g DW	SVI ml/mg	pH -	COD mg/g DW	NH_4-N mg/g DW	TKN mg/g DS	PO_4-P mg/L	P_{tot} mg/L	Thermal conductivity W/mK	Calorific value MJ/kg	Density kg/m^3
Dry VIP	83	381	0.57	0.43	0.11	7.6	680	13	40	0.73	3.86	0.54	14.06	1,356.5
Wet VIP	79	562	0.54	0.46	0.04	7.7	720	7	30	0.83	2.93	0.55	13.08	1,443.1
CAB VIP	77	139	0.49	0.51	0.51	7.4	650	3	30			0.60	14.31	1,350.1
UD	60	246	0.45	0.55	0.23	7.5	490	5	30	1.00	3.27	0.38	12.93	1,450.4

Pit sludge categorisation was carried out at WWTP site near Tongaat, north of Durban. Three sludge samples were selected from a dry VIP pit and a urine diverting (UD) toilet (from active and inactive vaults). Manual sorting by different material categories was carried out, the weight of each of them measured and then calculated as a per cent of total wet mass. An example of screening FS during a sampling event is presented in Figure 2.5.

Although different materials were identified during the manual sorting, over 85% by wet mass was determined to be organics which was FS. The second most prevalent category was 'paper', about 7-8% total wet mass, followed by textiles (1-2%). Feminine hygiene products were about 1% of the total mass. Further investigation of the particle size distribution between the different material categories is expected to provide information about the size range of those components and estimate their impact on the pit emptying devices.

2.9 PHYSICAL-CHEMICAL CONSTITUENTS

This section provides an overview of the commonly measured constituents for the characterisation of FS and the typical methods of quantification. Further information on the fundamentals of their transformations is presented in Chapter 3: Treatment Mechanisms. Methods need to be rigorourslу adapted for FS, but a detailed description of the analytical methods for wastewater can be found in references such as Standard Methods, e.g. APHA (2005).

2.9.1 Nutrients

Excreta contains nutrients that originate from food consumption. Of the total nitrogen, phosphorus and potassium that is consumed, 10-20% of nitrogen, 20-50% of phosphorus, and 10-20% potassium is excreted in the faeces, and 80-90% of nitrogen, 50-65% of phosphorus, and 50-80% of potassium in the urine (Berger, 1960; Lentner *et al.*, 1981; Guyton, 1992; Schouw *et al.*, 2002; Jönsson *et al.*, 2005; Vinnerås *et al.*, 2006). Ammonia (NH_3) is produced by deamination of organic nitrogen, and hydrolysis of urea ($CO(NH_2)_2$) in urine by urease. The majority of ammonia in raw FS comes from the urine

(Mitchell, 1989; Jönsson *et al.*, 2005). The nitrogen content in faeces is about 20% as ammonia, 17% as organic nitrogen in the cells of living bacteria, and the remainder as organic nitrogen (e.g. proteins, nucleic acid) (Lentner *et al.*, 1981).

Nitrogen

Nitrogen is an important parameter to consider in FS treatment, as the total nitrogen concentrations in FS is typically quite high (e.g. 10-100 times the concentration in domestic wastewater). Depending on factors such as pH, length of storage, the presence of oxygen, and the type of FS, nitrogen will be present in a combination of the following forms; ammonium (NH_4-N)/ammonia (NH_3-N), nitrate (NO_3-N)/nitrite (NO_2-N), and organic forms of nitrogen (e.g. amino acids and amines).

When quantifying ammonia concentrations in FS, a preliminary distillation step is required, followed by a titrimetric method, electrode, or a phenate method (APHA, 2005). To prevent volatilisation, samples must be refrigerated and analysed within 24 hours, or frozen or acidified to pH 2 for analysis within 28 days (APHA, 2005). Total Kjeldahl Nitrogen (TKN) is the sum of organic nitrogen and ammonia (NH_3-N)/ammonium (NH_4-N). TKN can be determined by the macro-kjeldahl method, the semi-micro-kjeldahl method, or block digestion and flow injection analysis. Nitrate and nitrite can be determined by ion chromatography, capillary ion electrophoresis, cadmium reduction, hydrazine reduction, cadmium reduction flow injection, or UV spectrophotometric method (APHA, 2005). There are also commercially available kits for these analyses, which are commonly used. Total nitrogen can be determined by oxidative digestion to nitrate followed by quantification of the nitrate, or by the sum of TKN and nitrate/nitrite results.

Figure 2.5 Municipal solid waste removed from faecal sludge with bar screens at treatment plant influent during a sampling event in Kampala, Uganda (photo: Daniel Ddiba).

Phosphorus

The concentration of phosphorus is also an important parameter to consider, as the total phosphorus concentration in FS is quite high (e.g. 2-50 times the concentration in domestic wastewater). Phosphorus in FS will be present as phosphate, the acid or base form of orthophosphoric acid (H_3PO_4 / PO_4-P), or as organically bound phosphate (e.g. nucleic acids, phospholipids and phosphorylated proteins). The fate of phosphorus in the various treatment processes will be based on factors such as sorption, precipitation, complexation, sedimentation, mineralisation, pH, plant uptake in planted drying beds, and redox potential.

Phosphate can be determined colorimetrically to determine 'reactive' phosphorus, or following hydrolysis or digestion to quantify total phosphorus, including particulate and organic fractions (APHA, 2005).

2.9.2 pH

Measurement of pH is essential for the understanding of water chemistry processes, such as acid-base chemistry, alkalinity, neutralisation, biological stabilisation, precipitation, coagulation, disinfection, and corrosion control (APHA, 2005). The pH of FS from septic tanks is normally in the range of 6.5 to 8.0 (Ingalinella *et al.*, 2002; Cofie *et al.*, 2006; Al-Sa'ed and Hithnawi, 2006), but can vary greatly from 1.5 to 12.6 (USEPA, 1994). A pH outside the range of 6 to9 indicates an upset in the biological process that will inhibit anaerobic digestion and methane production. This could result from a change in the hydraulic loadings, the presence of toxic substances, a large increase in organic loading, or that the systems are receiving industrial or commercial wastewater. The pH can be measured with electrodes and meters and pH papers (APHA, 2005).

2.9.3 Total solids

TS concentration of FS comes from a variety of organic (volatile) and inorganic (fixed) matter, and is comprised of floating material, settleable matter, colloidal material, and matter in solution. Grit, sand and municipal waste are discussed below in Section 2.9.6. Parameters that are typically measured include total solids, fractions of volatile or fixed solids, and settleable, suspended or dissolved solids.

The total solids (TS) are quantified as the material remaining after 24 hours of drying in an oven at 103-105 °C. Volatile solids (VS) are the fraction that are ignited and burned off at a temperature of 500 °C, and this fraction is also considered to be the organic portion. The fixed solids are the amount remaining after ignition, and are generally considered to be the inorganic portion. The ratio of VS to TS is used as an indicator of the relative amount of organic matter and the biochemical stabilty of FS. Total solid values are important as they are used to design and dimension FS treatment technologies such as planted and unplanted drying beds.

The suspended solids fraction of the FS are defined as the solids that are not able to pass through a filter, while the solids that do pass through are termed dissolved solids. Since these values are dependent on the pore size of the filter that is used it is important to report filter size with suspended solids data. A 0.45 μm filter is typically used for the analysis of wastewater effluents, but filters up to 2.0 μm can be used. If FS is too dense to pass through filters, then total solids concentrations are more commonly reported.

Solids that settle out of suspension after a certain period of time, for example, the solids that accumulate in the bottom of an Imhoff cone after 30 to 60 minutes, are termed settleable solids. This value is reported as the sludge volume index (SVI), and is used for designing settling tanks. More information on solids measurements can be found in Chapter 9.

2.9.4 Biochemical Oxygen Demand and Chemical Oxygen Demand

The oxygen demand of FS is an important parameter to monitor, as the discharge of FS into the environment can deplete or decrease the oxygen content of water bodies resulting in the possible death of aquatic fauna. The oxygen demand is reduced through stabilisation, and can be achieved by aerobic or anaerobic treatment. FS dewatering technologies do not necessarily decrease oxygen demand.

BOD is a measure of the oxygen used by microorganisms to degrade organic matter. The standard method for detecting BOD involves incubation at 20°C for 5 days, and is reported as BOD_5 in mgO_2/L. Wastewater is considered to be weak, medium, strong and very strong respectively at BOD_5 of 200, 350, 500, and >750 mg/L (Mara, 2004). As shown in Table 2.3, FS typically has a much higher BOD_5 than that of 'strong' wastewater. Non carbonaceous material can also consume oxygen, for example the oxidation of ammonia to nitrate, which can increase the reported BOD value if not taken into account. To prevent this, nitrification can be inhibited through the addition of chemicals. The particle size distribution also has an effect, as smaller and more soluble particles have faster BOD reaction rate coefficients. Other factors that can account for sample variability include sample filtration, dilutions, and sampling methodologies.

BOD only represents biodegradable organics, whereas COD represents the oxygen equivalent of the organic matter that can be oxidised chemically with dichromate, a powerful chemical oxidant. The laboratory analysis of COD is more convenient than that for BOD, taking between a few minutes to hours depending on the method.

COD concentrations will be higher than BOD for a number of reasons including:

- complex organic molecules like lignin which are resistant to biodegradation, being oxidised by COD;
- some inorganic substances also being oxidised by COD;
- inhibition of bacteria in the BOD test.

COD is determined in the laboratory with an open or closed reflux method, and commercial kits are also readily available (APHA/AWWA/WEF, 2005). The ratio of BOD to COD can also be used as an indicator of the relative biodegradability of the organic matter in different waste streams. BOD, COD, and measurements of COD fractionation are explained in detail in Chapter 9. Other metrics of organic carbon not included here are measures of total organic carbon (TOC), and specific organic compounds of concern.

2.9.5 Oil and grease

Fats, oil and grease in FS comes from a wide variety of sources including lard, meats, seeds and nuts, kerosene, and lubricating oils. Oil and grease content is important to consider because it can reduce microbial degradation due to reduced solubility, increase the scum layer in settling tanks, cause maintenance problems, and create films in surface waters if discharged. Extraction with solvents can be used to determine concentrations of oil and grease, and is reported as total oil and grease soluble in the specific solvent that is used (APHA, 2005).

2.9.6 Grit and sand

Grit and sand concentrations are important to consider in the treatment of FS, as their presence influences the required size and filling rates of tanks used for storage and treatment, and can increase the frequency of clogging in pipes and pumps. Sources of grit and sand include unlined pit latrines, cleaning and washing of utensils and vegetables, cleansing (e.g. sand tracked into house), and flooding. In areas where sand is pervasive in the FS, it is a significant treatment and design concern (e.g. Dakar,

Senegal; Case Study 2.4). In order to reduce the sand and grit content, sand traps should be installed at entry points to pipes and sinks. Concentrations of sand can be determined by first drying the sample at 105 °C in an oven, then at 550 °C to obtain the total fixed solids. The ashes are then treated with a hot mixture of nitric acid and hydrochloric acid. The amount of sand is obtained after filtration following calcination at 1,000 °C.

2.9.7 Municipal solid waste

Municipal solid wastes are deposited in sanitation containment systems for a number of reasons, such as the lack of a functional system for the collection and management of municipal solid waste. In addition, menstrual hygiene products and baby diapers are commonly thrown into sanitation systems. As shown in Figure 2.6 the accumulation of these solid wastes can be significant and should be strongly discouraged through educational campaigns. Solid wastes can cause problems in the collection and transport of FS (Chapter 4), result in clogged pipes and pumps, increase required storage and treatment volumes, and affect the end quality of treatment products.

In order to prevent clogging at treatment plants, the installation of a bar screen is essential at the influent. It has been found that organic decomposable wastes are the largest constituent in the screenings from FS (Troschinetz and Mihelcic, 2009) and typically accounts for 48% of the total waste. Other constituents include pebbles, rubble, sand and fine particles (29%), iron, wood and textiles (20%), and plastics (3%) in mass (dry weight) percentage (Rouyat *et al.*, 2006). Similar results were observed in Dakar, and are presented in Case Study 2.4.

Figure 2.6 Municipal solid waste removed from faecal sludge with bar screens at treatment plant influent, Dakar, Senegal (photo: Linda Strande).

Case Study 2.4: Sand and solid waste content in faecal sludge in Dakar, Senegal; increased collection costs in Kampala, Uganda

Dakar

A study conducted in Dakar, Senegal found the average sand content in FS from septic tanks to be 935 g/m^3 (mg/L), with a range of 90 g/ m^3 to 4,000 g/m^3. In poor areas of Dakar, yards are typically bare soil and access roads are unpaved. A large amount of sand is therefore tracked into latrines by bare feet or shoes. This amount increases further in the rainy season, when wet sand adheres more readily to shoes and feet. Showers are also commonly located in the latrine superstructure. In addition, the predominance of squat toilets facilitates the direct introduction of sand in the system (M'Voubou, 2004).

When passing FS through a 2 cm diameter grid, the following composition was found for the solid waste content (mass dry weight percentage after 2 days of drying in the sun): sponges, bones, wood 1% each; textiles 2.5%; plant seeds 3%; stones 11%; plastics 12%; sand 25% and decaying (organic) matter 43%.

Kampala

Next to creating problems in pit emptying and FS treatment, additional solid waste in pits leads to higher emptying fees. In Kampala, the pit emptying fees are based on the volume of removed material and the travelling distance between the emptied system and the discharge location. The presence of accumulated solid wastes in a sanitation system results in a 10-50% increment on top of the emptying cost. Table 2.6 provides the typical costs for pit emptying in a 5 km travel radius and the increased costs due to this penalty.

Table 2.6 Typical costs for pit emptying in a 5 km travel radius plus increased costs due to a rubbish penalty

Truck Capacity (m^3)	Standard Costs (USD)	Penalty for rubbish (10 - 50%) (USD)	Range of total costs including rubbish fine (10 - 50%) (USD)
≤ 1.8	28	2.80 - 14	30.80 - 42
2.0 - 2.7	32	3.20 - 16	35.20 - 48
3.6 - 4.0	40	4.00 - 20	44.00 - 60
4.5 - 7.2	48	4.80 - 24	52.80 - 72
8.0 - 11	64	6.40 - 32	70.40 - 96

2.10 PATHOGENS IN FAECAL SLUDGE

Exposure to untreated FS should always be considered as a pathogenic health risk. Adequate reductions in pathogens need to be determined based on the intended enduse or disposal option for treated sludge and liquid effluents. This is described in more detail in Chapter 10, Enduse of Treatment Products. Some common pathogens of concern that may be excreted in faeces, and their importance in disease transmission, are presented in Table 2.7.

Table 2.7 Selected pathogens that may be excreted in faeces and related disease symptoms (adapted from Schönning and Stenström, 2004)

Group	Pathogen	Disease symptoms
Bacteria		
	Aeromonas spp.	Enteritis
	Campylobacter jejuni/coli	Campylobacteriosis - diarrhoea, cramping, abdominal pain, fever, nausea, arthritis, Guillain-Barré syndrome
	Escherichia coli (EIEC, EPEC, ETEC, EHEC)	Enteritis. For EHEC there are also internal haemorrhages that can be lethal
	Salmonella typhi/paratyphi	Typhoid/paratyphoid fever – headache, fever, malaise, anorexia, bradycardia, splenomegaly, cough
	Salmonella spp.	Salmonellosis – diarrhoea, fever, abdominal cramps
	Shigella spp.	Shigellosis – dysentery (bloody diarrhoea), vomiting, cramps, fever; Reiters syndrome
	Vibrio cholera	Cholera – watery diarrhoea, lethal if severe and untreated
Virus		
	Adenovirus	Various; respiratory illness, here added due to enteric types (see below)
	Enteric adenovirus types 40 and 41	Enteritis
	Enterovirus types 68-71	Meningitis; encephalitis; paralysis
	Hepatitis A	Hepatitis – fever, malaise, anorexia, nausea, abdominal discomfort, jaundice
	Hepatitis E	Hepatitis
	Poliovirus	Poliomyelitis – often asymptomatic, fever, nausea, vomiting, headache, paralysis
	Rotavirus	Enteritis
Parasitic protozoa		
	Cryptosporidium parvum	Cryptosporidiosis – watery diarrhoea, abdominal cramps and pain
	Cyclospora histolytica	Often asymptomatic; diarrhoea; abdominal pain
	Entamoeba histolytica	Amoebiasis – often asymptomatic, dysentery, abdominal discomfort, fever, chills
	Giardia intestinalis	Giardiasis – diarrhoea, abdominal cramps, malaise, weight loss
Helminths		
	Ascaris lumbricoides	Generally no or few symptoms; wheezing; coughing; fever; enteritis; pulmonary eosinophilia
	Taenia solium/saginata	Taeniasis
	Trichuris trichura	Trichuriasis - Unapparent through to vague digestive tract distress to emaciation with dry skin and diarrhoea
	Hookworm	Itch; rash; cough; anaemia; protein deficiency
	Schistosoma Spp. (blood fluke)	Schistosomiasis, bilharzias

2.10.1 The use of indicators

When measuring the treatment efficiency of FS treatment processes, it is too expensive and labour intensive to measure all types of pathogens. Instead, it is common practice to select indicators of pathogenic activity which are measured to provide an indication of the level of pathogen removal during treatment. These indicators can be either pathogenic, or non-pathogenic, but the organisms need to be carefully selected in order to provide adequate information on the inactivation of pathogens. The following requirements should be met when selecting an indicator (Mara, 2004):

- be exclusively of faecal origin;
- be in numbers greater than those of the pathogens of concern;
- have a removal that mimics, and is close to, that of pathogens of concern; and
- be simple, inexpensive, accurate, and reliable to measure.

In addition, indicator organisms should have the ability to survive longer than the pathogen of concern. Some of the typical indicator organisms for wastewater, FS and environmental contamination are the coliform bacteria, helminths and baceteriophage (as an indicator for viruses). Other indicators of faecal contamination that have been used in pollution control and pathogen die-off studies include faecal streptococci, Klebsiella, *Clostridium perfringens, Bacteroides*, Enterococci, and *Bifidobacterium* (Feachem *et al.*, 1983; WHO, 1984).

2.10.2 Coliform bacteria

Coliform bacteria are bacteria that populate the intestinal tract, and are pervasive in faeces. Their presence in the environment is therefore used as an indicator of faecal contamination. *Escherichia coli* (*E. coli*) is the target organism that has traditionally been used to identify faecal contamination in the environment (Feachem *et al.*, 1983). However, there are complicating factors such as bacteria from the genus Escherichia that can grow in the environment, and the test is therefore not 100% indicative of faecal contamination. Tests have been developed to enable the quantification of total coliforms, faecal coliforms, and *E. coli*. However since these bacteria are not indicators of viral or protozoan contamination, and although they are used as indicators of faecal pollution in the environment, they do not necessarily provide a good indication of pathogen reduction in FS treatment processes.

Total and faecal coliforms are not good indicators in tropical and sub-tropical climates. The standard method for coliforms analysis relies on the production of acid and gas in medium incubated at a temperature equal to that of the human body (37 °C). The standard method of analysing thermal tolerant faecal coliforms relies on their production of acid and gas from lactose when incubated at 44 °C. However, under tropical and sub-tropical conditions, it has been found that coliforms, which are not necessarily faecal, also grow and produce acid and gas from lactose at 44 °C (Mara, 2004). Enzymatic and biochemical assays have been developed for the detection of *E. coli* (APHA, 2005), and commercial kits are also available for total and faecal coliforms and *E. coli*.

Helminths

When analysing FS, helminths are most commonly used as an indicator of the effectiveness of pathogen reduction due to their prevalence in low- and middle-income countries, and their persistence following treatment. Helminths ('parasitic worms') are eukaryotic parasites, which are prevalent in about one third of the world's population. Helminths include nematodes (round worms), cestodes (flat worms) and trematodes (flukes). These are important pathogens to monitor, as eggs from one infected person can infect hundreds of people. *Ascaris lumbricoides*, a type of round worm, is the most commonly used indicator, as the eggs are one of the pathogens most resistant to inactivation in treatment processes, and can be identified relatively easily (Feachem *et al.*, 1983). The ability of *Ascaris lumbricoides* eggs to remain viable stems from a highly impermeable eggshell, which is considered to be one of the most resistant biological structures. The shell allows the passage of essential respiratory gases while protecting the eggs from a wide array of chemicals and extreme pH conditions (Nordin *et al.*, 2009). The monitoring of the removal of Ascaris eggs therefore provides an indication that less resistant pathogens have also been inactivated (Fgure 2.7).

To use Ascaris eggs as a metric of pathogen removal, the number of eggs can be enumerated, but a more sensitive metric is to measure the viability of eggs. A viable egg is one that still has the potential to develop, versus eggs which are no longer viable and have no risk of pathogen transmission. The enumeration of viable Helminth eggs employs the coproscopic method, involving sedimentation, desorption, centrifugation and floatation. During development, this method was estimated to achieve an efficiency of 30-70% in the enumeration of viable eggs (Gaspard and Schartzbroad, 1995). However, further work has increased this to 100% based on a sensitivity of 0.4 ppm, by calculating the number of helminth eggs in the sample, and incorporating the estimated efficiency of the procedure (Malicki *et al.*, 2001). In addition, improvements in the method have significantly reduced the number of replications that are necessary based on previous methods (USEPA 1995).

The improved USEPA standard methods for the enumeration of helminths (2003) are based on 4 steps that include:

- parasite dissociation from organic matter;
- flotation with natrium nitrate solution;
- sedimentation; and
- concentration and microscopic examination using a Sedguick-Rafter counting chamber.

The South African Water Research Commission has also developed and published the 'Standard Methods for the Recovery and Enumeration of Helminth Ova in Wastewater, Sludge, Compost and Urine Diversion Waste in South Africa', which is available free of charge via the internet (Moodley *et al.*, 2008).

Figure 2.7 Analysis of viable Helminth eggs in Dakar, Senegal (photo: Linda Strande).

Viruses

Quantification of total viruses in FS can be undertaken using an electron microscope, but the easiest method of evaluation is to measure their effects on hosts (Madigan and Martinko, 2006). The bacteriophage are commonly used as viral indicators. Examples include *Salmonella typhimurium* bacteriophage 28B, enterobacteria phage MS2 and coliphage Øx174 using host strains respectively as *Salmonella typhimurium* phage type 5, *Salmonella typhimurium* WG 49 (ATCC 700730) and *E. coli* ATCC 1370. The standard double-layer agar method is used, with viruses quantified by the plaque assay with the agar overlay technique (Adams, 1959; Madigan and Martinko, 2006).

2.11 CONCLUSION

Based on the current state of knowledge, it is evident that caution must be exercised when making assumptions for quantifying and characterising FS for the design of treatment systems. Fortunately, FSM is a rapidly growing field, and there is currently a great deal of research being conducted into the characterisation of FS. In the next few years, as FSM gains in acceptance, and research results are obtained, further information will become available, allowing for the more accurate design of treatment systems.

Previous research into the characterisation of FS has focused on parameters for environmental protection and agricultural use, e.g. BOD/COD, TS, TVS, nutrients, and pathogen indicators. Current research into innovative enduses has increased this field to include parameters such as COD fractionation, lipid content for biodiesel, structural properties, and calorific value. Examples include the use of FS treatment endproducts as a potential energy source, such as the biogas produced from anaerobic digestion systems, or direct use of dried FS in industrial boilers and kilns (Murray Muspratt *et al.*, 2014). Other possibilities include treatment with black soldier flies to produce protein, and incorporation in building materials (Diener *et al.*, 2014). Other research that will assist in improving collection and transport methods, and understanding dewatering behavior, includes the analysis of rheological properties and shear strength (AIT, 2012; Radford, 2012).

2.12 BIBLIOGRAPHY

AIT (2012). Assessment of FS rheological properties. Final Research Report. Environmental Engineering Programme. School of Environment, Resources and Development, Asian Institute of Technology (AIT).

Al-Sa'ed, R. M. Y., Hithnawi, T.M. (2006). Domestic septage characteristics and co-treatment impacts of Albireh wastewater treatment plant efficiency. Dirasat Engineering Sciences 33(2), p.187-197.

APHA/AWWA/WEF (2005). Standard Methods for the examination of water and wastewater. 21st Edition. American Public Health Association, American Water Works Association, and Water and Environment Federation Publication. Washington D.C., USA. ISBN 0-87553-047-8.

Bassan, M., Tchonda, T., Yiougo, L., Zoellig, H., Mahamane, M., Mbéguéré, M., Strande, L. (2013a). Characterisation of faecal sludge during dry and rainy seasons in Ouagadougou, Burkina Faso. Refereed paper presented at the 36th WEDC International Conference, Nakuru, Kenya, 2013.

Bassan, M., Mbéguéré, M., Tchonda, T., Zabsonre, F., Strande, L. (2013b). Integrated faecal sludge management scheme for the cities of Burkina Faso. Journal of Water, Sanitation and Hygiene for Development 3(2), p.216–221.

Bentley, R.W. (2002). Global oil and gas depletion: An overview. Energy Policy 30(3), p.189-205.

Berger, E.Y. (1960). Intestinal absorption and excretion. In: Comar C. L. and Bronner F. (eds). Mineral Metabolism, p.249-286. Academic Press, New York.

Diener, S., Semiyaga, S., Niwagaba, C., Muspratt, A., Gning, J.B., Mbéguéré, M., Ennin, J.E., Zurbrugg, C., Strande, L. (2014). A value proposition: resource recovery from faecal sludge – can it be the driver for improved sanitation? Resources Conservation & Recycling (in press).

Diongue, S.T. (2006). Stratégies d'optimisation de la filière de vidange de boues de latrines à Dakar, Institut International du Génie de l'Eau et de l'Environnement (Burkina Faso)/Ecole Ploytechnique Fédérale de Lausanne (Suisse). Master: 81.

Feachem, R.G., Bradley, D.J., Garelick, H., Mara, D.D. (1983). Sanitation and Disease. Health aspects of excreta and wastewater management. World Bank studies in water supply and sanitation. John Wiley and Sons. New York.

Foxon, K., Brouckaert, C., Wood, K. (2012). How fast do pits fill up? Empirical evidence and mathematical models. FSM2, Durban, South Africa.

Gao, X.Zh., Shen, T., Zheng, Y., Sun, X., Huang, S., Ren, Q., Zhang, X., Tian, Y., Luan, G. (2002). Practical manure handbook. (In Chinese). Chinese Agricultural Publishing House. Beijing, China. In: WHO. 2006. Guidelines for the safe use of wastewater, excreta and greywater. Volume 4. Excreta and greywater use in agriculture. ISBN 92 4 154685 9.

Gaspard, P., Schwartzbrod, J. (1995). Helminth eggs in wastewater quantification technique. Water Science and Technology 31(5/6), p.443–446.

Guyton, A.C. (1992). Human physiology and mechanisms of disease. W. B. Saunders Co, Philadelphia, USA.

Heinss, U., Larmie, S. A., Strauss, M. (1998). Solids Separation and Pond Systems for the Treatment of Faecal Sludges in the Tropics. Lessons learnt and recommendations for preliminary design. SANDEC Report No. 5/98. Second Edition. Swiss Federal Institute for Environmental Science ad Technology (EAWAG) and Water Research Institute (WRI), Accra/Ghana.

Heinss, U., Larmie, S.A., Strauss, M. (1994). Sedimentation Tank Sludge Accumulation Study. EAWAG/SANDEC publications.

Ingallinella A.M., Sanguinetti G., Koottatep, T., Montangero, A., Strauss, M. (2002). The challenge of faecal sludge management in urban areas – strategies, regulations and treatment options. Water Science and Technology 46(10), p.285-294.

Jönsson, H., Baky, A., Jeppsoon, U., Hellström, D., Kärrman, E. (2005). Composition of urine, faeces, greywater and biowaste for utilization in the URWARE model. Urban water Report of the MISTRA Programme, Report 2005:6, Chalmers University of Technology, Gothenburg, Sweden. Available at: www.urbanwater.org.

Jönsson, H., Vinnerås, B., Höglund C., Stenström, T.-A. (1999). Source separation of urine. Wasser und Boden 51 (11), p.21-25.

Katukiza A.Y., Ronteltap, M., Niwagaba, C., Foppen, J.W, A., Kansiime, F., Lens, P.N.L. (2012). Sustainable sanitation technology options for urban slums. Biotechnology Advances 30(5), p.964-978.

Kengne, I.M., Kengne, S. E., Akoa, A., Bemmo, N., Dodane, P.-H., Koné, D. (2011). Vertical-flow constructed wetlands as an emerging solution for faecal sludge dewatering in developing countries. Journal of Water, Sanitation and Hygiene for Development 01(1), p.13-19.

Koanda, H. (2006). Vers un assainissement urbain durable en Afrique subsaharienne : Approche innovante de planification de la gestion des boues de vidange, École Polytechnique Fédérale de Lausanne. Thèse de doctorat: 351.

Koné, D., Strauss, M. (2004). Low-cost Options for Treating Faecal Sludges (FS) in Developing Countries – Challenges and Performance. Paper presented to the 9th International IWA Specialist group conference on wetlands systems for water pollution control; and to the 6th International IWA Specialist Group Conference on Waste Stabilisation Ponds, Avignon, France, 27th Sept. - 1st Oct. 2004.

Koottatep, T., Surinkul, N., Polprasert, C., Kamal, A.S.M., Koné, D., Montangero, A., Heinss, U., Strauss, M. (2005). Treatment of septage in constructed wetlands in tropical climate: lessons learnt from seven years of operation. Water Science & Technology 51(9), p.119-126.

Lentner, C., Lentner, C., Wink, A. (1981). Units of Measurement, Body Fluids, Composition of the Body, Nutrition. Geigy Scientific Tables. CIBA-GEIGY Ltd, Basle, Switzerland. ISBN 0-914168-50-9.

M'Voubou, K. F. (2004). Evaluation de la gestion communautaire des boues de vidange dans la Commune d'Arrondissement de Sahm-Notaire. Ville de Guédiawaye- Région de Dakar-Sénégal, Ecole Inter-états d'Ingénieurs de l'Equipement Rural / Ecole Polytechnique Fédérale de Lausanne. Mémoire de DESS: 73

Madigan, T.M., Martinko, M.J. (2006). Brock Biology of Microorganisms. 11th Edition. Pearson Prentice Hall. Pearson Education Inc., Upper Saddle River, United States of America. ISBN 0-13-196893-9.

Malicki, J., Montusiewicz, A., Bieganowski, A. (2001). Improvement of counting helminthes eggs with internal standard. Research Note. Water Research 35(9), p.2333-2335.

Mara, D. D. (2004). Domestic wastewater treatment in developing countries. Earthscan UK and US. ISBN 1-84407-019-0.

Moodley, P., Archer, C., Hawksworth, D. (2008). Standard Methods for the recovery and enumeration of helminth ova in wastewater, sludge, compost and urine diversion waste in South Africa. A report of Water Research Commission, SA.

Murray Muspratt, A., Nakato, T., Niwagabe, C., Dione, H., Stupin, L., Regulinski, I., Mbéguéré, M., Strande, L. (2014). Fuel potential of feacal sludge: calorific value results from Uganda, Ghana, and Senegal. Journal of Water, Sanitation and Hygiene in Developing Countries (in press).

Niwagaba, C. (2009). Treatment technologies for human faeces and urine. PhD Thesis, Doctoral Thesis No. 2009: 70. Swedish University of Agricultural Sciences, Uppsala, Sweden. ISBN 978-91-576-7417-3.

NWSC (National Water and Sewerage Corporation) (2008). Kampala Sanitation Program (KSP) - Feasibility study for sanitation master in Kampala, Uganda.

Radford J.T., Fenner, R.A. (2013). Characterisation and fluidisation of synthetic pit latrine sludge. Journal of Water, Sanitation and Hygiene for Development 3(3), p.375–382.

Rossi, L., Lienert, J., Larsen, T.A. (2009). Real-life efficiency of urine source separation. Journal of Environmental Management 90, p.1909-1917.

Rouyat, J., Broutin, C., Rachmuhl, V., Gueye, A., Torrasani, V., Ka, I. (2006). La gestion des ordures ménagères dans les villes secondaires du Sénégal. Vers des politiques municipales incluant les quartiers périphériques, Éditions du Gret.

Schönning, C., Stenström, T.-A. (2004). Guidelines for the safe Use of Urine and Faeces in Ecological Sanitation. Report 2004-1. Ecosanres, SEI. Sweden.

Schouw, N.L., Danteravanich, S., Mosbaek, H., Tjell, J.C. (2002). Composition of human excreta – a case study from Southern Thailand. Science of the Total Environment Journal 286(1-3), p.155-166.

Sonko, E.M. (2008). Traitement des boues de vidange de systèmes autonomes d'assainissement à Dakar (Sénégal): évaluation de l'efficacité de la séparation solide/liquide de lits de séchage non plantés soumis à différentes charges de boues de vidange et à divers apports. Mémoire de DEA : Institut des Sciences de l'Environnement, UCAD. 73 p.

Steen, I. (1998). Phosphorus recovery in the 21st Century: Management of a non-renewable resource. Phosphorus and Potassium Journal 217, p.25-31.

Troschinetz, A. M., Mihelcic, J. R. (2009). Sustainable recycling of municipal solid waste in developing countries. Waste Management 29, p.915-923.

United States Environmental Protection Agency (USEPA). (1994). Guide to Septage Treatment and Disposal. Document EP/625/R-94/002. Washington D.C. 20460.

United States Environmental Protection Agency (USEPA) (1999). Decentralized Systems Technology Fact Sheet, Septage Treatment/Disposal. Document EPA 932-F-99-068. Office of Water, Washington D.C.

Vinnerås, B., Palmquist, H., Balmér, P., Weglin, J., Jensen, A., Andersson, Å., Jönsson, H. (2006). The characteristics of household wastewater and biodegradable waste - a proposal for new Swedish norms. Urban Water 3, p.3-11.

WHO (1984). Guidelines for drinking water quality, volume 2: Health criteria and other supporting information. WHO publication. ISBN 92 41541695.

Yen-Phi, V.T., Rechenburg, A., Vinnerås, B., Clemens, J., Kistemann, T. (2010). Pathogens in septage in Vietnam. Science of the Total Environment 408 (9), p.2050-2053.

Additional Reading Material

Ayres, R.M., Stott, R., Lee, D.L., Mara, D.D., Silva, S.A. (1989). Comparison of techniques for the enumeration of human parasitic helminth eggs in treated wastewater. Environmental Technology 12, p.617–623.

Badji K. (2008). Traitement des boues de vidange: Eléments affectant la performance des lits de séchage non-plantés en taille réelle et les mécanismes de séchage, Ecole Supérieure Polytechnique de Dakar, Mémoire de fin d'étude, 167 p.

Bradley, R.M. (1981). Disposal of septic-tank contents in Cyprus, Water Pollution Control 80, p.131–137.

Butler, D., Payne, J. (1995). Septic Tanks: Problems and Practice. Building and the Environment 30(3), p.419-425.

Chan, L., Li, Y. (2008). Protocol evaluation of the total suspended solids and suspended sediment concentration methods: Solid recovery efficiency and application for stormwater analysis. Water Environment Research 80(9), p.796-805.

Chen, M. (1988). Pollution of groundwater by nutrients and fecal coliforms from lakeshore septic tank systems, Water, Air and Soil Pollution 37, p.407–417.

Collin J.J., Salem, G. (1989). Pollution des eaux souterraines par les nitrates dans les banlieues non assainies des pays en développement, le cas de Pikine (Sénégal). Symposium Internationale sur des solutions intégrées pour des problèmes de pollution de l'eau, Lisbonne.

Franceys R., Pickford J., Reed, R. (1995). Guide de l'assainissement individuel. OMS. Genève.

Radford J.T., Fenner, R.A. (2013). Characterisation and fluidisation of synthetic pit latrine sludge Journal of Water, Sanitation and Hygiene for Development (IWA), Available on-line 15 February 2013 doi:10.2166/washdev.2013.023.

Strauss, M., Heinss, U., Montangero, A. (2000). On-Site Sanitation: When the Pits are Full – Planning for Resource Protection in Faecal Sludge Management. In: Proceedings, Int. Conference, Bad Elster,20–24 Nov. 1998. Schriftenreihe des Vereins fuer Wasser-, Boden und Lufthygiene,2000;105:353-60: Water, Sanitation and Health – Resolving Conflicts between Drinking Water Demands and Pressures from Society's Wastes (I. Chorus, U. Ringelband, G. Schlag, and O. Schmoll, eds.). IWA Publishing House and WHO Water Series.

Strauss, M., Larmie, S. A. (1997). Treatment of sludges from on-site sanitation - low-cost options. Water Science and Technology 35(6) p.129-136.

Strauss, M., Larmie, S. A., Heinss, U. (1998). Solids separations and ponds systems for the treatment of faecal sludges in the tropics: lessons learnt and recommendations for preliminary design. 05/98, SANDEC.

WHO (2006). Guidelines for the safe use of wastewater, excreta and greywater. Volume 4. Excreta and greywater use in agriculture. ISBN 92 4 154685 9.

End of Chapter Study Questions

- List four parameters which are important for FS characterisation, explain how they are analysed and which ranges determine high, medium and low strength FS.
- Describe how each of the following operational factors impact the variability of FS: toilet usage, storage duration, inflow and infiltration and climate.
- Describe two theoretical methods that can be used in the quantification of FS, and the difficulties in obtaining the relevant data.
- Describe FS treatment objectives, why they are important, and how their effectiveness can be monitored.

Chapter 3

Treatment Mechanisms

Magalie Bassan, Pierre-Henri Dodane and Linda Strande

Technology

Learning Objectives

- Know the difference between physical, chemical and biological treatment mechanisms.
- Understand how combinations of physical, chemical and biological treatment mechanisms are responsible for faecal sludge treatment.
- Obtain a basic insight into how mechanisms affect the operation and maintenance needs of treatment technologies.
- Understand the mechanisms for key parameters that can be controlled to increase treatment efficiency and meet treatment objectives.

3.1 INTRODUCTION

This chapter presents an overview of the mechanisms on which faecal sludge (FS) treatment processes are based, and highlights those on which the treatment technologies discussed in subsequent chapters rely. Many FS treatment technologies are based on those developed for wastewater and wastewater sludge treatment, but it is important to remember that these technologies cannot be directly transferred. FS characteristics differ greatly from wastewater, and have a direct impact on the efficiency of treatment mechanisms (Spellman, 1997; Kopp and Dichtl, 2001). Important properties of the sludge to consider include stabilisation, organic load, particle size and density, dissolved oxygen, temperature, pH, water content and viscosity. The current understanding of physical, biological and chemical mechanisms in FS management (FSM) is limited and has been acquired via empirical observations over the years. It is important that this lack of knowledge is overcome in order to improve the design and operation of FS treatment technologies. For more detailed background information on wastewater treatment mechanisms, it is recommended that the reader refers to an engineering textbook specific to this topic.

This chapter is divided into three sections, presenting the physical, biological, and chemical mechanisms used for the treatment of FS. Physical mechanisms include dewatering, drying and volume reduction. These are the most widely employed mechanisms in current FS treatment methodologies, and are generally considered to be robust. Biological mechanisms allow the removal and transformation of organic constituents, nutrients and pathogens via the activity of microorganisms. Chemical mechanisms involve employing additives to optimise and control desired reactions, and are mainly used for disinfection and enhanced dewatering.

3.2 PHYSICAL MECHANISMS

One of the most important treatment mechanisms in FSM is dewatering. FS is mainly comprised of water, the proportion of which is dependent on the type of onsite technology. Water is heavy and expensive to transport, and discharging this polluted water to the environment has significant negative impacts. Dewatering is also necessary prior to resource recovery for applications such as composting, or combustion as a fuel. Dewatering is based on physical processes such as evaporation, evapotranspiration, filtration, gravity, surface charge attraction, centrifugal force and pressure.

Water in FS can be available in free or bound forms. This is an important differentiation in understanding treatment mechanisms because the free water is fairly easily removed, while removal of the bound water is much more difficult (Kopp and Dichtl, 2001). Free water (also referred to as bulk water) usually represents the majority of water in untreated sludge. It can be separated from the solid phase by dewatering technologies such as settling or filtration. It is not adsorbed, bound, or influenced by capillary forces. As shown in Figure 3.1, bound water includes interstitial, surface, and intracellular forms of bound water. Interstitial water (also referred to as capillary water) is in pore spaces, but bound to solids through capillary forces. Surface water (also referred to as colloidal water) is bound to solids and microorganisms by adsorption and adhesion. Intracellular water is contained within microorganisms, and is only removed by treatment mechanisms that result in the lysing of cells, thus releasing the liquid. When water is physically bound to solids, it is much more difficult to remove than free water; it requires the addition of chemicals or the use of centrifugation, pressure or evaporation.

3.2.1 Gravity separation

Gravity is probably the most commonly employed method of liquid – solid separation in FSM. It can achieve the separation of suspended particles and unbound water. Particles that are heavier than water settle out under quiescent conditions at rates based on size of particles, suspended solids concentration, and flocculation. These basic fundamentals are used in the design of settling-thickening tanks (Chapter 6) and grit chambers.

The four types of settling mechanisms include discrete particle, flocculent, hindered, and compression. Discrete particle settling occurs in lower concentration waste streams when particles settle out individually without reacting with other particles. Flocculent settling occurs when particles join together and merge, increasing their mass and settling velocity. This is important for smaller particles that are held together through Van der Waals force, resulting in increased settling velocities. Hindered settling occurs in highly concentrated waste streams, where the particles settle out together as a 'blanket'. Compression occurs at the bottom of a settling tank when the sludge blanket is 'squeezed' by the weight of the solids from above, removing more liquid.

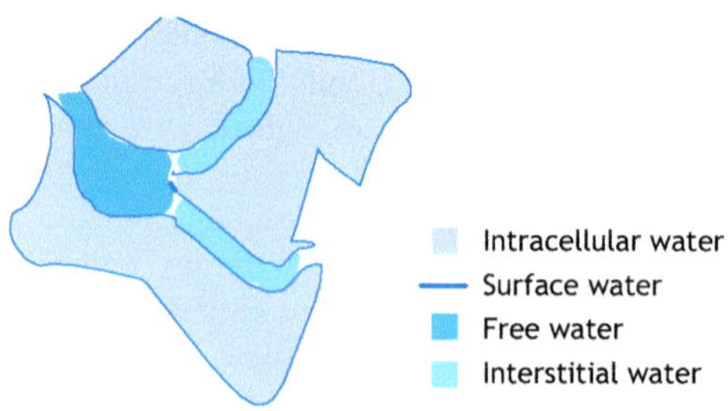

Figure 3.1 Water forms in a sludge floc (adapted from Kopp and Dichtl, 2001).

There are three main forces influencing the settling of a particle; gravity, buoyancy and drag force (or frictional resistance). The gravity force is dependent on the densities of the particle and the fluid, and the volume of the particle. The force due to gravity can be calculated as presented in Equation 3.1.

Equation 3.1: $F_g = \text{particle mass} \cdot g = \rho_p \cdot V_p \cdot g = \rho_p (1/6\, \pi d_p^{\,3})g$

Where:
F_g = Force due to gravity (N)
ρ_p = Particle density (kg/m^3)
V_p = Particle volume (m^3)
d_p = Diameter of the particle (m)
g = Gravitational constant (9.81 m/s^2)

The force due to buoyancy is in the opposite direction to that of gravity, (represented by the negative sign in Equation 3.2), and it is dependent on the density of the liquid.

Equation 3.2: $F_b = \text{liquid mass} \cdot g = -\rho_L \cdot V_p \cdot g = -\rho_L(1/6\, \pi d_p^{\,3})g$

Where:
F_b = Force due to buoyancy (N)
ρ_L = Density of liquid (kg/m^3)
V_p = Particle volume (m^3)
g = Gravitational constant (9.81 m/s^2)
d_p = Diameter of the particle (m)

The drag force is dependent on the particle velocity and diameter, the fluid density and viscosity, and a drag coefficient which is a function of the Reynolds number and the flow regime (laminar, transitional, and turbulent). The drag force is also in the opposite direction to that of gravity. For low Reynolds number (non-turbulent flows) and spherical particles, the drag force can be represented by the Stokes' law as shown in Equation 3.3:

Equation 3.3: $F_d = -3\pi\mu d_p$

Where:
μ = Water viscosity (N · s/m^2)
d_p = Diameter of the particle (m)

When the sum of the gravity, buoyancy and drag forces equals to 0, the particle is at its terminal settling velocity. The length of the tank needed for this particle to settle can be calculated based on this velocity, and the design parameter is the superficial area (width times length). Equation 3.4 is called Stokes' Law for settling, where the terms for $F_g + F_b + F_d = 0$ are substituted:

Equation 3.4: $$v = \frac{(\rho_p - \rho l)g d_p^2}{18\mu}$$

Flotation occurs when suspended solids have a similar or lower density to water, for example algal cells, fats, oils, and grease. Air bubbles can attach to particles, and if they have a similar density to water, this is sufficient to float them to the surface. The layer that forms on the surface of the liquid is referred to as 'scum'. In the design of settling tanks and stabilisation ponds for FS, it is important to address the scum layer as there is typically a significant accumulation (Figure3.2).

Figure 3.2 Settling tank at Niayes faecal sludge treatment plant in Dakar, Senegal (photo: Linda Strande).

3.2.2 Filtration

Filtration is also a commonly applied mechanism for liquid – solids separation in FSM. Several filtration media (e.g. membrane, granular) and types (e.g. slow, rapid, gravity driven or pressurised) are applied to water, wastewater and treated sludge (biosolids) processing. However, in FSM the most common types are unplanted and planted drying beds. These processes use filter media to trap solids on the surface of the filter bed, while the liquid percolates through the filter bed and is collected in a drain, or evaporates from the solids. In filter drying beds, slow filtration is occurring with filtration rates of 0.1-0.4 m/h, which requires less operations and maintenance than faster rates.

The parameters that have the greatest impact on slow filtration efficiency are the characteristics of the influent, the type of filtration media, and the filter loading rate (Metcalf and Eddy, 2003). For example, higher suspended solids concentrations in the influent can increase filter clogging, floc strength can impact the solids retained at the surface and the overall performance, and particle size distribution can affect performance as smaller particles are not as effectively removed by filtration.

Variable sizes of filter media can be used. Coarse media (e.g. gravel) has more pore spaces and allows the passage of more solids, whereas finer media provides a greater frictional resistance to the liquid flow and removes more solids. The need for solids removal must be balanced with the solids concentration of loaded FS and the potential for clogging. FS drying beds are usually designed with layers of increasing size media, from sand at the top to gravel at the bottom (Chapters 7 and 8). The flow rate of the liquid fraction that percolates (by gravity) through the bed is dependent on the resistance to flow exerted by the filter. The rate is calculated and reported as the volume passing through the filter in one hour, divided by the filter surface area. The depth of the filter determines the hydraulic retention time, and the head loss of the liquid, or the energy each unit volume needs to flow through the filter.

The main physical mechanisms that are responsible for filtration are shown in Figure 3.3. As these processes cannot be individually quantified, the design of drying beds relies on empirical calculations. Straining is the exclusion of particles based on size because they are larger than the pore size and not able to pass through. Sedimentation onto the media is a result of gravity settling. Interception is when particles come into contact with media as a result of the path of the liquid flow. Adhesion occurs when particles are removed from the liquid flow when they stick to the filter media. Flocculation is the joining of particles to form larger particles which are then removed by one of the above four mechanisms.

Diverse models have been developed based on filtration mechanisms to explain observed behaviors. Darcy's law can be used in the modeling of slow sand filtration, as the flow is considered to be slow enough to ensure laminar conditions. The resistance of the filtering media is best determined in the laboratory. The resistance to flow exerted by a 'clean' filter can be given by Equation 3.5 (Huisman and Wood, 1974):

Equation 3.5: $$H = \frac{v_f}{k} \cdot h$$

Where:
H = Resistance of the clean filter bed or pressure drop (m)
v_f = Filtration rate per unit area of the bed (m/s)
h = Thickness of the bed (m)
k = Permeability coefficient (m/s).

During operation of the filter, the effective pore sizes will become smaller as particles are trapped in the filter, and the growth of a biological biofilm develops on the media. This phenomena, known as 'ripening', results in increased efficiency of the filter with the retention of greater amounts of finer particles. With slow sand filtration, most of the solids are trapped on the surface of the filter. This increases the resistance for liquids to pass the surface of the filter, and result in reduced flow rates as head losses increase. This can result in clogging of the filter and a rapid decrease of filtration efficiency. To prevent this from occurring, care needs to be taken when designing, building and operating the filter (Chapters 7 and 8). It is also important to use washed sand and gravel when building a filter bed, to ensure that fine soil particles do not result in filter clogging.

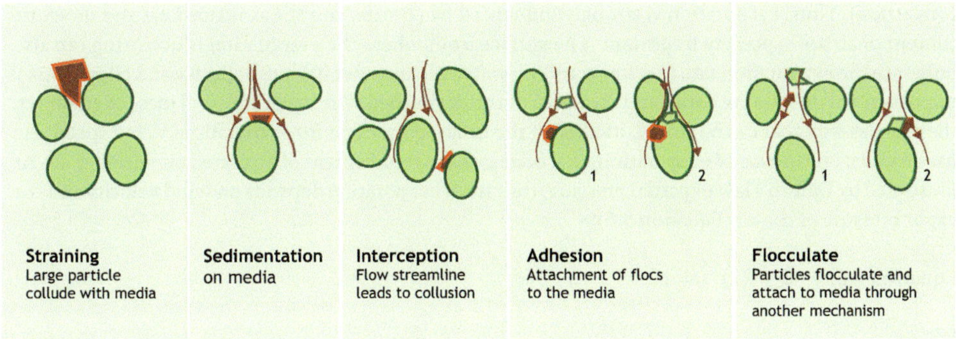

Figure 3.3 Schematic presentation of the mechanisms affecting flow in filter media (modified from Metcalf and Eddy, 2003).

Figure 3.4 Drying beds at Niayes faecal sludge treatment plant, Dakar, Senegal (photo: Linda Strande).

In addition to physical mechanisms, chemical and biological processes also occur within the filter. Chemical processes include attraction processes that result in flocculation or adhesion to filter surfaces. Biological growth happens throughout the filter, but tends to be more intense near the surface, depending on the presence of oxygen, carbon sources and nutrient availability. This can also result in biological removal of nutrients and BOD occurring within the filter (Panuvatvanich *et al.*, 2009).

3.2.3 Evaporation and evapotranspiration

Evaporation occurs when water is released into the air as a vapour, and transpiration is the process by which plants release water vapour to the air as a part of their metabolic processes. Evapotranspiration is the combination of these two mechanisms. In addition to filtration, dewatering in drying beds is also occurring through evaporation, and with planted drying beds through evapotranspiration. For both mechanisms to occur, the surrounding environment needs to have an evaporative demand, which means that the air is not saturated.

The energy required for evaporation to occur is provided by solar energy (with losses due to convection). Thus, evaporation is strongly influenced by climate, and the available heat and moisture content of air are especially important. The surface from where the evaporation is occurring can also influence the evaporation rate (e.g. free standing water versus water in sludge) (Musy and Higy, 2004). Important parameters are depth and total area of the drying bed. The larger the total mass of an object, the more energy that can be stored, increasing the heat requirement for evaporation. Wind speed also has an effect on the rate of evaporation, as it increases the replacement of saturated air with dry air. As illustrated by Dalton's law of partial pressure, the rate of evaporation depends on wind velocity and the vapor pressure of the air (Equation 3.6):

Equation 3.6: $$E_a = f(u) \cdot (e'_a - e_s)$$

Where:

E_a = Contribution of mass transfer to evaporation (mm/day)

$f(u)$ = Proportionality constant of the wind velocity

e'_a = Vapor pressure of water at saturation at the temperature of the surface (mm of mercury)

e_s = Effective vapor pressure (mm of mercury)

The Penman formula shown in Equation 3.7 utilises Dalton's law and incorporates empirical factors to calculate evaporation based on local climatic data. This type of information can typically be found on websites such as the Food and Agriculture Organisation (FAO) of the United Nations (www.fao.org), and in documents such as 'Crop evapotranspiration – Guidelines for computing crop water requirements' (Allen *et al.*, 1998).

Equation 3.7: $$E = \frac{+2\gamma}{+\gamma} \cdot E_c - \left(\frac{\gamma^{(2-\lambda)} \cdot E_2}{+2\gamma}\right)$$

Where:
E = Evaporation in (mm)
Δ = Slope of the saturation vapor pressure curve (kPa/°C)
γ = Psychrometric constant (kPa/°C), $\gamma = 0.00163 \cdot P/\lambda$, where P = Atmospheric pressure
E_c = Evaporation measured on a Colorado basin (mm)
λ = Latent heat of vaporisation

As with evaporation, rates of transpiration are influenced by heat, moisture, and wind, but are also dependent on additional factors such as plant species, growth phases, plant density, leaf shape and color, and water availability in the root zone (Stefanakis and Tsihrintzis, 2011). During transpiration, water is transported through the internal circulation system of the plant, and then released by the stoma, pores on the surface of leaves. The rate of evapotranspiration is always greater than the rate of evaporation alone (Musy and Higy, 2004). It has also been observed that temperature variations have a greater influence on evapotranspiration than on evaporation (Stefanakis and Tsihrintzis, 2011). For optimal evapotranspiration to occur with planted drying beds, the sludge load and rainfall data need to be considered to allow maximum biomass production of plants.

There are a few accepted methods for measuring rates of evapotranspiration (Musy and Higy 2004). Potential Evapotranspiration accounts for the theoretical water loss through evapotranspiration assuming that adequate water is available, there is dense plant coverage, and plants are in an intensive-growth phase. Maximal Evapotranspiration can be determined for individual plant species and for each growth phase, assuming optimal growing conditions. Actual Evapotranspiration utilises the recorded evaporation rate, relative humidity, and growth state of plants. This value is always smaller than the Maximal Evapotranspiration. To measure transpiration, water loss is measured in vegetated reference sites. This is more complex than measuring evaporation as the vegetation type and comparison to standard reference types needs to be considered. Calculating evapotranspiration therefore relies on values that are experimentally determined for a defined context and location. Therefore, the extrapolation involves control experiments and adjustments. The Penman-Monteith equation, derived from the Penman equation (as presented in Equation 3.7) is used to evaluate the potential evapotranspiration rate (Allen *et al.*, 1998; Uggetti *et al.*, 2012). This equation (Equation 3.8) allows the comparison of evapotranspiration at different periods of the year, in different locations, and between several types of plants.

Equation 3.8: $$PET = \frac{0.408 \cdot \Delta \cdot (R_n - G) + \gamma \frac{C_n}{T+273} u_2 \cdot (e_s - e_a)}{\Delta + \gamma \cdot (1 + C_d \cdot u_2)}$$

Where:
PET = Reference evapotranspiration in (mm/day)
Δ = Slope of the saturation vapor pressure curve (kPa/°C)
R_n = Net radiation at the crop surface (MJ/m^2/day)
G = Soil heat flux density (MJ/m^2/day)
γ = Psychrometric constant (kPa/°C), $\gamma = 0.00163 \cdot P/\gamma$, where P = Atmospheric pressure

Figure 3.5 Planted drying beds in the garden of a school, Bangkok, Thailand. The large leafs contribute to the evapo-transpiration and dewatering of the sludge (photo: Linda Strande).

C_n = Coefficient: 900 for short vegetation, and 1600 for tall vegetation
T = Mean daily air temperature at 2m height (°C)
u_2 = Daily wind speed at 2m height (m/s)
e_s= Vapor pressure at saturation at the temperature of the surface (kPa)
e_a= Effective vapor pressure (kPa)
C_d = Coefficient: 0.34 for short vegetation, and 0.38 for tall vegetation (mm)

3.2.4 Centrifugation

Centrifugation is mainly used for liquid – solid separation of wastewater sludge, but can also be employed for FS for the partial removal of bound water. Sludge is placed inside the centrifuge while it rotates at a high speed, and the centrifugal forces accelerate the sedimentation process. Solids settle out at the centrifuge walls, where they are pressed and concentrated. The liquid and solid fractions are then collected separately.

This process relies on the fact that when a particle in movement is forced to change its direction, it exerts a force against any obstacle toward its initial movement. The centrifugal force driving the movement from the center of a cylinder to its surface can be calculated by Equation 3.9 (Spellman, 1997):

Equation 3.9: $F_c = Wr(\rho_s - \rho)V$

Where:
Fc = Centrifugal force (N)
W = Angular velocity (radian per second)
r = Radius from the center of rotation to the particle (m)
ρ_s = Density of the particle (kg/m^3)
ρ = Density of the liquid (kg/m^3)
V = Volume of the particle (m)

The parameters influencing the efficiency of the centrifugation are not completely understood, but three important characteristics that were identified to be important with sludge from wastewater treatment plants are the settleability, the scrollability and the floc strength (Kopp and Dichtl, 2001).

3.2.5 Heat drying

Heat drying is used to evaporate and dewater wastewater sludge (biosolids) beyond what can be achieved with other more passive, or conventional methods. Currently, heat drying is applied more for wastewater sludge processing than for FS, but this technology should be transferable, and further information can be obtained from manufacturers and pilot studies.

Heat drying achieves both weight and volume reduction, as water is lost in the form of vapour. The temperature of the sludge is increased through energy transferred from an external heat source, which allows the evaporation of free water at the sludge surface, at a rate that depends on the ambient air temperature, humidity, flow and pressure, and the exposed sludge surface. As heat is continuing to be transferred, internal moisture is also being transferred to the surface and evaporated at a rate that depends on the physical characteristics of the sludge, the temperature and the moisture content. Heat drying involves convection, conduction, radiation, or a combination of these processes. Convection is employed in direct drying systems, conduction in indirect drying systems, and radiation in infrared drying systems (Chapter 5).

The amount of heat required depends on the specific heat capacity of the FS, which is the amount of energy required to raise the temperature of a unit mass by 1 °C. For example, the specific heat capacity of water at 25 °C is 4.18 kJ/kg/°C, which means that 4kJ are needed to raise the temperature of 1kg of water by 1 °C. No literature values were found for the heat capacity of FS, but the heat capacity of solids in wastewater sludge is reported to be 1.95 kJ/kg/°C (Kim and Parker, 2008).

3.2.6 Screening

Screening is another important physical mechanism in FSM. Bar screens at the influent of a FSTP are imperative to remove municipal waste and large solid objects from the FS, thereby preventing clogging and pump failures, and enhancing the value of treatment endproducts. Bar screens installed in a vertical or inclined position against the incoming flow make a physical barrier that retains coarse solids (Figure 3.6). The distance between the bars are set such that the liquid and small solid particles can flow through while the larger solids are trapped.

The velocity of the flow of FS through the bars influences the screen performance. A low velocity allows an increased removal of solids, but involves a greater solids deposition in the channel leading up to it, which should be avoided. Therefore, the flow velocity should reach, at a minimum, the self-cleansing velocity (greater than 0.3 m/s for wastewater). The flow should also not exceed 1 m/s in order to avoid coarse wastes being pulled through the bars due to the strength of the flow (Mara, 1976). The bars create a head loss that depends on the quantity and type of solid wastes retained.

Figure 3.6 Bar screen at Niayes faecal sludge treatment plant, Dakar, Senegal (photo: Linda Strande).

3.3 BIOLOGICAL MECHANISMS

In FSM, biology is essential in the achievement of treatment objectives through transformation of organic matter and nutrients. Biology is also important in understanding mechanisms of pathogen reduction. Pathogens of concern are presented in Chapter 2, while mechanisms of inactivation are covered in the following section.

Biological treatment harnesses the metabolism and growth rate of microorganisms in naturally occurring processes, and employs them in controlled situations to optimise the desired outcomes. Treatment systems usually rely on complex populations of microorganisms. As the microbes grow, they are dynamically altering the system, by modifying forms of organic matter, and releasing and binding up nutrients. They also release gases and other byproducts that can affect the environment.

The biodegradable organic matter in FS varies depending on the source, but usually needs to be stabilised prior to final enduse or disposal. Stabilisation involves the degradation of putrifiable, readily degradable material, leaving behind more stable, less degradable organics. This is important in order to reduce the oxygen demand, produce stable and predictable characteristics, reduce odors, and allow for easy storage and manipulation (Vesilind, 2001). 'Stabilised' organic matter does not have an exact agreed upon scientific definition, but in general refers to resistance to further biodegradation. Stabilised sludge consists of particles like cellulose, lignin, inorganic matter, and cellular matter of microorganisms that consumed readily degradable organics, whereas unstabilised sludge contains easily degradable compounds such as carbohydrates, proteins, and sugars. Volatile solids are used as a measure for stabilisation, as they are considered to be composed of readily degradable organic matter. Equation 3.10 can be used to assess the level of degradation that has occurred during FS treatment. For 'raw' or 'fresh' FS (unstabilised FS) $\rho = 0$; when fully digested and stabilised, $\rho = 1$ (Kopp and Dichtl, 2001).

Equation 3.10: $\rho_{VSS} = (1-(VSS_1/VSS_0)) \cdot 100$

Where:
ρ = Percentage of degradation
VSS_1 = Volatile Suspended Solids (g/L) at time 1
VSS_0 = Volatile Suspended Solids (g/L) at time 0

3.3.1 Metabolism

For growth to occur, microorganisms need energy and carbon sources. As illustrated in Figure 3.7, bacteria are grouped together based on their metabolic properties, including energy source, carbon source, and electron receptors (e.g. aerobic or anoxic). Energy can be provided through solar energy or chemical forms (i.e. phototrophs, and chemotroph organisms), the chemical forms can be either organic or inorganic (i.e. chemoorganotrophs and chemolithotrophs). The carbon source used for the synthesis of new cells can be obtained from organic matter or carbon dioxide. Essential nutrients for growth include nitrogen, phosphorus, sulfur, potassium, magnesium, iron and calcium.

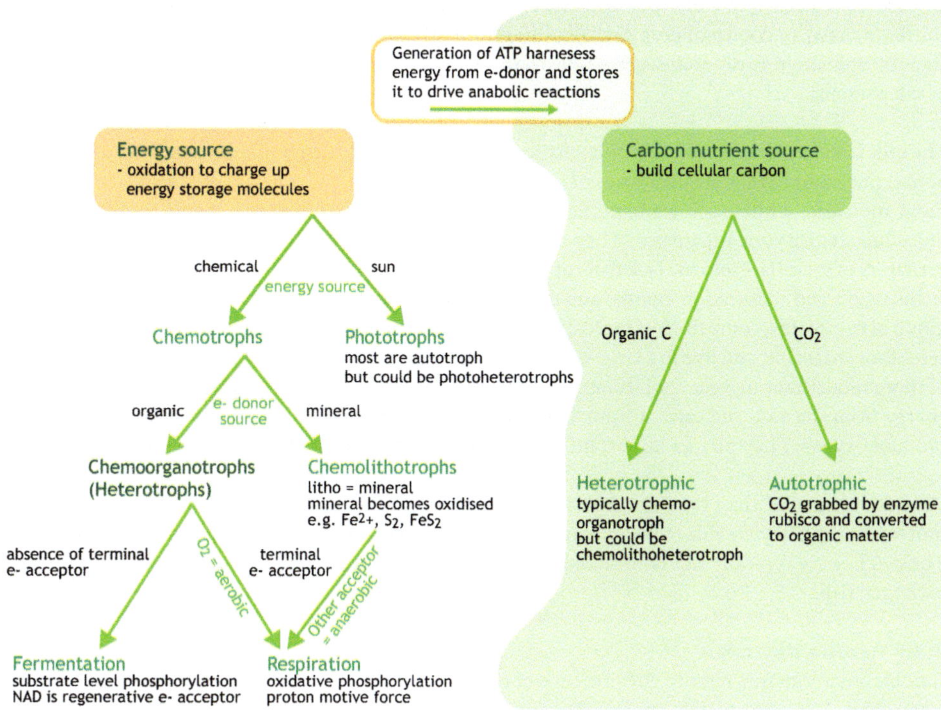

Figure 3.7 Nomenclature of microorganisms based on their energy and carbon requirements (figure: Linda Strande).

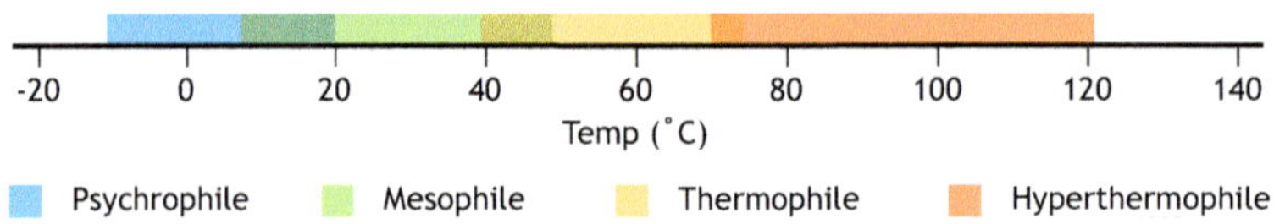

Figure 3.8 Partition of the different types of organisms based on their optimum temperature.

3.3.2 Temperature

The growth rate of microorganisms is also heavily reliant on temperature. Biological activity often doubles for every 10˚C increase in temperature within a given growth range for each organism. Each organism has a minimum temperature, defined as the point below which the organism cannot grow; an optimum temperature range, where enzymatic reactions happen at their greatest possible rates; and a maximum temperature, above which microorganisms can no longer grow due to denaturation of proteins. As shown in Figure 3.8, there are four types of organisms which can be defined depending on their optimal temperature range, namely: psychrophilic (optimal temperature at 15°C or lower), mesophilic (optimal temperature 20-45°C), thermophilic (optimal temperature 45-80°C) and hyperthermophilic (optimal temperature at 80°C or greater).

3.3.3 Types of microorganisms

All living organisms have a cellular structure that is either prokaryotic or eukaryotic. Prokaryotes consist of bacteria and archaea, are single celled, structurally less complex than eukaryotes, and their DNA is not enclosed in a nucleus. Bacteria are 0.5-1 μm in size, and have the shape of bacilli (rods), spirilla (spirals), or cocci (sphere). Archaea differ from bacteria in their evolutionary history. They are all chemotrophic, and many archea grow in extreme environmental environments (e.g. high temperature or salt content).

The cells of eukaryotes contain complex structures enclosed within membranes, and have a membrane bound nuclei. In FSM, the eukaryotes of greatest importance for treatment mechanisms are protozoa, fungi and algae, while pathogenic protozoa and helminthes determine pathogenic risk. Protozoa are unicellular, eukaryotic organisms. They are often motile and larger than bacteria, and are frequently predators of bacteria. Protozoa lack chlorophyll and cell walls. They also play an important role in waste stabilisation and maturation ponds, and can reduce pathogens through predation. Fungi are a large group of organisms, comprised of molds, yeasts, and mushrooms. They are chemoorganotrophs, can be aerobic or anaerobic, and live in a large variety of environments. Fungi are important in the stabilisation of more recalcitrant organic molecules (e.g. in composting). Algae are photoautotrophic, getting their energy from the sun, and carbon from CO_2. Algae use chlorophyll in a similar fashion to plants, and produce oxygen. They play an important role in waste stabilisation and maturation ponds.

Viruses are smaller than bacteria (20-300 nanometers), and consist of a core of RNA or DNA in a protein capsid. They are able to infect plants, animals, and bacteria, and are not able to replicate without a host. They are typically not considered living organisms. In FSM, they are primarily a concern as a pathogen risk.

3.3.4 Aerobic treatment

Aerobic environments refer to the presence of oxygen, and aerobic organisms rely on oxygen for their respiration. Microorganisms can be either purely aerobic requiring oxygen to grow, or facultative meaning they are also able to survive in anaerobic conditions. Typical aerobic treatment processes in wastewater treatment include activated sludge, sequencing batch reactors, trickling filters and facultative or maturation ponds. Aerobic processes occur in any solid or liquid treatment process where oxygen is present, including FS drying beds and composting.

Aerobic growth phases include oxidation and synthesis during rapid growth, and endogenous respiration. As shown in Equation 3.11 during oxidation organic matter is consumed, CO_2 released, and new cells are synthesised:

Equation 3.11: Organic matter + O_2 + nutrients $\rightarrow CO_2 + H_2O + C_5H_7NO_2$ + other products

Where $C_5H_7NO_2$ are the newly synthesised cells.

Endogenous respiration corresponds to the period where readily degradable organic matter is depleted, and microorganisms consume cellular content to maintain metabolic reaction. This is shown in Equation 3.12:

Equation 3.12: $C_5H_7NO_2 + 5O_2 \rightarrow 5CO_2 + 2H_2O + NH_3$ + energy

The dissolved oxygen content in FS is very low due to microbial activity which rapidly depletes available oxygen, and the low solubility of oxygen in water. For processes to remain aerobic, they typically rely on aeration or mixing, which can be energy intensive.

3.3.5 Composting

Composting is the controlled process by which biological decomposition of organic matter occurs by the same organisms that naturally degrade organic matter in the soil. The resulting endproduct is a dark, rich, humus-like matter that can be used as a soil amendment. Humus is defined as the stable fraction of the soil organic matter remaining after the major portions of added plant and animal residues have decomposed. Important mechanisms that govern this process are the oxidation of organic compounds, the release and immobilisation of nutrients, and the microbial synthesis of new compounds.

In thermophilic composting, the system goes through a three phase process. During the first phase, bacteria are growing rapidly while consuming readily degradable compounds (e.g. sugar, starch, protein). During this period, the temperature is also increasing due to the rapid rate of growth (due to exothermic catabolic reactions), which is faster than the rate at which heat can escape. In the second phase, thermophilic temperatures of 50-75 °C are achieved and thermophilic bacteria become active, further decomposing the organic matter. During this phase pathogen reduction and inactivation of plant seeds (e.g. weeds) occurs as a result of the high temperatures. In the third phase, stabilisation is being reached as the last of the readily degradable substrates are depleted, bacterial activity slows down, and the temperature lowers. During this phase actinomycetes and fungi are further degrading more recalcitrant organic molecules such as cellulose and lignin.

The composting process is controlled through the optimisation of the carbon to nitrogen mass ratio (C:N), moisture content, and oxygen supply. The empirically observed optimal C:N is between 20 and 30, based on the ratio that microbes utilise carbon and nitrogen during their growth. There needs to be a balance of enough carbon for cell synthesis and energy extraction, with nitrogen for synthesis of amino acids, enzymes and DNA. If the C:N is lower than 20 there will be excess nitrogen in the system that will be lost following mineralisation due to nitrate leaching or ammonia volatilisation. If the C:N is greater than 30, then the nitrogen is 'locked up' in organic matter and not bio-available.

The optimal moisture content is 40-60% by weight. Moisture is necessary for biological growth to occur, and to transport nutrients throughout the compost pile. If the moisture content is greater than 60%, then it can impede microbial growth by creating anaerobic conditions.

The free pore space should represent at least 20% of the volume, with an optimal oxygen content of the air greater than 10% to ensure that aerobic decomposition is occurring. To achieve this, it is important to have a mix of different textures of materials to allow for oxygen to pass through the pile, and to turn the pile at certain intervals to introduce oxygen, and redistribute partially decomposed matter. Turning the pile lowers the temperature as ambient air is introduced, but is followed by a rapid increase in temperature as biological activity speeds back up again. This process continues until the third stabilisation phase of the composting process is reached.

3.3.6 Anaerobic treatment

Anaerobic conditions are characterised by the lack of oxygen. Anaerobic degradation occurs anywhere in FSM systems where oxygen has been depleted, for example anaerobic and facultative waste stabilisation ponds, septic tanks, and settling tanks. Anaerobic fermentation can also be employed for the treatment of sludge. Anaerobic digesters can provide a beneficial method of stabilising FS, as it also results in the production of biogas that can be used for energy generation. Biogas is a mixture of mainly methane (55-75%) and carbon dioxide (30-45%) (Arthur *et al.*, 2011). Due to the less energetically favorable nature of anaerobic metabolisms, less sludge (i.e. microbial biomass) is produced during anaerobic digestion.

Anaerobic digestion is a complex process characterised by hydrolysis, fermentation, acidogenesis, acetogenesis and methanogenesis. Hydolysis is an enzymatic process through which particulate matter and more complex organic compounds are degraded and become more bioavailable. At the same time, proteins, lipids, and polysaccharides are converted into amino acids, fatty acids, and monosaccharides. During fermentation (or acidogenesis) acidogenic microorganisms further degrade amino acids, sugars, and fatty acids to methogenic substrates (e.g. H_2, CO_2, formate, methanol, methylamines, and acetate). Organic molecules are used as both electron donors and accepters. Therefore, methanogen archea can be characterised as chemoorganotrophs (Figure 3.7). During methanogenesis, one group of archea split acetate into methane and carbon dioxide, while another group produces methane through the use of hydrogen and carbon dioxide. Methanogenesis occurs more readily at mesophilic (30- 38 °C) and thermophilic (49-57 °C) temperatures. Methanogenic processes are presented in Equation 3.13 to Equation 3.15 (Madigan *et al.*, 2003):

Equation 3.13: $4H_2 + CO_2 \rightarrow CH_4 + 2H_2O$

Equation 3.14: $CH_3OH + H_2 \rightarrow CH_4 + H_2O$

Equation 3.15: $CH_3COO^- + H_2O \rightarrow CH_4 + HCO_3^-$

Acidogenic and methanogenic microorganisms have a syntrophic relationship. The methanogens use the hydrogen produced by acidogens, maintaining an optimal partial pressure for the acidogens. Hence, the slow growth rate of methanogens is the limiting step in the process. If this process slows down, then the volatile fatty acids produced by acidogens will build up in the reactor, resulting in a lowered pH, and a further disruption of the methanogenic activity. When this occurs, it is referred to as the digester going 'sour'. Because of this carefully balanced microbial relationship, it is important to ensure that consistent operation and monitoring are occurring, with pH monitoring being the most convenient and useful method. Methanogens are also strongly inhibited by the presence of oxygen, free ammonia, heavy metals, and sulfides.

Figure 3.9 Biogas reactors at 2iE in Ouagadougou, Burkina Faso (photo: Linda Strande).

3.3.7 Nitrogen cycling

Biological nitrogen cycling is an important aspect of FSM, as FS tends to be very high in ammonia nitrogen. Nitrogen is an essential nutrient that can be captured for beneficial enduse, and is also a potential pollutant that should not be indiscriminately discharged into the environment. Inorganic forms of nitrogen are available for microorganisms to use during growth. As shown in Figure 3.10, once it is utilised, nitrogen is immobilised and no longer bioavailable as it is bound up into organic molecules such as microbial cellular components and structures. Nitrogen is then mineralised and released back into bioavailable forms as organisms die off and the organic matter is degraded. The majority of nitrogen in FS is present as ammonia that is released during this hydrolysis process.

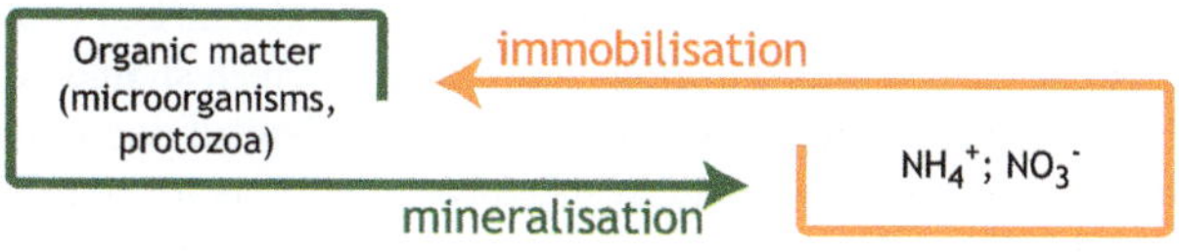

Figure 3.10 The process of mineralisation and immobilisation of nitrogen in the environment.

Nitrification

Ammonia nitrogen that is released during mineralisation, can be oxidised to nitrate through biological nitrification, which is an aerobic, autotrophic process. Ammonia oxidising bacteria oxidise ammonia to nitrite, rapidly followed by nitrite oxidising bacteria oxidising nitrite to nitrate, as shown in Equation 3.16 and Equation 3.17. This is a sensitive biological process and it is thus important to consider total nitrogen concentration, biochemical oxygen demand (BOD) concentration, alkalinity, pH, temperature, and potential for toxic compounds when designing systems that rely on nitrification (Metcalf and Eddy, 2003). The optimal temperature for nitrification is 28°C, with the process becoming inefficient below 10°C. The optimal pH range is between 7.5 and 8.0. Reasonable rates of nitrification occur at neutral pH (7.0), but become restricted below pH 6.8. Since nitrification is an aerobic process, it should be ensured that the dissolved oxygen concentration is greater than 1 mg/L. The nitrification process requires 7.14 g of alkalinity as calcium carbonate ($CaCO_3$) for each gram of ammonia nitrogen (as nitrogen) converted to nitrate (Metcalf and Eddy, 2003). Free ammonia concentrations above 100 mg/L at pH 7 can also inhibit the nitrification process (Metcalf and Eddy, 2003).

Equation 3.16: $$2NH_2 + 3O_2 + \text{Nitrosomonas} \rightarrow 2NO_2^- + 2H_2O + 2H^+$$

Equation 3.17: $$2NO_2^- + O_2 + \text{Nitrobacter} \rightarrow 2NO_3^-$$

Denitrification

Biological nitrogen removal happens in anoxic environments with the reduction of nitrate to nitrogen gas thereby releasing nitrogen to the air. Anoxic environments are low in oxygen, and nitrate is used as an electron receptor. Dissolved oxygen concentrations greater than 0.1-0.5 mg/L inhibit the anoxic process and the optimal pH range is 7.0-8.0. The process occurs with both heterotrophic and autotrophic bacteria, many of which are facultative aerobes. The process happens through a series of intermediate gaseous nitrogen oxide products. Denitrification generally proceeds through some combination of the intermediate forms shown in Equation 3.18.

Equation 3.18: $$NO_3^- \rightarrow NO_2^- \rightarrow NO \rightarrow N_2O \rightarrow N_2$$

Where:
NO_3^- = nitrate
NO_2^- = nitrite
NO = nitrogen oxide
N_2O = nitrous oxide
N_2 = nitrogen gas.

When designing a system that includes both nitrification and denitrification, it is important to ensure there is adequate BOD for denitrification to occur. More precise values can be calculated, but it is estimated that 4 g of BOD is needed per g of nitrate reduced (Metcalf and Eddy, 2003). In addtion 3.57 g of alkalinity as $CaCO_3$ is regained during denitrification which should be taken into account when calculating the total alkalinity requirement for nitrifying – denitrifying systems. Simultaneous nitrification and denitrification can also occur in anaerobic conditions without BOD with anammox bacteria that can oxidise NH_4^+ to N_2, utilising NO_2^- as an electron acceptor, which also gets reduced to N_2.

Phosphorus cycling

As with nitrogen, phosphorus is an essential nutrient that can be captured for beneficial enduse, and is also a potential pollutant that should not be indiscriminately discharged into the environment. Phosphorus in FS and excreta is mostly present as phosphates; molecules comprised of the acid or base form of orthophosphoric acid (H_3PO_4) or phosphate (PO_4^{3-}), or as organically bound phosphorus (e.g. nucleic acids, phospholipids and phosphorylated proteins).

The fate of phosphorus in treatment processes depends on factors such as sorption, precipitation, complexation, sedimentation, mineralisation, pH, plant uptake, and redox potential. During degradation of organic material, bound phosphates are mineralised and released. Phosphate is not lost due to off gassing or leaching like nitrogen, as the soluble inorganic form is adsorbed in the sludge. During biological treatment processes, about 10-30% of phosphorus is taken up by microorganisms. This can be increased through biological dephosphatation, or through chemical precipitation with $FeCl_3$ or $Al_2(SO_4)_3$ or $FeSO_4$ which are used for wastewater treatment. The greatest loss of phosphorus during FS treatment is due to removal by plants in planted drying beds.

3.3.8 Pathogen reduction

In this section, the mechanisms that result in biological die-off of pathogens from physical, biological, and chemical mechanisms are covered. Types of pathogens are presented in more detail in Chapter 2. It is important to have an understanding of all of these interrelated mechanisms, to ensure that pathogen reduction is achieved during FS treatment. They affect all biological processes, which need to be considered to ensure that treatment processes function as designed.

Temperature

Most pathogens are inactivated above temperatures of 60°C when cell proteins and nucleic acids are denatured. This is achieved in processes such as thermophilic co-composting as shown in Figure 3.11, as well as lime treatment. As the temperature increases, less time is needed for pathogen inactivation.

Figure 3.11 Pilot scale co-composting facility with faecal sludge and municipal waste in Bangalore, India (photo: Chris Zurbrügg).

Time

The duration of treatment (e.g. planted drying beds) or the storage of treated sludge can result in pathogen reduction, as they have a limited survival time in adverse conditions. In faeces, most bacteria can only survive between 1 week and 2 months. For example, Salmonella spp. survives on average for 30 days and Faecal coliforms for 50 days (Feachem *et al.*, 1983). Helminth eggs however are very persistent, and can maintain viability for several months to years. The required storage duration for pathogen reduction also depends on the ambient temperature. For example, Niwagaba (2009) recommends storage time of FS for up to one year at an ambient temperature of 35°C, and two years at 20°C. Storage at temperatures less than 10°C does not result in adequate inactivation (Weemaes and Verstraete, 2001).

Sorption

Helminth eggs tend to sorb or settle, and hence partition with the solids fraction in FS systems. In settling and thickening tanks, about 50% of the helminth eggs are separated from the liquid fraction due to settling (Heinss *et al.*, 1998). In filtration that occurs with drying beds, the majority of Helminth eggs remain with the solid fraction, as does 90% of indicator bacteria (Pepper *et al.*, 2008). However, indicator bacteria are not necessarily representative of all possible pathogens (e.g. viruses, different types of bacteria, protozoan cysts). Although the majority of Helminth eggs partition with the solids fraction, the fate of all pathogens must be considered.

Desiccation

Evaporation resulting in desiccation or dehydration reduces active pathogens, as microorganisms need water for survival. Water activity is represented by the ratio of water vapour pressure of the sludge to the water vapour pressure of pure water under the same conditions. Pure water has a water activity of 1, and most pathogens cannot survive under a water activity of 0.9, while some yeast and eggs survive in much drier conditions (Carrington, 2001). All dewatering technologies therefore contribute to the die-off of pathogens (e.g. drying beds) if the water content gets below a certain point where desiccation has an effect. Further storage also contributes to disinfection due to the reduction of the available water.

UV

Solar/UV radiation in the range of 300-400 nm effectively inactivates pathogens by denaturing DNA molecules via photochemical reactions (Borrely *et al.*, 1998). UV light has been shown to effectively inactivate *E. coli* in waste stabilisation ponds (Maïga *et al.*, 2009). However, it is important to remember that for this mechanism to be effective, the light rays must be able to penetrate the FS during treatment. This mechanism is therefore most likely only occurring at the surface, as the high organic matter and turbidity prevents penetration of UV radiation.

pH

Most microorganisms can only survive and grow within a range of 2-3 pH units, and very few can survive below pH 3 and above pH 10. In this way, chemical addition for pH control can result in pathogen reduction. However, the pH can also upset composting and anaerobic digestion processes, and it is therefore important to consider downstream treatment steps when employing pH control for pathogen reduction.

3.4 CHEMICAL MECHANISMS

Chemicals can be mixed with FS to improve the performance of other physical mechanisms (e.g. addition of a cationic polymer to increase the flocculation and the settling efficiency), or to inactivate pathogens and stabilise FS. The addition of chemicals can represent a significant increase in the overall cost of treatment, and the benefits therefore need to be carefully weighed.

3.4.1 Alkaline stabilisation

Alkaline additives, such as lime, can be used for the stabilisation of FS, either pre- or post-dewatering. If carried out prior to dewatering, more additives will be required. Lime is also used to precipitate phosphorus from liquid streams in wastewater treatment plants, and to polish effluents. The addition of adequate quantities of lime to FS raises the pH to 12, which stops the microbial activity. This results in an odor reduction due to factors that cause putrification, and in a reduction of pathogens. This chemical reaction also hydrolyses fats, carbohydrates, and proteins, as well as ammonia from amino acids. If quick lime (CaO) is used, it will also result in an exothermic reaction that can increase the temperature up to 60°C (Andreasen, 2001), thereby increasing pathogen reduction, and inactivating Helminth eggs. The reaction has also been documented to increase settling efficiency. However, after the initial reaction, the pH will lower again, therefore requiring that lime is added in excess. It should also be noted that regrowth of bacterial pathogens can occur over time. Concerns with this process include ammonia odors and lime scaling.

3.4.2 Ammonia treatment

It is well established that aqueous ammonia is effective at inactivating microorganisms, but the exact mechanisms are not yet fully understood. As described in Vinnerås (2013), possible mechanisms for bacterial inactivation are that NH_3 denaturates proteins, destroys membrane potentials, or causes rapid alkalinisation of the cytoplasm resulting in a critical loss of potassium (K). Viral inactivation is possibly due to the cleavage of RNA, but for larger organisms, such as helminths, the mechanisms are still not fully understood. Ammonia disinfection has been shown to be effective in urine (Vinnerås *et al.*, 2008), sewage sludge (Pecson *et al.*, 2007), and compost (Adamtey *et al.*, 2009), but applications for FS are still in the research phase of development.

It is aqueous NH_3 that is responsible for microbial inactivation, not the ammonium ion (NH_4^+). The pK_a of ammonia is 9.25 (the pH where 50% is NH_3 and 50% is NH_4^+), and the percent NH_3 concentration based on pH can be determined by Equation 3.19.

Equation 3.19: $$NH_3,\% = \frac{100}{1+[H^+]/K_a}$$

The total aqueous NH_3 concentration will also depend on temperature and total ammoniacal nitrogen concentration (NH_3+NH_4^+). For NH_3 disinfection to be effective, the pH has to be above 8.5 (Vinnerås, 2013). Ammonia can be added as as aqueous NH_3 solution, or urea ($CO\ (NH_2)_2$), which is rapidly enzymatically transformed to NH_3. The treatment needs to be carried out in a confined space to avoid loss of gaseous NH_3. At this stage, the time for inactivation of microorganisms needs to be determined empirically for each specific organism of concern. As long as the pH remains stable, the aqueous NH_3 will remain constant and pathogen regrowth will not occur (Vinnerås, 2013). If the treated FS is applied to the soil it will result in a decrease of pH, thereby increasing the concentration of NH_4^+, which is beneficial as a fertiliser.

3.4.3 Coagulation and flocculation

Colloidal particles that are not removed through gravity settling tend to be negatively charged, making them stable in suspension. In coagulation and flocculation additives are added that destabilise particles, allowing them to come in contact with each other, form larger flocs and settle, thereby achieving enhanced sedimentation. These additives are chosen based on the hydrophobic or hydrophilic characteristics of the particles, together with their surface charge.

Coagulation and flocculation are achieved by adding polymers that form a bridge between particles, or by adding potential determining ions (strong acid or base) that reduce the total surface charge. Polymers

Technology

can be natural or synthetic based chemicals. They work by either forming a bridge between the anionic and non-ionic ends of the polymer to particles, or by forming a bridge with high molecular weight polymers that are adsorbed to particles.

3.4.4 Conditioning

Chemical conditioning is based on the same physical properties as coagulation and flocculation, and can be carried out prior to physical forms of dewatering as described in Section 3.4.3 to enhance performance. Common additives include ferric chloride, lime, alum, and organic polymers. Iron salts and lime can increase the total solids of dried sludge (increasing bulk), whereas polymers do not increase the total solids. To select the appropriate chemical to use, important aspects to consider are sludge age, pH, source, solids concentration, and alkalinity. In general, the dosage is determined in the laboratory with simple settling jar tests. The information that is currently available relates to wastewater sludge treatment. To transfer this technology to FS, information from manufacturers, laboratory, and pilot-scale testing is necessary.

3.4.5 Disinfection of liquid effluents

Disinfection of liquid effluents is not covered in detail in this book, as it is not specific to FS, and is covered in detail in wastewater and water treatment references (e.g. ponds, wastewater treatment). Liquid effluents from settling tanks or drying beds typically require further treatment prior to disinfection. Disinfection should be considered as a 'polishing' step to achieve a final reduction in pathogens and is not a primary form of FS treatment. Liquid effluent treatment also needs to be considered in the context of adequate and appropriate levels of treatment for the intended enduse, as discussed in Chapter 10. Disinfection refers to a reduction of pathogens, not a total elimination (which is referred to as sterilisation). Chemical forms of disinfection include chlorination, ozonation, and UV, but it is also achieved through mechanical means such as filters or membranes.

Chlorination is the most widely used method of disinfection, and can be applied in either a solid or liquid form. Important design parameters include contact time, chlorine concentration, pathogen load, temperature, and other constituents in effluent (e.g. remaining organic load). Chlorine is toxic to microorganisms, as it has a high capacity for oxidation, which damages cell membranes. The oxidation process is not specific to microbes, and it is therefore important to consider the total organic load. Chlorination is not effective for disinfecting FS, or liquid effluents that contain high organics, as the chlorine will be used up in the oxidation of these other constituents.

3.5 BIBLIOGRAPHY

Adamtey N., Cofie O., Ofosu-Budu G.K., Danso S.K.A., Forster, D. (2009). Production and storage of N-enriched co-compost. Waste Management 29, p.2429-2436.

Arthur, R., Baidoo, M.F. (2011). Biogas generation from sewage in four public universities in Ghana: A solution to potential health risk. Biomass and Bioenergy 35(7), p.3086-3093.

Allen, R.G. (2000). Using the FAO-56 dual crop coefficient method over an irrigated region as part of an evapotranspiration intercomparison study. Journal of Hydrology 229 (1), p.27-41.

Allen, R.G., Pereira, L.S., Raes, D., Smith, M. (1998). Crop evapotranspiration - Guidelines for computing crop water requirements. FAO Irrigation and Drainage Papers. FAO Irrigation and Drainage Paper 56, FAO, Rome.

Andreasen, P. (2001). Chemical stabilization. Sludge into Biosolids: Processing, Disposal, Utilization. Edited by L. Spinosa and P. A. Vesilind. IWA Publishing, United Kingdom.

Borrely, S.I., Cruz, A.C., Del Mastro, N.L., Sampa, M.H.O., Somessari, E.S. (1998). Radiation processing of sewage and sludge. A review. Progress in Nuclear Energy 33(1–2), p.3-21.

Carrington, E.G. (2001). Evaluation of sludge treatments for pathogen reduction – final report. E. Communities. Luxembourg.

Feachem, R. G., Bradley, D. J. (1983). Sanitation and Disease – Health Aspects of Excreta and Wastewater Management. Washigton, D.C, USA, The World Bank.

Heinss, U., Larmie, S.A. (1998). Solid separation and pond systems for the systems for the treatment of faecal sludges in the tropics. EAWAG. Dübendorf, Switzerland, EAWAG.

Huisman, L., Wood, W. E. (1974). Slow sand filtration (Vol. 16). World Health Organization, Geneva.

Kim Y., Parker W. (2008). A technical and economic evaluation of the pyrolysis of sewage sludge for the production of bio-oil. Bioresoures Technology 99(5), p.1409-1416.

Kopp, J., Dichtl, N. (2001). Characterization. Sludge into Biosolids – Processing, Disposal, Utilization. Edited by L. Spinosa and P. A. Vesilind. IWA Publishing, United Kingdom.

Madigan, M.T., Martinko, J.M., Parker, J., Fernândez, C.R., Pérez, M. S. (2004). Brock's microbiology. Pearson Education, Inc., New Jersey, United States.

Mara, D. D. (1976). Sewage Treatment in Hot Climates. Wiley, London, United Kingdom.

Metcalf and Eddy (2003). Wastewater Engineering: treatment, disposal, reuse. Tchobanoglous, G., Burton, F.L. eds. McGraw-Hill Book Company.

Musy, A., Higy, C. (2004). Hydrologie, une science de la nature. Presses Polytechniques et Universitaires Romandes, Lausanne, Switzerland.

Niwagaba, C.B. (2009). Treatment Technologies for Human faeces and Urine. PhD Thesis, Swedish University of Agricultural Science, Uppsala, Sweden.

Panuvatvanich, A., Koottatep, T., Koné, D. (2009). Influence of sand layer depth and percolate impounding regime on nitrogen transformation in vertical-flow constructed wetlands treating faecal sludge. Water Research 43 (10), p.2623-2630.

Pecson, B.M., Barrios, J.A., Jimenez, B.E., Nelson, K.L. (2007). The effects of temperature, pH, and ammonia concentration on the inactivation of Ascaris eggs in sewage sludge. Water Research 41 (13), p.2893-2902.

Pepper, I., Gerba, C., Gentry, T., Maier, R. (2008). Environmental Microbiology, Elsevier.

Spellman, F.R. (1997). Dewatering Biosolids. Technomic Publishing, Lancaster, United States.

Stefanakis, A.I., Tsihrintzis, V. A. (2011). Dewatering mechanisms in pilot-scale Sludge Drying Reed Beds: Effect of design and operational parameters. Chemical Engineering Journal 172(1), p.430-443.

Uggetti, E., Argilaga, A., Ferrer, I., García, J. (2012). Dewatering model for optimal operation of sludge treatment wetlands. Water Research 46(2), p.335-344.

Vesilind, P.A. (2001). Introduction to stabilization. Sludge into Biosolids – Processing, Disposal and Utilization. Edited by L. Spinosa and P. A. Vesilind. IWA Publishing, United Kingdom.

Vinnerås, B., Nordin, A., Niwagaba, C., Nyberg, K. (2008). Inactivation of bacteria and viruses in human urine depending on temperature and dilution rate. Water Research 42(15), p.4067-4074.

Vinnerås, B. (2013). Sanitation and hygiene in manure management. In: Sommer, S.G., Jensen, L.S., Christensen, M.L., Schmidt, T. (eds). Animal Waste – Recycling, Treatment and Management. Wiley-Blackwell, Oxford, UK (In Press).

Weemaes, M., Verstraete, W. (2001). Other treatment techniques. Sludge into Biosolids – Processing, Disposal, Utilization. Edited by L. Spinosa and P. A. Vesilind. IWA Publishing, United Kingdom.

End of Chapter Study Questions

1. Give two examples each of physical, chemical and biological mechanisms, and what treatment technologies they are relevant to.

2. Does composting rely on physical, chemical or biological treatment mechanisms? What are three conditions that are required for efficient composting?

3. Which mechanisms are responsible for pathogen reduction and in which treatment technologies do they exist?

Chapter 4

Methods and Means for Collection and Transport of Faecal Sludge

Georges Mikhael, David M. Robbins, James E. Ramsay and Mbaye Mbéguéré

Technology

Learning Objectives

- Understand the social, procedural and technical aspects related to the collection and transport of faecal sludge from onsite sanitation technologies, and of its range of magnitude.
- Know which types of equipment can be applied for different types of onsite sanitation technologies.
- Be aware of the issues and the variables at play in the transport of faecal sludge to a treatment point or transfer station.
- Know what faecal sludge transfer stations are, how they are operated and factors that influence their location.
- Be aware of the health and safety issues regarding faecal sludge collection and transport.

4.1 INTRODUCTION

This chapter presents best practices in faecal sludge (FS) collection and transport that are applicable to a range of service providers, from sole proprietors with a bicycle driven cart, to large companies with multiple (sometimes hundreds) of transport vehicles operating in densely populated urban areas. Consequently, it presents a variety of techniques from the most basic manual sludge removal methods to those required for the most sophisticated vacuum truck operations. Often, due to the complexity of different types of onsite technologies, economic status and access, a variety of service providers can be found operating simultaneously in any given geographical region, and even within the same company.

People and companies that collect and transport FS from onsite sanitation technologies such as septic tanks and pit latrines perform a valuable service for residents, neighbourhoods and the cities where they are located (Figure 4.1). They provide a critical link in the service chain that makes access to sanitation a reality. Without collection and transport companies to remove FS, onsite systems will not function properly. This chapter is focused on the procedural and technical aspects related to the collection of FS from onsite technologies, transport to a location where treatment occurs, and the way in which service providers accomplish these tasks.

The objective of this chapter is to highlight aspects of an ideal, professional and safe sludge removal service. Effective service providers rely on trained personnel, functional equipment, and procedures for conducting the work safely and with minimal impact to the environment. Sanitation authorities should encourage measures such as worker training and certification, and licensing of collection vehicles. Activities should be adapted to suit the local context within which the services are implemented, while keeping these overall objectives in mind.

4.2 TYPICAL DUTIES AND RESPONSIBILITIES

Typical duties and responsibilities of FS collection and transport service providers include those that occur prior to the FS removal, the FS collection itself, and the subsequent transportation of the FS to the treatment facility. As will be explained in Chapter 12, service providers could be sole proprietors, different size companies, or a municipality.

Figure 4.1 Empyting operation of faecal sludge from a septic tank using vacuum equipment. Job safety would be improved by better protective equipment (photo: David M. Robbins).

When emptying the FS from onsite systems, a number of tasks are performed in accomplishing the job. Ideally, a typical job requires the service provider to:

- interact with customers prior to removing FS to arrange logistics and inform them of procedures;
- share the standardised fee or negotiate one, depending on the business model;
- locate onsite sanitation systems that are to have sludge removed;
- determine the accessibility of the system once it is located;
- open the system to facilitate the process;
- collect the FS;
- evaluate the condition of the system post-collection;
- close and secure the system once the FS removal has been completed;
- clean up after the process is completed; and
- perform the final inspection and report any issues with the system to the customers after the service is completed.

This section provides more in depth information on some of those activities that would ideally be performed prior to sludge removal tasks. The collection and transportation of FS are covered in more detail in later sections.

4.2.1 Interfacing with clients

The operator who comes to collect the FS is often the only person that a resident will interact with regarding their onsite system. As such, the operator has a responsibility not only to perform the tasks properly, but to be knowledgeable about the onsite system, and to be able to communicate why sludge removal is necessary and beneficial to the client and their community. The operator is also the only person who will be able to observe the onsite storage system both when it is full, and when it is empty. They should use this opportunity to assess how well it is functioning, identify repair needs and issues related to proper operation that might increase the life span of the system. As such, they can also troubleshoot and be a source of valuable information about FS management (FSM) in the community in which they work. This is also a good opportunity for service providers to work in conjunction with local governments to disseminate information, such as pamphlets on the proper care of septic tanks, or information on how unimproved latrines might be updated or improved to provide better service.

Case Study 4.1 : Customer Interaction in Marikina City, The Philippines

Marikina City, Philippines takes customer interaction a few steps further. Here the city, in conjunction with the water utility, has set up an organised sludge removal program that services neighborhood by neighborhood on a rotating 5 year cycle. They partner with the private sector service providers as follows:

- a few days before the service providers are in the neighborhood, they send out a truck with a loudspeaker to advise residents of the pending service;
- the day before sludge removal takes place, the city workers visit the homes and pass out informational brochures;
- they identify homes that require assistance in opening up their septic tanks and provide a list of people that can do this for a small fee; and
- on the day of sludge removal, they are present to answer questions, direct traffic and troubleshoot.

The result is 95% compliance with sludge removal requirements provided for in their local ordinance.

Figure 4.2 An example of an emptying operation in Dakar, Senegal where the system is located in an internal courtyard and gaining access requires cooperation with the residents (photo: Linda Strande).

Service providers work with residents at the household level to determine the location of the onsite system that requires emptying, to identify access ports and manholes (if they exist), to identify where to place their emptying equipment, and any other relevant issues. Frequently, onsite systems are located directly under the kitchen or bathroom, which makes entering the home with emptying equipment a necessity. Therefore cooperation and communication with residents at the household level is critical, and makes the emptying process more efficient. An example is shown in Figure 4.2, where an emptying operation is in process, with the service provider entering a courtyard to access the onsite system.

General rules that service providers should follow when interacting with household level residents include:

- being courteous and always obtaining explicit permission to perform the service;
- answering questions to the best of their ability and referring to the regulatory authority as needed;
- using caution when entering homes with hoses and other equipment, and protecting floors, walls and furniture from damage;
- communicating findings to the residents (preferably a written record of the service and any identified issues); and
- ensuring cleanliness during and after the service.

4.2.2 Locating the system to be emptied

Frequently, the location of the onsite sanitation system that needs to be emptied is not obvious. For example, septic tanks are typically buried and their location may not be known, and if latrines are grouped it is not always apparent for which one the service is being hired.

Methods to locate sanitation systems tanks include the following:

- asking the client the location of the tank;
- if not known, looking for obvious indicators such as manholes, tank lids or exposed concrete slabs, as shown in Figure 4.3;
- identifying sewer cleanouts outside or under the building. The direction of the cleanout might indicate the location of the tank;
- hammer a metal probe (e.g. one cm diameter metal rod) gently into the ground and determine by feel if rocks or tanks are encountered;
- looking for depressions in the yard around the house, which may signify the location of an underground tank;
- if the house is on piers, looking underneath to inspect the plumbing and determining if sewer lines or vent stacks are buried (scraping the earth around these pipes might indicate the location of the tank;
- if the house is constructed on top of a concrete slab, gently tapping with an iron bar on the floor to reveal hollow sounds.

Figure 4.3 Example of a septic tank access port designed for ready access, Vung Tau Province, Vietnam (photo: Linda Strande).

4.2.3 Determining accessibility

Determining the accessibility of septic tanks or pit latrines involves first determining if the site itself is accessible, and then assessing if each compartment of the system can be accessed to accommodate the FS emptying service. The following are typical factors that determine accessibility of a site:

Width of the road

- if using a truck, roads need to be wide enough to accommodate the truck or sludge emptying equipment.

Access to the site

- does neighboring property need to be accessed to reach the system;
- are there any weather related concerns regarding site access, such as stream crossings, or roads which are impassable during heavy rain event.

Location of the site

- if a truck or cart is being used, is the onsite system located close enough to the parking area to facilitate the emptying operation;
- is the client's location close enough to a FS treatment plant (FSTP) to accommodate transport.

The following questions can be used as a checklist to assist the service provider in determining if the system is accessible for emptying:

- can the system be opened to accommodate the sludge removal equipment (e.g. hose)?
- are there existing manholes over each compartment that can be opened?

- will the installation of new access ports be required? If so, is that a service that the residents have agreed to?
- will slabs, floors, or septic tank covers have to be rebuilt following emptying?
- will the pit collapse if emptied?

4.2.4 Tools of the trade

The sludge emptying process requires that the service provider has access to a number of tools and that the equipment is properly used and maintained. An example of inadequate maintenance is illustrated in Figure 4.4. The specific tools used by service providers vary based on the technology used and the availability on the local market.

Some tools common to all service providers include:

- shovels, pry bars and probes to locate tanks and manholes;
- screwdrivers and other hand tools to open manholes and access port lids;
- long handle shovels and buckets which may be necessary to remove solids that cannot otherwise be removed;
- hooks to remove non-biodegradable solids;
- hoses for FS pumping as well as for adding water to tanks if available; and
- safety equipment including:
  - wheel chocks to prevent the vehicle from moving when parked;
  - personal protective equipment such as hardhat, face protection, eye protection, boots and gloves;
  - disinfectants, barriers, sorbents and bags for cleaning up and collecting spilled material.

Employees of the collection and transport company should be responsible for maintaining tools and equipment in proper working order, and reporting to supervisors when repairs are required.

Figure 4.4 Hoses and fittings require frequent attention and repairs to keep them functioning properly. This picture illustrates inadequate maintenance, resulting in leaking of FS from the emptying hose (photo: David M. Robbins).

4.3 PROPERTIES OF FAECAL SLUDGE IN RELATION TO COLLECTION AND TRANSPORT

FS can be removed from septic tanks or latrines through the use of manual and mechanised techniques that may rely upon hand tools, vacuum trucks, pumping systems, or mechanical augers. The specific method utilised will be based on the type of onsite system, accessibility of the site, the type of equipment owned by the service provider, and the level of expertise.

Awareness of the properties of FS is necessary in order to understand the challenges faced in its collection and transport. These properties are primarily influenced by water content, sludge age, the presence of non-biodegradable material, and organic material. For example, within a pit latrine containment system, recently deposited FS found in the top portion typically has a higher water and organic content than that found in deeper layers, and consequently a lower density (Buckley, *et al.*, 2008). The top portion is therefore less viscous and relatively easy to collect. The absence of water and organic content in the deeper, older and more digested layers makes collection much more difficult, this condition is frequently referred to as 'thick'. Depending on the collection method, 'thick' FS often needs to have water added to facilitate pumping. This suggests that the deposition period could be used as a strong indicator of the ease with which FS could be collected. FS characteristics are covered in more detail in Chapter 2.

4.4 MANUAL COLLECTION

Service providers using manual collection methods generally come from low-income communities living in informal settlements. They are often given offensive titles such as scavengers, *vyura* (Swahili for frogmen), *baye pelle* (pick & shovel men) or *kaka bailers* (excreta bailers). In some regions, they are members of ostracised groups, such as southern Asia's *Dalit* (untouchables). However, regardless of how these social prejudices manifest themselves, the service providers often find themselves stigmatised within their own families and communities due to the nature of their work.

Manual sludge collection falls into two general categories, namely 'cartridge containment' and 'direct lift'. Cartridge containment and direct lift methods can be practiced safely when operators perform their tasks with the proper equipment following appropriate procedures. For instance, descending into pits as currently practiced in several Sub-Saharan, and south and southeast Asian countries is not safe. Dumping of FS directly into the environment rather than discharging at a transfer or treatment site must also be avoided.

National and local governments, such as those in Ghana and Bangladesh, are starting to recognise the unsafe and unhygienic practises commonly used in manual sludge collection, and are prohibiting these activities. In addition, local government, can help promote hygienic FS collection by highlighting best practices, imposing restrictions on unsafe practices, and providing incentives such as training, capacity building, and licensing. Formalising the informal sector through training and licensing will drive the demand for improved services, will improve hygiene, and enable business development and job creation.

4.4.1 Cartridge containment devices

One example of a cartridge containment device is the "Uniloo" (Figure 4.6), an innovative technology designed for hygienic manual collection. This system consists of a modular, mobile, urine-diverting toilet, which has a replaceable and sealable cartridge waste tank (IDEO, 2012). Capable of holding approximately 20 litres of waste, the "Uniloo" isolates both users and collectors from direct contact with FS. Collectors remove and seal a full cartridge and replace it with a pre-cleaned empty one on a regular basis, while the full cartridges are transported by low-cost transport and decanted at a transfer or treatment station using appropriate personal safety equipment and protection.

4.4.2 Direct lift

The direct lift method involves the collection of FS from latrines or tanks by using long handled buckets and long handled shovels. Filled buckets are hoisted to the ground surface, where they are emptied into tanks fitted onto carts which are then transported to transfer stations or treatment sites.

4.5 MANUALLY OPERATED MECHANICAL COLLECTION

Recent innovations in human powered mechanical devices are assisting service providers in servicing septic tanks and pit latrines more quickly, safely and efficiently. This section provides an overview of four (4) of the most common types of mechanical pumping equipment that has been developed and trialled; namely, the Sludge Gulper, the diaphragm pump, the Nibbler, and the Manual Pit Emptying Technology (MAPET).

4.5.1 Sludge Gulper

The Sludge Gulper, as shown in Figure 4.6, was developed in 2007 by the London School of Hygiene and Tropical Medicine (LSHTM). It is a low-cost manually driven positive displacement pump that operates along the same principles as that of direct-action water pumps.

The Gulper has a simple design, and can be built using locally available materials and fabrication techniques generally common in low-income countries. It consists of a PVC riser pipe containing two stainless steel 'non-return' butterfly valves. One valve, the 'foot' valve, is fixed in place at the bottom of the riser pipe and a second valve, the 'plunger' valve, is connected to a T-handle and puller rod assembly. As the handle is moved up and down, the two valves open and close in series and sludge is lifted up the riser pipe to exit the pump via a downward angled spout. A strainer is fitted to the bottom of the riser pipe to prevent non-biodegradable material from entering and blocking the pump.

Figure 4.5 The 'Uniloo's' cartridge developed by Unilever, Water and Sanitation for the Urban Poor (WSUP) and IDEO in Ghana (photo: Nyani Quarmyne).

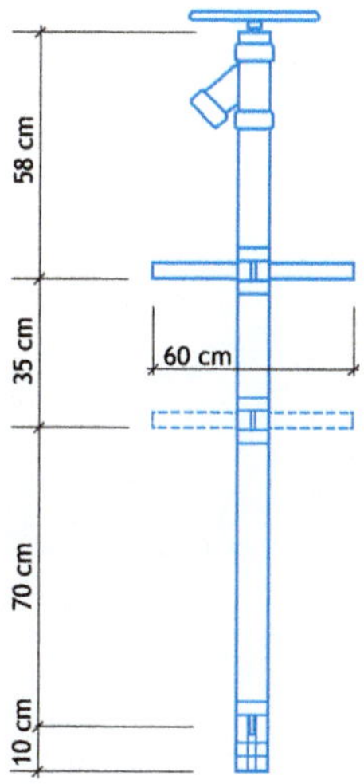

Figure 4.6 Schematic of the Sludge Gulper (Tilley *et al.*, 2014).

Since its initial development, the Gulper has undergone several modifications to make it more user friendly and better adapted to local conditions. Modifications that have been trialed include a lever-type handle to make pumping easier and a retractable riser to allow for the emptying of varying-depth containment systems. Other pumps have been developed that use similar principles, including the Poor Pump or the Manual Desludging Hand Pump (MDHP).

The Gulper performs well with less viscous sludge and is capable of pumping at a rate of approximately 30 L/min. The pumping head is fixed for each pump but is dependent on the configuration of the Gulper.

Depending on the design and materials used, the capital cost of the Gulper ranges from 40 – 1,400 USD (Boot, 2007; Godfrey, 2012; Still and Foxon, 2012).

Some challenges reported by developers and users of the pump include:

- difficulty setting up and operating the pump in toilets with a small superstructure (Godfrey, 2012);
- clogging of the pump by non-biodegradable material present in the sludge;
- cracking of the PVC riser pipe after long-term use; and
- splashing of wet sludge during operation (Godfrey, 2012).

Of the manually driven mechanical collection systems discussed in this section, the Gulper has reached the widest number of pit emptying service providers in Africa and Asia. However, independent uptake or production of the pump by service providers without the intervention of external organisations (e.g. funding, training, technical support) has not been reported.

4.5.2 Manually operated diaphragm pumps

Manually operated diaphragm pumps, as illustrated in Figure 4.7, are simple low-cost pumps capable of extracting low viscosity FS that contains little non-biodegradable materials. They typically consist of a rigid, disc shaped body clamped to a flexible rubber membrane called a diaphragm. An airtight seal between the diaphragm and the disc forms a cavity. To operate the pump, the diaphragm is alternately pushed and pulled causing it to deform into concave and convex shapes in the same way a rubber plunger is used to unblock a toilet. A strainer and non-returning foot valve fitted to the end of the inlet pipe prevents non-biodegradable material from entering the pump and stops backflow of sludge during operation respectively.

Figure 4.7 Manual diaphragm pump operation in Bangladesh (photo: Georges Mikhael).

While they are light enough to be transported by one or two persons, in some cases the pumps are mounted on wheels for ease of transport. Depending on the model, the capital cost of a pump can vary between 300-850 USD.

The following challenges have been reported with this technology:
- clogging when pumping sludge containing non-biodegradable material;
- difficulties in keeping air-tight seals at fittings resulting in air entrainment and consequently low functionality;
- cracking of the rubber diaphragm (Muller and Rijnsburger, 1992); and
- difficulty in locally sourcing or manufacturing the pumps and spare parts.

4.5.3 Nibbler

A continuous, rotary action, displacement sludge pump called the Nibbler was developed by the LSHTM at around the same time as the Gulper. It is capable of collecting medium viscosity sludge using a continuous roller chain loop enclosed in a PVC pipe. The pipe can be inserted into the access hole of a containment structure or a pit latrine without the need to break any part of the structure.

The chain is driven by manually rotating a double crank and sprocket located at the top of the pipe. Semi-circular metal discs loosely and horizontally attached to the chain at regular intervals scoop out the waste from the bottom of the pit and displace it to the top. Once at the top of the pipe, sludge is scrapped off the discs and into a Y-shaped connector, which guides the sludge into the container being used for onward transport. A vertical plate spanning the length of the pipe divides the downward and upward travel directions of the chain and discs. Due to limited success during trials, development of the nibbler was suspended.

4.5.4 MAPET

In Tanzania in 1992, WASTE developed and trialled a human-powered vacuum system for the collection and short-distance transport of sludge called the Manual Pit Emptying Technology (MAPET). Of the technologies in this chapter, he MAPET is both the earliest and the most technically advanced manually driven mechanical collection system. It has two separate components, a pump and a 200 litre vacuum tank, each mounted on a dedicated pushcart.

From a technical point of view, trials have proven that the MAPET works well and is able to pump sludge from a depth of 3 metres at a rate of 10 to 40 L/min depending on the depth and viscosity of the sludge (Brikké and Bredo, 2003). Trials also concluded that WASTE succeeded in tackling most of the technical challenges that they initially set out to address. However, eight years after their introduction, only one of the eight MAPETs in Tanzania was found to be operational, and after thirteen years none were operational (BPD, 2005). Some of the reasons for their lack of long-term sustainability include:
- a breakdown in institutional support on which the service providers using MAPET were highly dependent;
- a reliance on the importation of a key spare part (a leather piston ring) which could not be sourced locally ; and
- the inability of the MAPET service providers to recover maintenance and transport costs from emptying fees (WASTE Consultants, 1993).

4.5.5 Comparison of equipment

Table 4.1 provides a summary of the four manually operated mechanical equipment discussed in this section.

Table 4.1 Summary comparison table of manually operated mechanical equipment

Equipment type	Performance	Purchase/Operating cost (USD)	Challenges
Gulper	• Suitable for pumping low viscosity sludges • Average flow rates of 30 L/min • Maximum pumping head is dependent on design	• Capital Cost: 40 – 1,400 (depending on design)/ • Operating Cost: Unknown	• Difficulty in accessing toilets with a small superstructure • Clogging at high non-biodegradable material content • PVC riser pipe prone to cracking • Splashing of sludge between the spout of the pump and the receiving container
Manual diaphragm pump	• Suitable for pumping low viscosity sludges • Maximum flow rate of 100 L/min • Maximum pumping head of 3.5m – 4.5m	• 300 – 850 (depending on manufacturer and model) • Operating Cost: Unknown	• Clogging at high non-biodegradable content • Difficult to seal fittings at the pump inlet resulting in entrainment of air • Pumps and spare parts currently not locally available
Nibbler	• May be suitable for pumping higher viscosity sludges	• Capital Cost: Unknown • Operating Cost: Unknown	• May be unsuitable for dry sludge with high non-biodegradable material content
MAPET	• Maximum flow rates of between 10 and 40 L/min depending on the viscosity of the sludge and the pumping head • Maximum pumping head of 3.0m	• Capital Cost: 3,000 (1992) • Operating Cost: 175 per annum (maintenance costs only) (1992)	• Requires strong institutional support for MAPET service providers • A reliance on the importation of a key spare part • MAPET service providers unable to recover maintenance and transport costs from emptying fees

4.6 FULLY MECHANISED COLLECTION

Fully mechanised technologies are powered by electricity, fuel or pneumatic systems. They can be mounted on a frame or trolley for increased mobility, or mounted on vehicles for emptying and transporting large quantities of sludge over longer distances. This section introduces a range of fully mechanised technologies. It includes equipment that is widely available such as motorised diaphragm pumps, trash pumps and some types of vehicle-mounted vacuum equipment. It also details some less commonly used equipment that is either in early stages of development such as the motorised pit screw auger, or no longer being developed, such as the gobbler.

4.6.1 Motorised diaphragm pumps

Other than being powered by a motor, motorised diaphragm pumps operate using the same principles as the manual diaphragm pumps. Many different commercial brands and types exist covering multiple applications, one of which is the pumping of FS. Although they can be powered hydraulically, electrically or by compressed air, the most common type used for pumping FS is driven by a petrol or diesel engine. The pumps are typically mounted on a frame and moved by hand or using a trolley for increased mobility.

The performance of the petrol or diesel engine differs depending on the size and model. They are generally suited to pumping liquid sludge but can handle some solid particles (MSF, 2010). A typical 3-inch pump can pump solids ranging in size from 40 to 60 mm, with a maximum flow rate of 300 to 330 L/min, and maximum pumping head of 15 metres.

Motorised diaphragm pumps were used to empty VIP latrines in South Africa but frequently blocked due to the presence of large pieces of non-biodegradable material in the sludge (O'Riordan, 2009). A lack of spare parts for some components of the engine is also a constraint in some low-income countries.

The approximate purchase price of a 3-inch motorised diaphragm pump is 2,000 USD.

4.6.2 Trash pump

Trash pumps work in a similar way to centrifugal impeller water pumps, with some different features. The impeller of a trash pump typically has fewer solid blades, sometimes with cutting edges that can break up the material being pumped. The impellers' housing is usually simple and easy to remove allowing for rapid unblocking if and when required (MSF, 2010).

Trash pumps are suitable for pumping sludge with high liquid content. Similar to motorised diaphragm pumps, the performance of the pumps differs depending on the size and model. The 3-inch pumps can typically handle solid particles in the range of 20 to 30 mm, have maximum flow rates of approximately 1,200 L/min, and maximum pumping heads of 25 to 30 metres. The approximate purchase price of a 3-inch trash pump is 1,800 USD.

4.6.3 Motorised pit screw auger

Pit screw augers (SAS) are based on the Archimedean screw design. Trials were carried out using manually operated AS, however they were found to operate too slowly to be effective (Still and O'Riordan, 2012). Motorised SAS are currently under development with prototypes mimicking certain aspects of commercial motorised soil augers. They consist of an auger placed inside a plastic riser pipe and protruding by approximately 5 to 15 cm from the bottom end of the pipe. An electric motor is mounted on top of the riser pipe where it connects to the auger (Figure 4.8).

Figure 4.8 Mechanised pit screw auger in South Africa (photo: David M. Robbins).

To operate, the riser pipe is placed in the FS and as the auger turns, FS is picked up by cutting blades at the bottom of the auger and lifted up the riser pipe along the auger flights. A downward angled spout at the top of the riser pipe allows material to be discharged into a collection container. Weighing between 20 and 40 kg, motorised SASs can be operated by one person. Flow rates are estimated to vary between 40 to 50 L/min and it may be suitable for pumping high viscosity FS and semi-solids. They can handle a small amount of non-biodegradable waste (de los Reyes, 2012), with more recent prototypes being fitted with a reverse gear to facilitate dislodging of waste. It is reported that one of the prototypes costs in the region of 700 USD to build, however no data is currently available on operating costs.

Some of the challenges faced by the motorised PSAs include (Still and O'Riordan, 2012; Still and Foxon, 2012):

- complicated emptying process due to the fixed length and rigidity of the auger and riser pipe;
- unsuitability for use with dry sludge and large amounts of non-biodegradable waste;
- difficulties with cleaning after use; and
- difficulties manoeuvring due to weight and size.

4.6.4 Gobbler

The Gobbler was prototyped by the South African Water Research Commission (WRC) in 2009 as a more robust and efficient version of the Nibbler. Using the same operating principles as the Nibbler, the Gobbler is powered using an electric motor. The motor turns a double chain drive that rotates a heavier gauge chain that of the Nibbler. The metallic disks used to pull FS out of the pits in the Nibbler have been replaced by metal scoops, and a scraper is installed at the discharge point to remove the FS off the scoops (Still and O'Riordan, 2012).

During testing, sludge blocking the drive chains of the Gobbler was found to be a significant problem (Still and O'Riordan, 2012; Still and Foxon, 2012). Other issues encountered in the construction and operation include (Still and O'Riordan, 2012):

- a complex fabrication process requiring a high number of parts;
- difficulty moving and setting up the pump due to its heavy weight; and
- difficulty emptying containment structures of different depths as the length was not adjustable.

The estimated cost of the prototype Gobbler was approximately 1,200 USD. As with the Nibbler, and due to the significant challenges experienced, no further investments have been made in the development of the Gobbler (Still and Foxon, 2012).

4.6.5 Vehicle-mounted vacuum equipment

Pumping systems that utilise a vacuum have been shown to be effective at removing FS from onsite water-retaining systems. Vacuum pumps may be mounted on heavy duty trucks or trailers, on lighter duty carts or even on human powered carts when smaller volumes are being collected, or for use in dense urban settings not accessible by larger trucks. Vacuum pumps often utilise the truck's transmission to power the system, although independently powered, dedicated motors can also be used. Vacuum trucks are available in a wide variety of sizes and models to accommodate different needs, with the most commonly used having capacities ranging from 200 litres to 16,000 litres. Different types of vacuum systems are described in the following section.

Conventional vacuum trucks

Vacuum pumps are sized based on lift elevation, pumping distance, volume of sludge to be removed, and volume of the tank. When designing collection and transport systems, local manufacturers should be consulted in order to determine what equipment is available. Product specifications must be checked to verify that the proposed truck is adequate for the need.

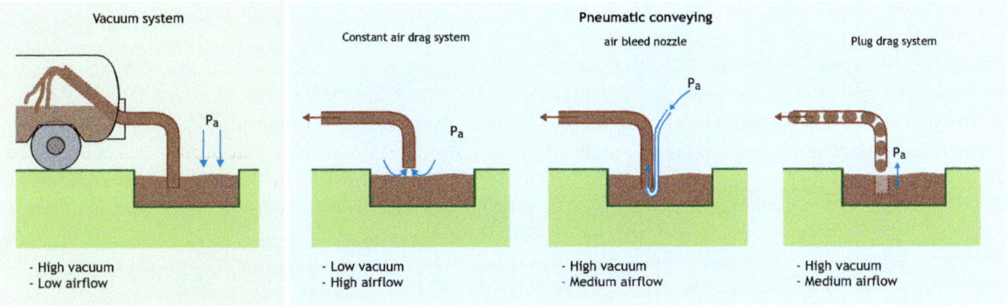

Figure 4.9 Four types of vacuum sludge removal techniques (adapted from Böesch and Schertenleib, 1985).

The typical volume of trucks used for the collection of FS ranges from 10,000 litres to 55,000 litres. Various factors influence the selection of a vacuum truck by a service provider, including:

- typical volume of the tanks or vaults that will be serviced;
- road widths and weight constraints;
- distance to the treatment plant;
- availability;
- budget; and
- skill level of the operators.

Conventional vacuum tankers are typically fitted with either a relatively low cost, low-volume sliding vane pump or a more expensive liquid ring pump. The former is more appropriate for low-capacity vacuum tankers where high vacuum and low airflow sludge removal techniques are used. Vacuum conveyance techniques work best for removing low-viscosity sludge such as that found in septic tanks (Böesch and Schertenleib, 1985).

Liquid ring pumps are more appropriate for high-capacity vessels and pneumatic conveying techniques. Three such techniques namely, constant air drag, air bleed, and plug drag, are briefly described in Figure 4.9. They are most suitable for emptying higher viscosity sludge typically found at the bottom of a septic tank or in a pit latrine.

Some conventional vacuum tankers are also fitted with dewatering capabilities in order to reduce the volume of the sludge transported and increase efficiency. Wastewater disposal points (typically a sewerage network) then become necessary to dispose of the untreated liquid effluent. The downfall of this more complex system is the expert maintenance needs, in terms of human capacity and spare parts.

BREVAC

In 1983, the International Reference Centre for Waste Disposal (IRCWD) undertook a series of field tests in Botswana using multiple conventional and specialist vacuum tankers as well as mechanical collection equipment. The BREVAC, developed by Building Research Establishment (BRE), was one of the specialist vacuum tankers being tested (Figure 4.10).

The equipment was designed to haul a double-compartmental vessel; the first being a 4.3 m^3 compartment for sludge, and the second a 1 m^3 compartment for service liquid (i.e. water) (Böesch and Schertenleib, 1985). It was fitted with a high performance liquid ring vacuum pump with a 0.8 bar suction capacity and 26 m^3/minute air flow rate. The tanker was also fitted with a hydraulic tipping cylinder to incline the vessel and facilitate cleaning after it had been emptied.

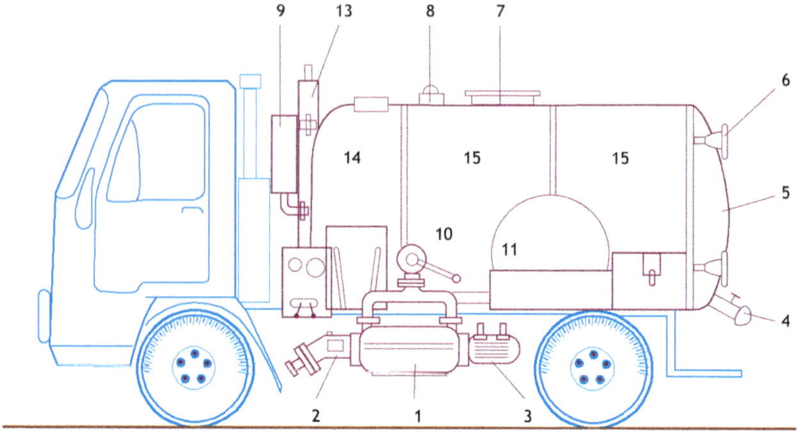

1 Liquid ring vacuum pomp
2 Hydraulic motor
3 Service liquid pomp
4 Suction and discharge valves
5 Swing-out rear door
6 Handwheel
7 Hatch
8 Pressure safety valve
9 Water separator for discharge air
10 Load/discharge control valve
11 Spare wheel
12 Tool locker
13 Hydraulic tank tipping cylinder
14 Service liquid (water tank)
15 Slurry tank

Figure 4.10 Schematic diagram of BREVAC (adapted from Böesch and Schertenleib, 1985).

During testing, the BREVAC proved to be capable of emptying high viscosity FS from pit latrines and manoeuvring in tight spaces and difficult terrain (Böesch and Schertenleib, 1985). Its ability to hydraulically break up sludge masses with pressurised water streams prevented the need for someone to get inside the septic tank or pit during cleaning. Some design features required reconfiguration (e.g. addition of a floating ball gauge to determine sludge level), and non-biodegradable materials caused clogging of the hoses. However, overall the technology was deemed to be technically sound and fit for purpose.

Due to the highly technical design and specialist parts necessary to operate it, and the high cost associated with such issues, the BREVAC failed to sustain demand and presence in the target market.

Figure 4.11 The Mark III (left) and Mark IV (right) models of the BREVAC collection and transport device (photo: Peter Edwards).

Vacutug

In 1995, lessons learned from the development of the BREVAC and MAPET by IRCWD were taken into account in the development of the UN-HABITAT Vacutug. The first version, Mark I, was developed in Ireland by Manus Coffey Associates (MCA), and trialled in Kenya by the Kenya Water and Health Organisation (KWAHO). Since then, a further four versions have been developed in Bangladesh, and several units of each sold. Examples are shown in Figure 4.11 and in Table 4.2.

Table 4.2 Different Vacutug versions and corresponding general properties

Version	Capacity (litres)	Relative Width	Travel Distance	Mounting & Propulsion	Cost (USD) (excluding shipping)
Mark I & II	500	Very Narrow	Short-Haul	Mounted on self-propelled chassis	10,000
Mark III	1,900	Average	Long-Haul	Mounted on trailer chassis and propelled by tractor or pick-up	20,000
Mark IV	700	Narrow	Medium-Haul	Mounted on chassis of motorised tricycle	15,000
Mark V	1,000	Narrow	Medium-Haul	Mounted on chassis of motorised tricycle	15,000

4.6.6 Delivering vehicle-mounted vacuum services

Vacuum units for sludge removal are complex mechanical systems that must be operated correctly, not only to accomplish sludge removal, but also to protect the equipment and health of the service providers.

The following steps are recommended for the operation of vacuum trucks:

1 Park the truck as close to the system as possible. The maximum distance is determined by the length of hose and elevation rise from the bottom of the pit or septic tank to the vacuum truck tank inlet. This should typically be no more than 25 meters in linear distance and 4 meters in elevation gain. Further distances or elevation differences may require intermediate pumps.
2 Inform the occupant of the pending service and note any concerns or issues.
3 Inspect the site for possible hazards, such as clearing the area of people, or identifying high groundwater that could cause a tank to 'float' if emptied.
4 Secure the truck using wheel chocks.
5 Lay out and connect the hoses from the truck to the tank or pit to be emptied.
6 Open the tank or pit by removing the access ports or covers over the storage system.
7 Engage the vacuum equipment by using a power take-off from the truck's transmission.
8 Increase the vacuum to the proper level with the valve closed by watching the vacuum gauge, then lowering the end of the hose into the storage system, and open the valve sufficiently such that the FS is drawn out of the tank or pit. Closing the valve periodically re-builds the vacuum to enable the removal of further FS.
9 Continue this process until the job is complete.
10 Break up FS that has agglomerated into a solid mass, either by making use of a long handle shovel and adding water when necessary to reduce the viscosity of the FS; or by reversing the direction of the flow and forcing the contents of the vacuum truck tank back through the hose and into the sanitation system in order to use the high pressure stream to break up the sludge. The direction of

the flow is then returned to normal and the contents removed. It is essential to ensure that the hose is in sound condition, and that the hose connections are locked into place prior to using this method;

11 Operators should remove between 90% and 95% of the contents. It is recommended that this is verified by management through periodic spot checks.

12 Identify any abnormal conditions, such as high concentration of non-biodegradable materials, oils and grease. The colour and odour of the FS can provide clues as to how the occupants are using the system, and if excessive chemicals are being discharged down the drain.

13 Inspect the system once empty. In the case of a septic tank, the following checks should be carried out by the operator:

a. Listen for water running back from the discharge pipe, which could indicate plugged leach lines (if present);
b. Check to make sure that inlet and outlet tees are properly in place. Frequently, these structures break off and can sometimes be found at the bottom of the tank;
c. Inspect the tank for cracks or damage;
d. Verify that the tank is properly vented;
e. Ensure that the tank lids are properly attached when the pumping is complete and that they are properly secured;
f. Prepare a written report indicating:
   - how much waste was removed;
   - the condition of the tank or pit;
   - any recommendations for repairs or maintenance;
   - any recommendations for proper use of the system.

14 Secure the tank lid and pack away the hoses;

15 Clean up any spillage using proper sorbent materials;

16 Inform the client that the work is complete, and give them the final report. In some instances, payment is received immediately for the service however, payment is often made directly to the service provider through some type of billing system. During this final interview, the operator informs the client of the findings and any recommendations;

17 Remove the wheel chocks and drive the truck to the next site or to the nearest approved disposal site.

Figure 4.12 Example of correct method where the hose is easy and quick to assemble (photo: Linda Strande).

Figure 4.13 Example of an improperly maintained hose held together with plastic bags and twine (photo: David M. Robbins).

4.6.7 Summary of fully mechanised systems

Table 4.3 summarises the main aspects of the mechanised mechanical sludge emptying equipment discussed in this section, highlighting the performance, cost and concerns.

Table 4.3 Summary table for mechanised mechanical sludge emptying equipment

Equipment Type	Performance	Cost (USD) Capital	Cost (USD) Operating	Challenges
Motorised diaphragm pump	• can handle liquid sludge and solid particles 40 to 60mm in size • maximum flow rate of 300 to 330 L/min • maximum pumping head of 15 m (can easily empty from variable depths)	2,000	Unknown	• blocking due to non-biodegradable waste in the sludge • spare parts not available locally
Trash pump	• can handle very liquid sludge and solid particles 20 to 30 mm in size • maximum flow rate of approximately 1,200 L/min. Maximum pumping head of 25 to 30 m (can easily empty from variable depths)	500 – 2,000	Unknown	• difficult to find spare parts • requires containment system • potential for clogging
Pit screw auger	• can handle liquid sludge and a small amount of non-biodegradable waste • flow rates of over 50 L/min. pumping head of at least 3m (difficulty emptying from variable depths)	700	Unknown	• the fixed length of the auger and riser pipe • unsuitable for use with dry sludge and large quantities of non-biodegradable waste • difficult to clean after use • difficult to manoeuver due to weight and size
Gobbler	• blocks easily due to sludge build up in the working parts • pumping head of at least 3 m • difficulty emptying from variable depths	1,200	Unknown	• complex fabrication process and a high number of parts • weight of the pump • length not adjustable
Vacutug	• can handle low-viscosity sludge well and some non-biodegradable waste • ideal for areas with limited access. • pumping head varies depending on model used	10,000 – 20,000	25 USD/ load[1]	• can be slow to transport • difficulty emptying high-viscosity sludge • small volume (500 to 1,900 litres) • not financially viable for long-haul transport
Conventional vacuum tanker	• can easily handle low-viscosity sludge well and some non-biodegradable waste • Ideal for transporting large quantities of sludge over long distances • Pumping head varies depending on pump model used	10,000 – 100,000[2]	Highly Variable	• difficulty accessing high-density areas • difficult to maintain in low-income contexts due to specialised parts • prohibitively expensive for some service providers

1 Assuming two loads emptied per day from an average distance of 10 kilometres from the disposal point and an average travel speed of 10 km/h (Mikhael and Parkinson, 2011)

2 The price range of conventional vacuum tankers varies significantly depending on whether the vehicle is brand new or used, capacity, extra capabilities (e.g. jetting), and shipping costs.

4.7 TRANSPORT OF FAECAL SLUDGE

Most of the manually operated small-scale equipment described in Section 4.5 and some of the mechanised equipment described previously in Section 4.6 are not capable of transporting sludge. Low-cost transport equipment, standardised or customised, is therefore often used for the transport of sludge to the transfer station or treatment facility.

This equipment can be categorised into two main forms: that which is manually propelled by human or animal power, and that which is motor-propelled using a fuel-powered engine. This section describes the different types of manual and motorised equipment currently in use and presents their advantages and disadvantages.

The aspects that need to be considered for the transportation of FS include:

- the type of vehicle to be used including its road worthiness, maintenance, licenses and permits, and where it is kept when it is not in service;
- the type of sludge removal equipment, including hoses, pumps, augers, and other tools of the trade;
- the spill management equipment to be used including shovels, disinfectants, sorbents, and collection bags;
- the skills of the operator including the training and certifications that might be required to perform the work;
- procedures that need to be followed including rules of the road and activities at the treatment plant; and
- other aspects such as the use of transfer stations, worker health and safety, and emerging technologies.

4.7.1 Manual transport

Today, both standard carts used for general transport of materials, as well as customised carts designed specifically for transport of FS, can be found in many low-income countries, an example of which is shown in Figure 4.14. Although designs vary widely, standardised carts typically consist of a load-bed mounted on a single axle with one or more wheels. Containers of sludge with capacities of up to 200 litres can be carried on or in a manually pulled or pushed cart (Still and Foxon, 2012, Strauss and Montonegero, 2002; Barreiro, *et al.*, 2003; Chowdhry and Koné, 2012). The carts are designed to be manoeuvrable in tight spaces and have an effective range of up to 3 km.

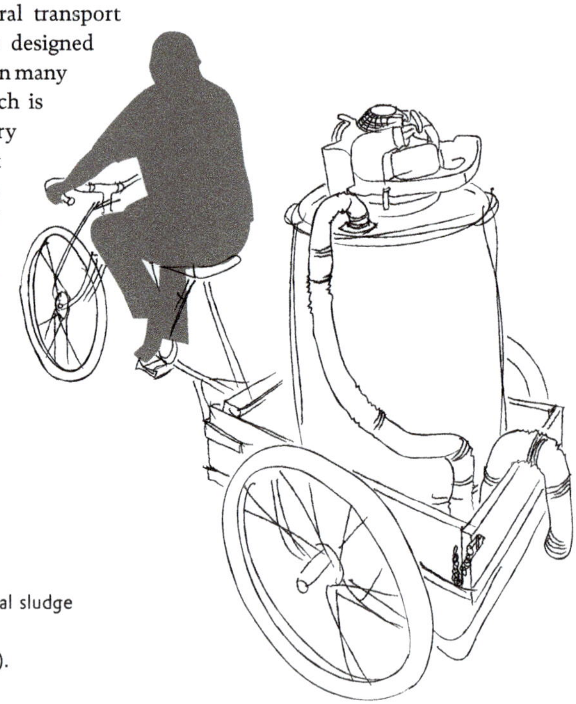

Figure 4.14 Schematic of human powered faecal sludge management transport technology (figure: Research Triangle Institute).

Figure 4.15 Sludge collection and transport vehicle navigating roads in the busy city of Manila, Philippines (photo: David M. Robbins).

Since manual low-cost transport equipment generally have a small load capacity, limited and a low travel range and low speed, they are not suitable for long distance transport.

4.7.2 Motorised transport

Motorised transport equipment offers the potential for larger load capacities and increased speed, leading to reduced travel times and a greater range as compared to manual transport. The operation and maintenance of motorised transport is generally more complex than that of manual transport, however many variations are widely used in low-income countries. Before selecting the type of transport system, it is important to verify that the knowledge and skills to carry out repairs are locally available.

Motorised tricycles are the smallest type of low-cost motorised transport used to move FS. They vary in size and power and are able to access narrower streets than the larger motorised vehicles. Some models are capable of carrying loads of up to 1,000 kg. Sludge can either be transported in drums on the load bed of a tricycle (O'Riordan, 2009) or in a tank fitted to the back (Figure 4.14).

More expensive motorised transport equipment has also been used for the collection of FS. Examples include pick-up trucks with load capacities ranging from 2,000 to 5,000 kg, but these are not always affordable for small-scale service providers (McBride, 2012; Bhagwan *et al.*, 2012), sometimes trucks can be fitted with additional options such as cranes with hook lifts (Losai Management Limited, 2011).

4.7.3 Delivering faecal sludge to the treatment plant or transfer station

It is becoming more common for larger FSTPs to make use of mechanised receiving stations as shown in Figure 4.16, where the operator connects the hose from the vacuum truck to the input port, electronically signs in, and discharges the load through the system provided. The receiving station will track the time and date of the load, the volume received, the operator's name, and any other relevant information as required. Mechanised receiving stations can therefore reduce human error and increase the accuracy and accountability of service providers. While mechanised receiving stations may be used for large operators, smaller systems and transfer stations rely on manual methods of vacuum truck emptying.

Figure 4.16 Automated faecal sludge receiving station at Manila Water's South Septage Treatment plant, Manila, Philippines (photo: WSUP, Sam Parker).

Independent of the delivery method of FS to the treatment plant or transfer station, operators should adhere to the following safety guidelines:

1 Check in with facility guard or operator.
2 Carefully following instructions regarding the sampling of FS. Some FSTPs have designated sites for residential septage, and others for commercial sludges. Plant operators may request samples of the FS prior to allowing discharge if it is suspected that the FS may contain materials hazardous to the plant.
3 Position the truck in the designated location for sludge removal, park and take the truck out of gear, apply the parking brake, and chock the wheels.
4 Remove the hose and make the connections.
5 Engage the power take-off or other mechanism for unloading the tank and complete the offloading process.
6 Obtain the necessary authorisation and access to the transfer station prior to transporting FS, as some transfer stations have locked inlets.
7 Ensure sufficient water is available for washing the solids as some transfer stations have screens to remove non-biodegradable solids.
8 Store any screened non-biodegradable solids in a safe location to drain and dry prior to containment and/or proper disposal either through incineration landfilling.
9 Use proper lifting techniques when discharging drums into a transfer station such as standing on a stable surface, and ensur all protective equipment is worn.
10 Clean up any spillage in the area around the inlet after completing the discharge of FS into the transfer station and re-seal the inlet.
11 Use personal protective equipment such as gloves and hard hats, and do not smoke during the entire collection and discharge operation.
12 Replace hoses and equipment, following adequate hygiene practices (e.g. hand washing), and completing the required paperwork.

4.8 TRANSFER STATIONS

4.8.1 Introduction

Earlier sections in this chapter introduced different options for the collection and transport of FS including small-scale equipment suitable for use in areas inaccessible to large vacuum tankers. While some of this equipment may be suitable for transporting sludge over short distances, the travel times are often too great for it to be economically viable for transporting FS to a final centralised treatment or disposal site.

In order to address this issue, decentralised transport approaches have been developed dividing the transport process into two main stages: primary and secondary. During the primary stage, carts or small vehicles are used to transport the FS from the point of collection to a nearby transfer station. In the secondary transport stage, the transfer station is emptied by means of large-capacity transport equipment such as a tanker truck, and transported to a final disposal point. It is essential that transfer stations be accessible to both primary and secondary transport equipment.

4.8.2 Types of transfer stations

There are two main types of transfer stations: 'fixed' and 'mobile'. These are described in more detail in the following sections.

Fixed transfer stations

Fixed transfer stations can be divided into four main categories, the first of which are 'permanent storage tanks'. Constructed as vault-like concrete structures, these tanks are designed to provide storage capacity for FS over a short period of time without capacity for treatment. An example of such tanks are the underground holding tank (UHT) reported by Boot (2007) in Accra, Ghana. With capacities of approximately 23 m^3, the UHTs were designed to provide access to pan latrine collectors (primary transport) and vacuum trucks (secondary transport). However, the natural solid-liquid separation and siltation that takes place when FS is stored over relatively long periods soon became an operational issue for local authorities. As a result, many UHTs fell into disuse as de-silting became a prohibitively costly and time-consuming process.

To mitigate the challenges of de-silting, a second category, the 'modular transfer station' has been developed using portable containers to replace the concrete vault. These come in various sizes such as:

- small sized (e.g. 200-litre metal drums, McBride, 2012);
- medium-sized (e.g. Intermediate Bulk Containers (IBCs) made of plastic liner and metallic frame, 500 – 3,000 litres);
- large-sized (e.g. customised metallic tanks or skips, >2,000 m^3 (Macleod, 2005; Strauss and Montangero, 2002).

The fixed transfer station essentially serves the role of a secure, safe, storage facility and can be designed according to the type of containers used. For example, in one project in Ghana a concrete-lined pit within a fenced compound was used to store the containers in order to avoid tampering, flooding or spillage. Once full the IBCs are emptied with a vacuum truck.

A third type of fixed transfer station is the 'multi-functional permanent tank'. In addition to providing storage capacity, it can also accept fresh FS from a public toilet, and/or provide partial sludge treatment. This latter design feature could include processes such as dewatering (settling tanks, drying beds, geotubes - ERE Consulting Group and Indah Water Konsortium, 2012) or anaerobic digestion (e.g. septic tanks, anaerobic baffled reactors, biogas digesters). The main advantage of stations providing both access to fresh FS and treatment capacity is easier siting due to acceptance by community and a reduction in secondary transport fees due to dewatering. Furthermore, treatment byproducts (e.g. liquid effluent or biogas) could be used if further treatment is provided. An example of a transfer station with capacity for receiving fresh FS is provided in Case Study 4.2.

The fourth type of fixed transfer station is the 'network-connected station'. These types of stations provide direct or indirect access to the sewerage network, if one exists, for the secondary transport of FS and/or its liquid effluent. Utilities and asset owners rightfully discourage the use of manholes for sludge disposal as it can lead to increased blockages in the network due to the low water content of the FS and also high BOD loading at the wastewater treatment plant (Chapter 9). However, illegal dumping into the network is not uncommon generally due to a lack of alternative facilities and the ease of access.

Mobile transfer stations

Mobile transfer stations consist of easily transportable containers providing temporary storage capacity at any point near the structure being emptied - essentially a tank fitted on a wheeled chassis. Examples of such transfer stations include motorised collection vehicles, or tanker trailers pulled via a truck or tractor.

The stations are sited in any area where multiple trips by small-scale transport equipment are required. The main advantage of these stations is that they sidestep the complex and often lengthy procedures required for siting fixed stations in high-density settlements. They can also double as secondary transport containers once full as they can be easily driven or towed to the final disposal site.

If towed, the motorised vehicle towing the container is capable of performing other related or unrelated duties thus allowing for cost savings and potential for increased revenue. Such systems have reportedly been used in places such as Maseru, Lesotho (Strauss and Montangero, 2002).

4.8.3 Siting of transfer stations

In order to establish a successful multi-staged transport system using transfer stations, the siting of the stations needs to be carefully planned. The following section highlights important aspects that should be considered in the planning stages.

Optimising coverage

The coverage of transfer stations needs to be sufficient to meet the demand generated by sludge collection using small-scale equipment, while at the same time minimising the overall cost of primary transport. In order to determine the appropriate coverage of transfer stations, their sizing, and their proximity to one another, the cost of primary and secondary transport methods being used need to be taken into account. The provisional use of mobile stations may assist in optimising coverage by allowing the evaluation of the suitability of potential locations over a period of time without committing to the construction of a fixed station.

Land availability

The process of finding suitable land space and obtaining the relevant permission for transfer stations can be difficult and time consuming. This can sometimes involve lengthy negotiations with multiple governmental agencies and land owners, particularly when siting within informal settlements. Due to their non-permanent nature, mobile stations could potentially mitigate such challenges, or alternatively consideration could be given for the use of modular transfer stations with small footprints. However, it should be noted that without adequate legal assurances, service providers could be required to remove such stations by dissatisfied landowners.

Acceptance

It is not uncommon for communities to reject the siting of a transfer station in close proximity to their homes. This so-called 'Not-In-My-Back-Yard'(NIMBY) effect can be challenging in densely populated informal settlements where there is little, if any, open land. Such rejections have been reported in Dar es Salam, Tanzania (Muller and Rijnsburger, 1992), Maputo, Mozambique (Godfrey, 2012) and Freetown, Sierra Leone. Early involvement of the communities in the siting process may therefore be necessary. Offering incentives such as combining transfer stations with other facilities like communal toilets and showers may help increase the level of acceptance.

Access

Depending on the type, access to a transfer station by primary and secondary transport vehicles or trunk sewers is necessary for proper operation. For instance, while siting a transfer station in the middle of a densely populated informal settlement would reduce the primary transport travel distance, it might not be accessible to larger vehicles used in secondary transport. It is thus necessary to ensure transfer stations are sited on roads large enough for access by secondary transport vehicles.

Case Study 4.2: Large capacity modular and multi-functional transfer station

A transfer station comprising a 6 m^3 shipping container placed atop a salvaged 6 m^3 tank container (ISO tank) was constructed by GOAL, Sierra Leone (Figure 4.16). The shipping container was customised to provide a disposal point for primary transport service providers, as well as one male and one female public sanitation facility. The disposal point was configured to receive sludge from 60-litre sealed drums through a screened chute connected to the tank container below. If and when necessary, the two containers can be disassembled, lifted with a crane and transported via a lorry to another location.

Water for the station was provided through a networked connection to a PVC water storage tank. This allowed for cleaning of the emptied 60-litre drums, as well as the installation of pour-flush toilets and hand washing facilities. The wastewater from the toilets was piped into the holding tank while the washing water was piped into a nearby soakaway. Access to the sanitation facilities was provided via a staircase and walkway, the latter being a necessity albeit one that limited access to those with reduced mobility.

Siting the station

A suitable site was identified on a piece of privately owned land, with the agreement that the station could be moved in the future if required. Obtaining the necessary permission from the local authorities was significantly easier than would have been for a fixed station, as the station was not legally classified as a permanent structure. Acceptance from neighbouring residents was also forthcoming as the station provided them with much needed toilet facilities. It has yet to be determined if the location is optimal for the primary and secondary transport service providers.

Performance

As of 2012 the transfer station was not yet operational and a management system had not been established. It is therefore difficult to evaluate performance, however some of the future potential challenges that have been identified include:

- the possibility that the land owner may require the demobilisation of the station, which would require a renewed siting process;

- the risk of public exposure to the sludge given the close proximity of the toilets to the disposal point;
- the difficulty of carrying the 60-litre drums up the stair case to the disposal point;
- the frequency of siltation and the need for de-silting the tank; and
- the cost implications of the potentially high volume of water entering the holding tank from the toilets.

Some of these issues might be addressed in the future, including providing a ramp dedicated for wheeling the drums to the disposal point, and providing dewatering capabilities.

Figure 4.17 Large capacity modular and multi-functional transfer station (top), and sanitation facility (bottom) (photo: GOAL).

4.9 OCCUPATIONAL HEALTH AND SAFETY

There are many specific health and safety concerns associated with the collection, transport, and discharge of FS. Unfortunately, the current situation is that the majority of FS collection and transport service providers in low-income countries do not practice adequate health and safety protection, particularly when employing manual emptying techniques or using small-scale equipment. As a result, service providers are at a high risk of exposure to physical, chemical and biological hazards. This section provides a brief summary of some of these hazards, and methods for reducing exposure. Additional health and safety information can be found in Chapter 11.

4.9.1 Physical hazards

The following physical hazards can exist in the handling of FS:

- low bearing capacity of the soil surrounding an unlined pit can lead to the collapse of its sidewalls during emptying (in particular for manual emptying);
- slips, trips and falls;
- exposure to sharp objects contained in the sludge (e.g. glass, metals);
- carrying heavy loads (e.g. containment structure cover or sludge-filled containers); and
- traffic (particularly relevant during transport).

4.9.2 Chemical hazards

The following chemical hazards are known to exist:

- direct and indirect oral, nasal and dermal exposure to chemicals (e.g. hydrocarbons that are introduced as odour suppressants, although this practise is not recommended);
- working in confined spaces in the presence of harmful gases (e.g. methane, ammonia, sulphur dioxide), in an oxygen depleted environment (in particular during manual emptying).

4.9.3 Biological hazards

The following biological hazards can exist in the handling of FS:

- direct and indirect oral, nasal and dermal exposure to multiple types of pathogens in FS (e.g. bacteria, viruses, protozoa, and helminthes – for more information refer to Chapter 2).

4.9.4 Other hazards

- pervasiveness of alcohol consumption in the FS collection profession (Godfrey 2012, Mikhael 2011).

4.9.5 Mitigating risks

Preventative measures for mitigating risks can be adopted voluntarily or, assuming effective enforcement measures are in place, through the introduction of regulation.

The first and best line of defence for mitigating risks is by limiting exposure to the previously defined hazards. This includes:

- provide and wear the appropriate personal protective equipment (PPE) to avoid direct and indirect exposure to FS (e.g. gloves, coveralls, rubber boots with a metal sole, safety glasses and safety masks);
- develop and provide training on the use of tools customised for local conditions and local containment systems in order to avoid direct contact with FS; and
- provide a training programme on standard operating procedures including the proper use of PPE, tools and equipment.

Preventative measures related to personal health care are recommended including immunisation, and a deworming program. The latter is recommended particularly for service providers transitioning from unsafe to safe practices.

4.10 CONCLUSION

Much has been achieved in the field of FSM to improve collection and transport practices. However, there are still significant gaps that will require innovative and practical solutions.

Many of the challenges faced with emptying onsite sanitation systems are related to or influenced by their design. For instance, a septic tank placed far from the road makes access by vacuum trucks more difficult. Sludge containment structures for toilets with water seals have much less non-biodegradable material (e.g. household solid waste, menstrual hygiene products, rags, and cloth) mitigating clogging during collection and thus making the process easier and more efficient.

Innovations for safer, more effective and more efficient collection and transport services need to consider improvements to existing containment structures, as well as developing future designs. For the implementation of new designs to be effective, a multidisciplinary approach is required, involving, a range of specialists including sanitation specialists, design engineers, promotion and marketing personnel, masons, and legal advisors.

From a technical perspective, some recommendations have been made for improvements to existing sanitation facilities to enable easier collection. One suggestion is the installation of a built-in suction pipe extending to the bottom of a pit latrine (Coffey, 2007). The pipe could be used to inject air and water at low pressures into the pit to fluidise the sludge, thereby allowing collection to start from the most dense and difficult-to-pump sludge at the bottom of the pit, to the least dense and easy-to-pump sludge at the top. This recommendation is supported by studies from Hawkins (1982) where it was suggested that the addition of 2% water to sludge could increase its fluidity by 30 to 300 times. While initial trials of this design were reportedly promising, there is no evidence of its large-scale promotion or adoption.

A further innovative method currently under investigation is the Omni-Ingester, being developed with funding by the Bill & Melinda Gates Foundation. The Omni-Ingester will be designed to separate water, sand, and trash from sludge followed by sanitation of byproducts in the collection truck, thereby reducing the need for transport to treatment plants, often the most expensive part of the service. Such technologies, when used in conjunction with community based FSM programs can provide transformative opportunities for wide scale sanitation improvement while considering the health and safety of the service provider.

4.11 BIBLIOGRAPHY

Barreiro, W.C., Strauss, M., Steiner, M., Mensah, A., Jeuland, M., Bolomey, S., Koné, D. (2003). Urban excreta management - Situation, challenges, and promising solutions. In IWA Asia-Pacific Regional Conference.

Bhagwan, J., Wall, K., Kirwan, F., Ive, O.M., Birkholtz, W., Shaylor, E., Lupuwana, N. (2012). Demonstrating the Effectivenes of Social Franchising Principles: The Emptying of Household VIPs, a Case Study from Govan Mbeki Village.

Boesch, A., Schertenleib, R. (1985). Emptying On-Site Excreta Disposal Systems: Field Tests with Mechanized Equipment in Gaborone (Botswana). Duebendorf, Switzerland: International Reference Centre for Waste Disposal (IRCWD).

Boot, N.L.D. (2007). Talking Crap: Faecal Sludge Management in Accra, Ghana. Water, Engineering and Development Centre (WEDC). Loughborough, UK: Loughborough University.

BPD (2005). Sanitation Partnerships: Dar es Salaam Case Study. Dar es Salaam, Tanzania: Building Partnerships for Development.

Brikké, F., Bredero, M. (2003). Linking Technology Choice with Operation and Maintenance in the Context of Community Water Supply and Sanitation. Report for World Health Organization and IRC Water and Sanitation Centre, Geneva, Switzerland.

Buckley, C., Foxon, K.M., Brouckaert, C.J., Rodda, N., Nwaneri, C., Balboni, E., Couderc, A., Magagna, D. (2008). Scientific Support for Design and Operation of Ventilated Improved Pit Latrines and the Efficacy of Pit Latrine Additives. University of KwaZulu-Natal, Pollution Research Group, School of Chemical Engineering. Gezina: Water Research Commission.

Chowdhry, S., Koné, D. (2012). Business Analysis of Faecal Sludge Management: Emptying and Transportation Services in Africa and Asia. Seattle: The Bill & Melinda Gates Foundation.

Coffey, M. (2007). Propose and Test Pit Latrine Designs for Effective Emptying by Mechanical Exhauster. Newtownmountkennedy, Ireland: Manus Coffey Associates.

De los Reyes, F. (2012). Hygienic Pit Emptying with Low Cost Auger Pump. Available from Sustainable Sanitation Alliance: http://www.susana.org/images/documents/07-cap-dev/b-conferences/12-FSM2/c8.3-fsm2-de-los-reyes-north-carolina-state-university-usa.pdf, accessed January 2013.

ERE Consulting Group and Indah Water Konsortium (2012). Landscape Analysis and Business Model Assessment in Faecal Sludge Management: Extraction and Transportation Model in Malaysia. Seattle: The Bill & Melinda Gates Foundation.

Godfrey, A. (2012). Faecal Sludge Management Demonstration Project in Maxaquene A and B, Maputo, Mozambique. Maputo: WSUP.

GRET Cambodia. (2011). Landscape Analysis and Business Model Assessment in Faecal Sludge Management: Extraction & Transportation Models - Cambodia, Volume 1 – Main Report. Phnom Penh, Cambodia: Bill & Melinda Gates Foundation.

Hawkins, P. M. (1982). Emptying On-Site Excreta Disposal Systems in Developing Countries: An Evaluation of the Problems. IRCWD News (17), p.1-9.

IDEO (2012). Clean Team Brand Guide. San Francisco, California: IDEO.org.

Losai Management Limited (2011). Landscape Analysis and Business Model Assessment in Fecal Sludge Management: Extraction and Transportation Models in Africa. Webster: The Bill & Melinda Gates Foundation.

Macleod, N.A. (2005). The Provision of Sustainable Sanitation Services to Peri-urban and Rural Communities in the ethekwini (Durban) Municipality.

MCA (2007). Operating and Maintenance Manual for UN-HABITAT MK III Vacutug Latrine Emptying Vehcile. Dublin, Ireland: Manus Coffey and Associates.

McBride, A. (2012a). A Portable Pit Latrine Emptying Machine - The eVac. Pietermaritzburg, South Africa: PID, EWB-UK, WfP.

McBride, A. (2012b). The eVac in Malawi. Blantyre, Malawi: PID, EWB-UK.

Mikhael, G. (2011). Assessment of Faeca Sludge Emptying Services – Freetown, Sierra Leone. Freetown, Sierra Leone: GOAL Sierra Leone.

Mikhael, G., Parkinson, J. (2011). Assessment of Financial Costs of Sludge Emptying Vacutug Services in Dhaka, Bangladeh (Unpublished). London, UK: Water and Sanitation for the Urban Poor (WSUP) and International Water Association (IWA).

MSF (2010). Public Health Engineering in Precarious Situations. Médecins Sans Frontières (MSF).

Muller, M. S., Rijnsburger, J. (1992). MAPET: A Neighbourhood-based pit emptying service with locally manufactured handpump equipment in Dar es Salaam, Tanzania. Gouda, The Netherlands: WASTE Consultants.

O'Riordan, M. (2009). WRC PROJECT 1745 Management of sludge accumulation in VIP latrines Investigation into Methods of Pit Latrine Emptying. Durban: Partners in Development (Pty) Ltd.

Still, D., Foxon, K. (2012). Tackling The Challenges of Full Pit Latrines Volume 1: Understanding sludge accumulation in VIPs and strategies for emptying full pits. Gezina: Water Research Commission.

Still, D., O'Riordan, M. (2012). Tackling the Challenges of Full Pit Latrines Volume 3: The Development of Pit Emptying Technologies. Gezina: Water Research Commission.

Strauss, M., Montangero, A. (2002). Faecal Sludge Management – Review of Practices, Problems and Inititatives. Dubendorf: EAWAG / SANDEC.

Tilley, E., Lüthi, C., Morel, A., Zurbrügg, C., Schertenleib, R. (2008). Compendium of Sanitation Systems and Technologies. Dübendorf, Switzerland: Swiss Federal Institute of Aquatic Science and Technology (EAWAG).

WASTE Consultants. (1993). Summary of the Comparitive Study on Pit Emptying Technologies (COMPET). Dar es Salaam, Tanzania: WASTE Consultants.

Additional Reading Material

Boesch, A., Schertenleib, R. (1985). Emptying onsite excreta disposal system – Field Tests with Mechanized Equipment in Gaborone (Botswana). International Reference Centre for Waste Disposal (IRCWD). (IRCWD report; no. 03/85). Duebendorf, Switzerland: International Reference Centre for Waste Disposal.

O'Riordan, M. (2009). Investigation into methods of pit latrine emptying – Management of sludge accumulation in VIP latrines. Water Research Commission (WRC), South Africa

Still, D., Foxon, K., O'Riordan, M. (2012). Tackling the challenges of full pit latrines – Volumes 1 to 3. WRC Report No. 1745/1/12, Water Research Commission, South Africa.

End of Chapter Study Questions

1. List four manually operated mechanical collection technologies and describe how these systems work.
2. What technical challenges are often encountered in the operation of the Gulper System?
3. Explain what transfer stations are, and describe two types of transfer stations.
4. Name three types of hazards in the collection and transport of FS, and how to mitigate them.

Chapter 5

Overview of Treatment Technologies

Mariska Ronteltap, Pierre-Henri Dodane and Magalie Bassan

Technology

Learning Objectives

- Have an overview of existing faecal sludge treatment technologies.
- Have an overview of potential future treatment technologies and what information is currently lacking for their reliable implementation. Understand the advantages, constraints and field of application for treatment technologies.
- Be able to compare and contrast different treatment technologies based on performance and appropriateness for local contexts.
- Understand the importance of finding a context-adapted combination of technologies

5.1 INTRODUCTION

In the preceding chapters the characteristics of faecal sludge (FS) were discussed, an introduction to treatment mechanisms was given, and it was described how FS can be collected from the point of storage and delivered to treatment facilities.

In this chapter, an overview and introduction of FS treatment technologies is provided. Each technology has different fields of application. They can be used for the treatment or co-treatment of undigested FS (e.g. from public toilets), or digested or pretreated FS. Given the high content of coarse wastes such as plastics, tissues and paper in the FS discharged by collection and transport trucks, a preliminary screening is needed for most treatment technologies. Also, the characteristics of FS collected at industrial and commercial facilities should be checked as they can be contaminated with metals, have high concentrations of fats, oil and grease, or other concerns as described in Chapter 2. They should be segregated and treated separately. After treatment, three types of endproducts can be distinguished:

- screenings;
- treated sludge; and
- liquid effluents.

Examples of potential combinations of FS and endproducts are presented in Figure 5.1, and are further detailed in the Technology Selection Scheme (see Chapter 17). Selection of the technologies, or combinations thereof, should be done taking into account the local context, existing regulations and the enduse goals (see Chapter 10). Treatment will be comprised of combinations of:

- the treatment of FS directly transported from the onsite sanitation systems - this treatment can be done in one or several steps, and produces solid and liquid endproducts; and
- the further treatment of the resulting endproducts (either the solid part of the treatment endproducts (treated FS) or liquid effluents) before enduse or final disposal.

Additionally, technologies have different fields of application: some can be used either for the treatment or co-treatment of "fresh" FS (e.g. from public toilets), others are better suitable for treating digested FS (e.g. from septic tanks) or pretreated FS. This is related to the fact that fresh FS, derived from bucket or public latrines with a high emptying frequency (once per month or more often), is more difficult to dewater (Chapter 2 and 3) and may produce odors during digestion (Heinss *et al.*, 1998). In order to overcome these problems, fresh FS can be exposed to a chain of treatment technologies, with first a digestion step. Another option is to mix fresh with digested FS. Cofie *et al.* (2006) successfully conducted experiments on unplanted drying beds with a mix of fresh and digested FS at a ratio of 1:2.

In addition to treatment goals, a choice of technologies cannot be made without some form of cost comparison. However, a complete overview of the costs is difficult to obtain as many factors play a role and costs are strongly context specific; moreover, the lack of long-term experience of operational FS treatment implementations further complicates reliable cost estimation. To aid in the selection of appropriate technologies, the last section of this chapter offers insight in the process of cost calculation over the expected life time of technologies, and how costs of faecal sludge management (FSM) and sewer based technologies can be compared.

There are a number of technologies available for the treatment of FS; however, the same level of operational information is not available for all of them based on their varying degrees of implementation and current level of research. In this book, the FS treatment technologies that are covered in an operational level of detail are settling tanks, unplanted drying beds, planted drying beds and co-treatment with wastewater, which are described in detail in Chapter 6, 7, 8 and 9 respectively. Technologies that have also been implemented and are relatively well established are co-composting of FS together with municipal solid waste (MSW); co-treatment of FS in waste stabilisation ponds; and deep row entrenchment. These are described in this chapter under Section 5.2.

Several other technologies have been adapted from activated sludge wastewater treatment and sludge management practises. This category includes anaerobic digestion of FS, sludge incineration, mechanical sludge drying processes such as centrifugation, and chemical treatment through lime addition. These are covered in Section 5.3. Finally, there is a whole range of technologies that are currently under development. Much of this current research has the goal to increase the financial value of treated endproducts, and some of these technologies are also introduced in Section 5.4 and in Chapter 10.

5.2 TREATMENT TECHNOLOGY OVERVIEW

An overview of treatment technologies, together with their treatment objectives and functionality, are presented in Figure 5.1. It is important to realise that for the conversion of FS into a product that is safe for enduse or disposal, several processes need to take place. FS typically contains large volumes of water and hence needs to be dewatered, which can be achieved on its own, or in combination with solid / liquid separation. Depending on the endgoal, further treatment needs could include converting organic matter into a stabilised form and/or pathogen reduction. The underlying mechanisms are described in

Chapter 3. One of the key elements in designing any particular series of technologies is to keep the final goal in mind (see Chapter 10). If the final goal is to make a dry product that can be reused in agriculture, then particular care has to be paid to dewatering and pathogen reduction. If the goal is to incinerate the sludge for energy production, then dryness is very important while pathogens do not play a role (outside of worker protection).

The technology selection scheme (Figure 17.10) provides assistance for selecting technologies that are appropriate for a particular setting, taking into account all role playing factors. Other key elements in selecting technologies, such as economic feasibility are covered at the end of the chapter, and local regulations and local context are discussed throughout the Planning section of the book.

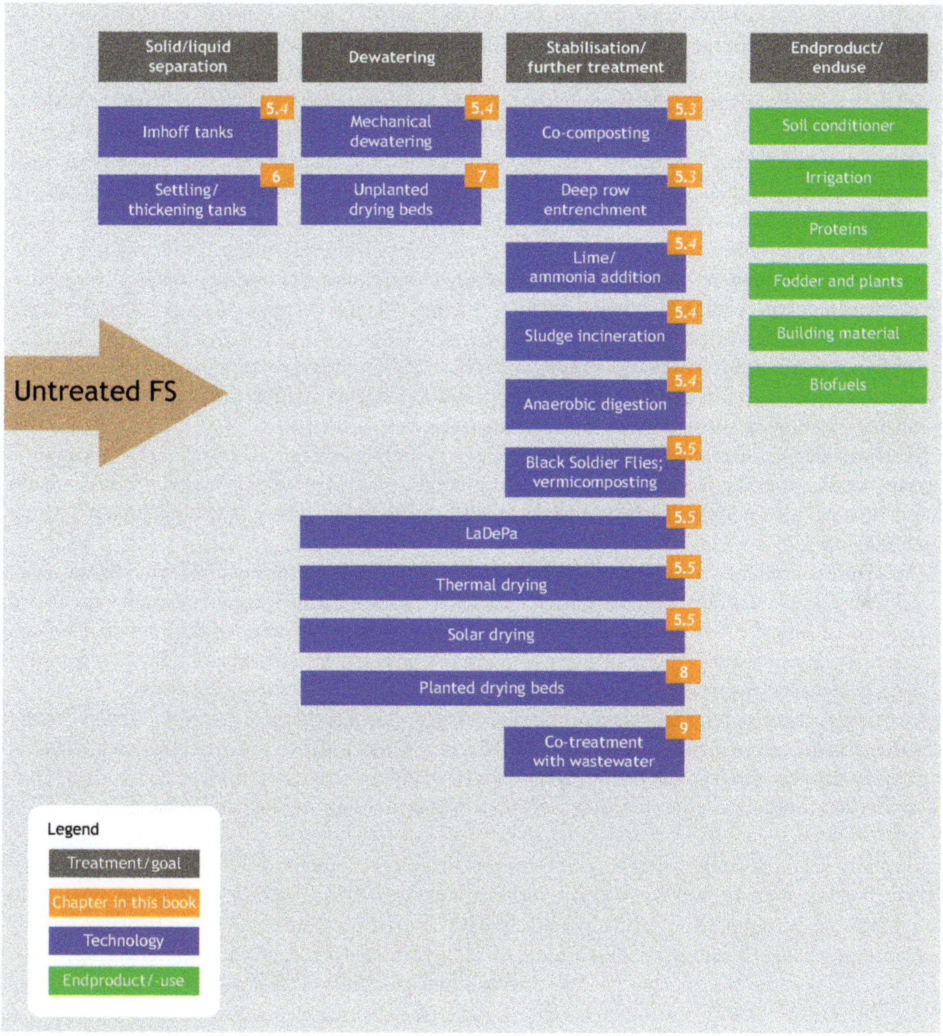

Figure 5.1 Grouping of treatment technologies according to their treatment goal. Endproducts are discussed in Chapter 10. Specific flows from one technology into the next is presented in the Technology Selection Scheme of Chapter 17.

5.3 ESTABLISHED FAECAL SLUDGE TREATMENT TECHNOLOGIES

5.3.1 Co-composting of faecal sludge

Composting is a biological process that involves microorganisms that decompose organic matter under controlled predominantly aerobic conditions. The resulting endproduct is stabilised organic matter that can be used as a soil conditioner. It also contains nutrients which can have a benefit as a long-term organic fertiliser. There are two types of composting systems, open and closed, of which open systems are lower in capital and operating costs but typically require more space. In an open composting system, raw organic matter is piled up into heaps (called windrows) and left for aerobic decomposition. To increase space efficiency, the heaps of waste can also put into walled enclosures which is called box composting. If untreated waste feedstock is placed in a closed container this is called in-vessel or closed drum composting and is considered in the category of closed systems.

To ensure an optimal composting process, the following parameters need to be controlled (EAWAG and IWMI, 2003):

- A carbon to nitrogen ratio (C:N) between 20-30:1 to ensure biological availability; as the organisms degrading organic matter need carbon as a source of energy and nitrogen for building cell structure. High nitrogen enhances ammonia loss due to volatilisation. Higher C:N ratio hinders optimal growth of the microbial populations due to insufficient nitrogen. The compost heap will remain cool and degradation will proceed slowly. High carbon in the final compost product can create problems as microbial activity in the soil may use any available soil nitrogen to make use of still available carbon, thereby "robbing" the soil of nitrogen and thus hindering its availability for plants. During composting carbon is converted to CO and the C:N ratio decreases to a ratio of around 10:1 when the compost is stabilised.
- An oxygen concentration of 5-10% to ensure aerobic microbiological decomposition and oxidation. Aeration can be ensured by either providing passive ventilation structures (air tunnels) or can be enhanced by blowing or sucking air through the waste heap (called active or forced aeration). With forced aeration external energy is required. In open systems mechanically or manually turning the heaps can also contribute to better aeration, although the main objective of this turning is to ensure that material on the outside of the heap is moved to the centre where it will be subject to high temperatures.
- A moisture content between 40 and 60 % by weight to ensure adequate moisture for biodegradation, and that piles are not saturated creating anaerobic conditions. Turning removes water vapour and thus the turning frequency depends primarily on the moisture content of the material, as high moisture content reduces the availability of air in the pore space (Cooperband, 2002). If compost heaps become too dry, water must be added to ensure continuous biological activity.
- A particle diameter of less than five centimetres for static piles. Smaller particles degrade more rapidly as they provide more surface area for microbial decomposition. But on the other hand with smaller particles size aeration through the pile is hindered if structural strength cannot be maintained. Thus particles size influences pore structure and aeration as well as surface area for degradation.

In a properly operated composting heap the temperature rises rapidly to 60-70 °C as heat is released when carbon bonds are broken down in an exothermic process. Pathogen die-off is highest during this time of high temperature. After approximately 30 days, the temperature drops down to 50 °C. During the maturation phase the temperature is around 40 °C, and the process ends once ambient temperature is reached. The whole composting process (including maturation) takes a minimum of six to eight weeks (Klingel *et al.*, 2002).

Optimal composting conditions with appropriate C:N ratio and moisture content can be achieved by mixing different wastestreams together. For example, mixing of FS and MSW for co-composting can be advantageous as excreta and urine are relatively high in nitrogen (see Chapter 2) and moisture, whereas municipal waste can be relatively high in carbon and low in moisture. Concentrations of high lignin materials should be limited as they are resistant to biological degradation. Corn stalks and straw made of a tough form of cellulose are also more resistant to degradation. Although all these materials can be used a higher initial C:N ratio to compensate for lower bioavailability must be considered. A moisture content between 40 and 60% is considered ideal which corresponds to the feel of a damp sponge. Higher moisture content limits air supply, creates anaerobic conditions and results in odor emissions.

Co-composting of FS with MSW is best implemented with sludge that has undergone dewatering (e.g. settling-thickening tanks or drying beds). Although untreated FS can also be used and sprayed over the compost heaps, its high water content will only allow the use of very little volume before the compost heap is too wet and is therefore not practical. Organic MSW usually already has a moisture of 40-60% so typically not much additional moisture can be added before the system gets too wet. In the case of dewatered sludge, FS with a total solids (TS) content higher than 20% is mixed together with MSW in compost piles (Koné *et al.*, 2007). For further guidance on ensuring optimal carbon, nitrogen and moisture content, refer to the Sandec website (www.sandec.ch), including the publications: Co-composting of Faecal Sludge and Municipal Organic Waste and Marketing Compost (EAWAG and IWMI 2003) Rouse *et al.* (2008) and Strauss *et al.* (2003).

Potential advantages and constraints of co-composting

The main advantage of co-composting is formed by the thermophilic conditions and the resulting pathogen inactivation. The output of co-composting is a good soil conditioner which provides potential for income generation depending on the demand for compost (see Chapter 10). However, operating a co-composting plant and generating a safe product with value requires technical and managerial skills, which can be limiting if not available.

Case Study 5.1 : Co-composting of faecal sludge and organic solid waste in Kumasi, Ghana

(Adapted from Cofie and Koné, 2009)

In February 2002 a research pilot plant was opened in Kumasi, Ghana to study sludge drying in an unplanted drying bed and the subsequent operation of co-composting with organic MSW. For the location the facility of the Buobai treatment plant was chosen. FS was delivered to the plant in trucks, collected from unsewered public toilets and household septic tanks in Kumasi. On the project site, sludge was first dried in unplanted drying beds, as due to its high moisture content, fresh FS is unsuitable for direct aerobic composting. The FS was applied to the beds in a mixture of public toilet and septic tank FS ratio of 1:2. The dried FS was removed from the drying beds once it became spadable (about 10 days) and stored prior to co-composting. For organic waste, the organic fraction of municipal solid waste from markets or residential areas, collected and delivered by trucks to the composting site, was used. The organic fraction of the MSW and dried FS were mixed in a ratio of 3:1 and composted using an open windrow system (aeration by manual turning). During a composting cycle, the following activities were carried out: turning, watering, temperature measurement, weighing and sampling (followed by laboratory analysis). The matured compost was sieved, packed in 50 kg bags and stored prior to reuse - more details are provided in Table 5.1. The compost was tested for its impact on the germination capacity of selected vegetables and the germination capacity varied between 70-100%, which is an acceptable range. A large number of interviewed farmers (83%) were willing to use excreta-based compost.

With regard to helminth eggs, an optimum composting period of at least 2 months was necessary to produce compost that complied with the WHO guidelines of 1 Ascaris egg/gTS. High Ascaris inactivation efficiency (90–100%) was reached after 80 days due to heat generation during the composting process, thus exposing the helminth eggs for more than one month to temperatures over 45°C. Note that if these conditions are not met, pathogen reduction will not be adequate to meet the WHO guidelines. In that case, one possibility is extended storage prior to enduse.

From the above described investigation, it can be concluded that composting of dewatered FS is technically possible; it results in a soil conditioner safe for use in agriculture. Whether the technology is economically viable, however, depends on various local conditions (see Chapter 10 for issues regarding economic feasibility of endproducts).

Table 5.1 Design criteria and assumptions, used for the pilot-scale unplanted sludge drying bed and co-composting plant in Kumasi, Ghana (Cofie and Koné, 2009)

Faecal sludge dewatering		Co-composting	
Volume of FS treated	45 m^3/month = 1.5 m^3/day	**Ratio MSW:dried FS**	3:1 (by volume)
Dewatering cycles	3 per month	**Composting time**	1 month thermophilic; 1-2 months maturation
FS sludge truck loads	3 per cycle (1 truck ~ 5 m^3)	**Composting cycle**	One starting each month
Ratio public toilet sludge: septic tank sludge	1:2	**Required volume MSW**	3 · 4.5 = 13.5 m^3/month
Surface sludge drying beds	50 m^2	**Raw compost produced**	4.5 + 13.5 = 18 m^3/month
Hydraulic load on beds	30 cm / cycle	**Volume reduction through co-composting**	50%
Dried sludge produced	1.5 m^3 / cycle	**Mature compost produced**	9 m^3/month = 4.5 tonnes/month (density 0.5t/m^3)

5.3.2 Co-treatment in waste stabilisation ponds

Waste stabilisation ponds (WSPs) are widely used for the treatment of municipal wastewater. The mechanisms for stabilisation are based on natural processes that occur in aquatic ecosystems. WSPs are considered to be good options for wastewater treatment in low-income countries when adequate land is available, particularly in tropical climates (Mara, 2004). WSPs consist of several ponds having different depths and retention times.

A combination of three types of ponds in series is frequently implemented in wastewater treatment (Figure 5.2):

1 Anaerobic ponds that are two to four meters deep are used for settling of suspended solids and subsequent anaerobic digestion. The effluent flows to the facultative pond.
2 Facultative ponds that are 1 to 1.8m deep allow for remaining suspended solids to settle. In the top layer of the pond dissolved organic pollution is aerobically digested, while anaerobic conditions are prevalent at the bottom.

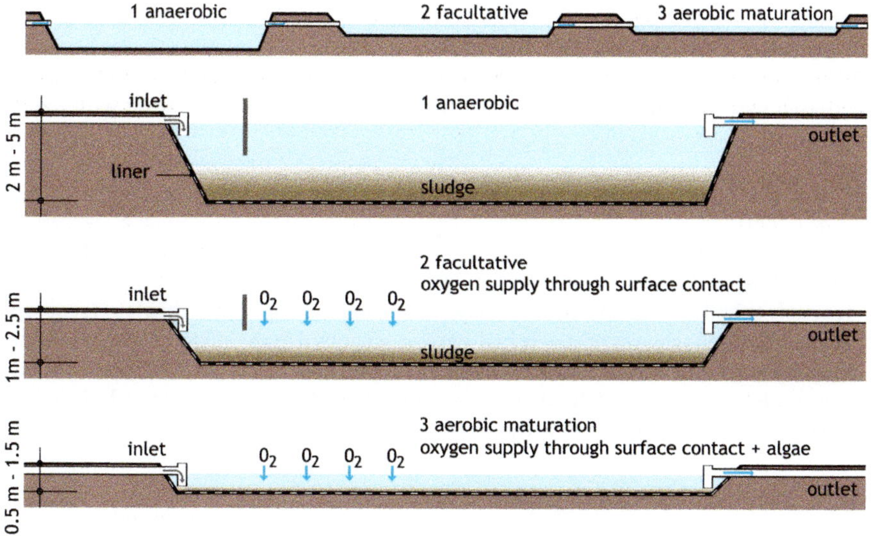

Figure 5.2 Design and principles of the three types of ponds constituting waste stabilisation ponds (Tilley *et al.*, 2014).

3 Maturation ponds that are 1 to 1.5 m deep allow for pathogen reduction and stabilisation. The ponds are mainly aerobic. Oxygen is supplied through algae and diffusion from the air. Pathogen reduction occurs via UV rays from the sun.

With the addition of FS to WSPs, ammonia quickly becomes a limiting factor. Cascade or mechanical aeration can be implemented to ensure adequate oxygen, which also helps to lower the ammonia concentration (Strauss *et al.*, 2000).

Stabilisation ponds are designed for organic loading rates. Anaerobic ponds have 2-3 m depth, remove 60-70 % of BOD and produce limited odor when loaded with 250–350 gBOD/m^3/day. Facultative ponds are 1-2 m deep and loaded with 350 kgBOD/ha/day (35 gBOD/m^2/day; Klingel *et al.*, 2002).

Waste stabilisation ponds can be used for the co-treatment of wastewater with the effluent following solid-liquid separation of FS in settling-thickening tanks. Papadopoulos *et al.* (2007) also applied FS directly after screening, but only in small volumes compared to the wastewater influent. However, problems were observed when dosing FS to the anaerobic pond and dosing was discontinued. Typically, due to the high ammonia concentration and high organic loads and solid content, treating solely FS in WSPs is not recommended, nor is the addition of large quantities (Strauss *et al.*, 2000). WSPs can also be used for the co-treatment of FS with landfill leachate (Kurup *et al.*, 2002), and can treat liquid by-products of other FS treatment technologies, including:

- Leachate from unplanted and planted drying beds. Leachate is low in organic matter compared to domestic wastewater and direct discharge into the facultative pond might be possible as the solid fraction is relatively low. However, the ammonia concentration can still present a problem, and algae and methanogenic inhibition by free ammonia can also occur (as discussed above).

- Effluent from settling-thickening tanks. This was implemented in Argentina as co-treatment with the influent of anaerobic ponds, where tests were conducted for the treatment of the effluent from settling ponds (Fernández *et al.*, 2004; Ingallinella *et al.*, 2002). This solution has also been adopted in Dakar, Senegal, where preliminary solid/liquid separation is done by settling tanks, the effluent is co-treated with wastewater in a WSP, and the thickened sludge is dewatered with unplanted drying beds.

More detailed reading on WSPs, including design, can be found in Mara (2004), Mara *et al.* (1992) and Strauss *et al.* (1999).

Potential advantages and constraints of waste stabilisation ponds

WSPs are simple to build and require relatively low O&M requirements. The technology is appropriate for tropical climates, and achieves relatively high pathogen removal in the effluent. Constraints include land availability, high rate of solids accumulation if preliminary solids separation is not performed, and potential inhibition due to high salt and ammonia concentrations. The removal of sludge that accumulates in the anaerobic ponds may require heavy mechanical equipment (Strauss *et al.*, 2000).

5.3.3 Deep row entrenchment

Deep row entrenchment is a technology that can be considered as both a treatment and enduse option, and is therefore also covered in Chapter 10. Deep row entrenchment was implemented for wastewater sludge in the US in the 1980s and has been adapted for FS in Durban, South Africa (Still *et al.*, 2012). Deep row entrenchment consists of digging deep trenches, filling them with sludge and covering them with soil. Trees are then planted on top, which benefit from the organic matter and nutrients that are slowly released from the FS. In areas where there is adequate land available, deep row entrenchment can present a solution that is simple, low cost, has limited O&M issues and produces no visible or olfactory nuisances. Benefits are also gained from the increased production of trees. However, the availability of land is a major constraint with deep row entrenchment, as is the distance/depth to clean groundwater bodies. In the application in Durban, limited nitrate leaching was found in the soil and tests conducted in the area showed that surrounding groundwater bodies remained free from pollution. It also appeared that the fast growing trees took up the additional nutrients (Still *et al.*, 2012). Deep row entrenchment is considered most feasible in areas where the water supply is not directly obtained from the groundwater source and where sufficient land is available, which means the sludge would have to be transportable to rural and peri-urban areas. In many countries legislation is still lacking for this option; in South Africa for example, environmental regulations will only allow deep row entrenchment for pit sludge disposal at the pilot scale in the foreseeable future.

Potential advantages and constraints of deep row entrenchment

The main advantage of deep row entrenchment is that very little is needed for it: no expensive infrastructure or pumps that are very susceptible to poor maintenance. In addition, growing trees has many benefits such as extra CO_2 fixation, erosion protection, or potential economic benefits. Constraints are that sufficient land has to be available in an area with a low enough groundwater table and, moreover, legislation still needs to catch up in many countries to allow for this technology.

Case Study 5.2: Deep row entrenchment in Durban, South Africa

(Adapted from Still *et al.*, 2002)

The water and sanitation unit (EWS) of the eThekweni municipality in Durban has been pursuing deep row entrenchment for disposal and treatment of both sludge from municipal wastewater treatment and FS derived from ventilated improved pit latrines (VIPs). The EWS project in Umlazi, south of Durban, started operation in 2009. Pit latrine sludge was buried at different loading rates in sandy soils (Figure 5.4; Still *et al.*, 2012). Positive effects were seen on the trees that were planted, however, there where substantial differences depending on the species and experimental conditions.

Figure 5.4 The Umlazi Deep Row Entrenchment Test Site – top the burial of faecal sludge from pit latrines in 1m deep trenches; below an overview picture of the filled trenches with trees planted on top. Groundwater wells were mapped to follow the fate of nutrients, organics and pathogens (photos: Jay Bhagwan, Water Research Council, South Africa).

At a second testing site near Durban it was observed that the relative difference in growth between trees grown with sludge and controls was reduced over time. After one year a 300% increase was observed for the trees growing with FS while at the end of the nine year growth cycle only a 30% to 40% more biomass was obtained, which is still a substantial increase. In addition to nutrients, Ascaris were also monitored. The South African researchers found that while a significant number of helminth ova were found in freshly exhumed pit latrine sludge, after 2.8 years of entrenchment less than 0.1% were found to still be viable (capable of growth or infection, Still *et al.*, 2012).

5.4 TRANSFERRED SLUDGE TREATMENT TECHNOLOGIES

Activated sludge wastewater treatment produces waste sludge that needs treatment. Technologies that are typically applied there may be transferable to application in FSM. The benefit of these technologies is that they have generally been applied for many years and much knowledge is present regarding design, operation and maintenance. The difficulty is however that the application of these technologies to FS has not been researched in much detail yet, which is key for succesful long-term implementation. More details are given per technology.

5.4.1 Anaerobic digestion

During anaerobic digestion, organic matter is converted into biogas and the remaining sludge is referred to as slurry or digestate. Biogas is a mixture of mainly methane and carbon dioxide and the digestate is relatively biologically stable and can be used as a soil conditioner. More information on fundamentals of biogas and its use are provided in Chapters 3 and 10.

Anaerobic digestion treats organic waste in airtight chambers to ensure anaerobic conditions. Anaerobic digestion has been widely applied in centralised wastewater treatment facilities for the digestion of primary sludge and waste activated sludge, typically with plug flow (PFR) or continuously stirred reactors (CSTRs). Anaerobic treatment technologies also include upflow anaerobic sludge blanket (UASB) reactors, anaerobic baffled reactors (ABRs) and anaerobic filters. Anaerobic treatment is also well known and developed for industrial wastes and highly loaded wastewater treatment plants (e.g. agro-industries, Arthur *et al.*, 2011). Throughout Asia, the onsite anaerobic digestion of animal manure with or without the addition of FS is widely practised (Koottatep *et al.*, 2004). However, the potential for semi-centralised to centralised treatment of FS in urban areas still remains untapped. There is great potential for the future development of anaerobic digestion of FS.

The main design parameters for anaerobic digesters are the hydraulic retention time (HRT), the temperature and the loading pattern. Operating conditions that play an important role in the design and operation of anaerobic digesters include:

- solids retention time (SRT);
- HRT;
- temperature;
- alkalinity;
- pH;
- toxic / inhibiting substances; and
- bioavailability of nutrients and trace elements.

When designing an anaerobic reactor, it is important to know the organic load that can be expected, in order to allow for a long enough HRT for degradation to occur. For systems without recycling, the SRT is equal to the HRT (e.g. plug flow reactors). The anaerobic reaction steps are directly related to the length of the HRT: an increase or decrease in the HRT leads to an increase or decrease in the degree of hydrolysis, acidification, fermentation and methanogenesis (Metcalf and Eddy, 2003). It is therefore important to keep track of the HRT to prevent reactor failures. The temperature also plays an important role, especially on the degree and rate of hydrolysis and methane formation. At the same time, temperature also affects physical and chemical parameters in reactors such as gas exchange and solubility of salts, and inactivation of pathogens.

Experience with faecal sludge

There have been a few studies evaluating the anaerobic digestion of FS and excreta. Arthur *et al.* (2011) and Klingel *et al.* (2002) recommend preliminary thickening to reduce the sludge volume and, consequently, the digester size. For the anaerobic digestion of fresh excreta, Daisy and Kamaraj (2011) report in their review that reduction of bacteria and viruses is very well possible, if a long enough hydraulic retention time is chosen. Song *et al.* (2012) found that between 15 and 90 mL biogas/g FS can be produced over 15 to 30°C respectively. However, the gas production only represented a 30% reduction of volatile solids, whereas the theoretical reduction is 50-60%. This indicates that gas production could be higher if operating conditions were optimised.

Potential advantages and constraints of anaerobic digestion for faecal sludge management

Anaerobic digestion has the potential to produce biogas while stabilising FS, reducing sludge volume and odors. However, operation and maintenance (O&M) of anaerobic digesters requires a relatively high level of skilled operation. Inhibition of digestion needs to be considered due to the inconsistent nature of FS, and also detergents and heavy metals should be addressed at the household level. A constraint of anaerobic digestion as a technology for FS treatment is that, despite the vast amount of knowledge on anaerobic digestion, it has not yet been proven for FS alone in semi-centralised to centralised treatment in urban areas. Hence, further research is needed, and pilot scale facilities need to be installed prior to full scale implementation in order to learn more about the safe and sustainable application of this technology.

5.4.2 Imhoff tank

An Imhoff tank is a compact sized tank that combines the effect of a settler and an anaerobic digestion system in one (Figure 5.5). It is a compact system which is well-known for wastewater treatment and has been implemented in Indonesia for FS treatment. Imhoff tanks are most often used as a primary treatment technology in wastewater treatment where it serves as a solid-liquid separation system including partial digestion for the settled sludge. The same considerations for sludge characteristics that were described under anaerobic digestion apply here.

The Imhoff tank is a high-rised tank (up to nine meters for wastewater sludge) where sludge settles at the bottom and biogas produced by the anaerobic digestion process rises to the top. The settling compartment has inclined walls (45° or more) and a slot at the bottom, which allows the sludge to slide down to the centre into the digestion compartment. The gas transports sludge particles to the water surface, creating a scum layer. T-shaped pipes or baffles are used at the inlet and the outlet to reduce velocity and prevent scum from leaving the system. The sludge accumulates in the sludge digestion

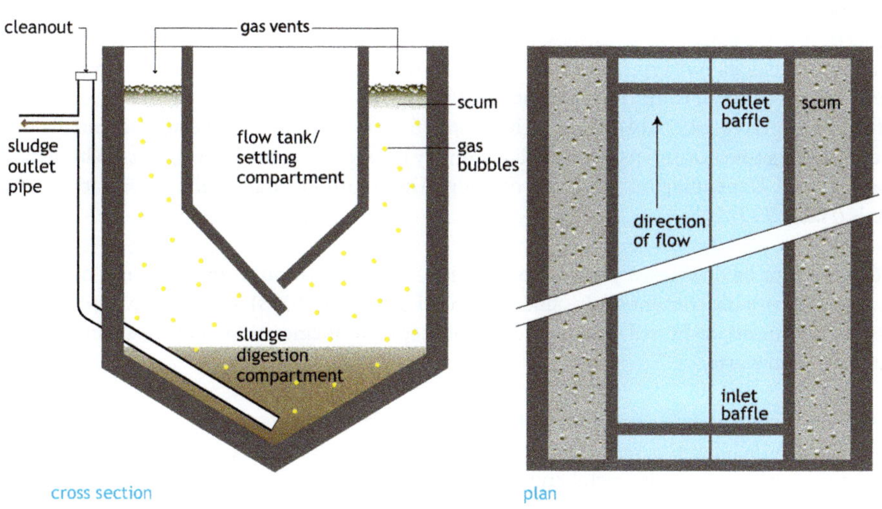

Figure 5.5 Schematic representation of an Imhoff tank (Tilley *et al.*, 2014).

chamber, and is compacted and partially stabilised through anaerobic digestion. The liquid fraction has a short retention time (2 - 4 hours) while the solids can remain up to several years in the digestion chamber. Both the supernatant and the settled sludge need further treatment before final disposal or enduse. The sludge can be further treated in a settling - thickening tank or on a sludge drying bed; the liquid can be treated in for example a constructed wetland. An Imhoff tank can be used when conditions are not favorable for biogas digesters or space for stabilisation ponds is not available.

Dimensioning of the anaerobic digestion compartment depends mainly on the temperature and sludge accumulation and the targeted degree of sludge stabilisation, which are linked to the desludging frequency. The digestion chamber is usually designed for four to 12 months sludge storage capacity to allow for sufficient anaerobic digestion. In colder climates longer sludge retention time and, therefore, a greater volume is needed. The Imhoff tank is usually built underground with reinforced concrete. It can, however, also be built above ground, which makes sludge removal easier because it can be done by gravity. For desludging, a pipe and pump have to be installed or access provided for vacuum trucks and mobile pumps (see Chapter 4). A minimum clearance of 50 cm between the sludge blanket and the slot of the settling chamber has to be ensured at all times. Scum and gas vent chambers are located at the sides of the tank; an outlet for desludging can be added (WSP, 2007; see Figure 5.5). A bar screen or grit chamber is recommended prior to the Imhoff tank to prevent coarse material from disturbing the system.

Potential advantages and constraints of Imhoff tanks

The main advantages of Imhoff tanks compared to settling-thickening tanks are the small land requirement, the possibility of operating only one tank (Klingel *et al.*, 2002), and the physical separation between the settled sludge and the liquid fraction. The main constraints compared to settling thickening tanks are the increased operational complexity, slightly higher costs as the Imhoff tanks require an additional elevation to accommodate the inclined baffles, and the risk of damage to the sludge draw-off pipe in case of an inadequate draw-off frequency. Operation and maintenance of an Imhoff system is not as complex as some technologies, but it requires skilled operators. Cleaning of flow paths, the sides of the tank as well as the removal of scum is very important. Stabilised sludge from the bottom of the digestion compartment should be removed according to the design (EAWAG *et al.*, 2010).

5.4.3 Sludge incineration

Incineration of sludge is a form of disposal which involves the burning of sludge at temperatures between 850-900 °C. It does not typically take advantage of the potential for resource recovery, however, energy can be captured from the incineration of sludge, for example in cement kilns (see Chapter 10 – Murray Muspratt *et al.*, 2014). The ash that is produced from incineration could potentially be used, for example as a cover material for urine diversion dry toilets or in construction, or it can be disposed of in landfill sites. Depending on the source of sludge, the ash may contain high concentrations of heavy metals (Hall, 1999).

Sludge needs to be dewatered prior to combustion, but stabilisation treatment is not necessary as it decreases the volatile content of the sludge (Metcalf and Eddy, 2003). Commonly used incineration systems are multiple-hearth incineration, fluidised-bed incineration and co-incineration with municipal solid waste.

Potential advantages and constraints of sludge incineration

Disadvantages include: the potential emission of pollutants; the need for highly skilled operating and maintenance staff, high capital and O&M costs; and residual ashes (Metcalf and Eddy, 2003). Advantages are that the sludge volume is substantially reduced and all pathogens are removed.

5.4.4 Mechanical sludge treatment

Mechanical dewatering or thickening can be carried out prior to, or following other treatment steps. Dewatering and thickening are important for reducing the volume of sludge that needs to be further treated or managed. After the sludge thickening process, additional reduction of the water content is often necessary and this can be done either naturally or by machine processes such as centrifugation or pressing.

Four technologies that are widely used for dewatering WWTP sludge are the belt filter, the centrifuge, the frame filter press, and the screw press. Only few examples are available in the literature for the implementation of these technologies to FS, but theoretically the technology is transferable. In Malaysia, centrifugation is used to dewater FS after screening and addition of flocculants where there is no space available for more land intensive technologies.

The following technologies are well recognised for wastewater management, and preliminary addition of flocculant is recommend for all of them to facilitate the separation of liquid from the solid particles. Although they are widely used for treating wastewater sludge, further experiments are required before recommendations can be made on design and operation of such systems for FS treatment.

Belt filter press: This allows the water to be squeezed out of the sludge as it is compressed between two belts. The main disadvantages of a belt filter press compared to other mechanical dewatering techniques are the need for skilled maintenance and the difficulty in controlling odors. The system consists of:

- a gravity drainage zone where the flocculated sludge is deposited and conveyed on a porous and mobile belt;
- a compression zone where a second belt is applied on the upper layer of the sludge, and compresses it to a pressure that can reach 7 bars; and
- a zone where the belts are separated and the dewatered sludge is released.

Centrifuge: This technology dries the FS as it is squeezed outwards on the surface of a cylinder rotating around its horizontal axis, due to the centrifugal force. The flocculated sludge is injected into the middle of this cylinder, and the particles are pushed outward against the surface. An Archimedean screw transports the released liquid to the side where the sludge entered, while another transports the sludge to the other end. The main disadvantage of the centrifuge is the high energy requirements.

Frame filter press: This system consists of porous vertical frames fixed in two walls that are positioned in front one of the other to create a chamber. This is a batch process in which the sludge is filled into the chamber at high pressure (up to 15 bars resulting in the leachate being released through the porous frames and the dewatered sludge being released through the opening of the lower wall).

Screw press: A screw press consists of a rotational screw placed in a perforated cylinder. The sludge is loaded at one end, it is pressurised due to a diminishing distance between the screw and the cylinder, and the liquid that is squeezed out is removed through the pores in the cylinder. The dewatered sludge is discharged at the other end. Screw presses provide dewatering at relatively low equipment and operational costs, and minimal maintenance skills are required. However, the dewatering is comparatively lower than other mechanical dewatering technologies.

Potential advantages and constraints of mechanical sludge treatment

The main constraints of these technologies in comparison to non-mechanical options are the investment costs, the O&M requirement, the need to add flocculants and the dependency on electricity. The general advantages are the compactness, and the speed of the process. To transfer these types of technologies to treat FS, information from manufacturers, laboratories, and pilot-scale tests is necessary.

Technology

5.4.5 Lime addition

Lime is used for wastewater sludge treatment to achieve the reduction of pathogens, odours, degradable organic matter, and also as sludge conditioner to precipitate metals and phosphorus (Méndez *et al.*, 2002), and has been implemented in the Philippines for FS treatment (Case Study 5.3). The process of pathogen reduction during alkaline stabilisation is based on an increase of pH, temperature (through exothermic oxidation reactions) and ammonia concentration (Pescon and Nelson, 2005). Its effect is enhanced by a longer contact time and higher dosing amount. All chemical compounds which have highly alkaline properties are generally termed as lime. Its most common forms are quick lime (CaO) and slaked lime $Ca(OH)_2$. Quicklime is derived from lime stone by a high temperature calcination process; quicklime is then hydrated to get slaked lime, also known as hydrated lime, or calcium hydroxide (Equation 5.1; Biosolids Technology Factsheet).

Equation 5.1: $CaO + H_2O \rightarrow Ca(OH)_2$

Equation 5.2: $Ca(OH)_2 + CO_2 \rightarrow CaHCO_3 + H_2O$

The high pH due to the formation of $CaHCO_3$ (Equation 5.2) creates an environment that halts or retards microbial degradation of organic matter (Turovskiy and Mathai, 2006). However, it is important to consider a number of design parameters like sludge characteristics, lime dose, contact time and pH in order to achieve optimum results from lime stabilisation in the most economic way possible (Turovskiy and Mathai, 2006).

An added benefit of lime is that heavy metals can precipitate with the lime. However, the pathogen removing effect of lime also affects desired microbial processes such as composting and other soil processes. Moreover, safety is very important: as lime is corrosive to the skin, eyes and lungs and proper personal protection equipment (PPE) is crucial (see also Case Study 5.3). Furthermore, protection from fire and moisture must be ensured.

Potential advantages and constraints of lime treatment

The main disadvantages of this technique are the requirement of consumables (lime), and a dry storage area. Pathogen regrowth is also a concern. Lime is an alkaline material which reacts strongly with moisture and high risks of hazard to the eyes, skin and respiratory system are observed. Therefore, skilled staff is required who must follow health and safety procedures and make use of good protective equipment.

Case Study 5.3: Lime stabilisation in the Philippines, San Fernando Valley

(Adapted from Robbins, 2009)

In June 2008, engineers and staff from the San Fernando City Health, Planning, Environment and Engineering offices engaged in the first stage of a pilot demonstration project to determine the effectiveness of lime stabilisation as a method of treating domestic FS from septic tanks in the Philippines (Robbins, 2009). Lime treatment was also mentioned as an appropriate technology by the Philippine's Department of Health (DOH) but it was not tested before in the field.

In the start-up of the study, one load (approximately $5m^3$ of FS) was used in a test run. First, a proper site had to be selected, using the following criteria: 1) the site is at a sufficient distance from residential areas; 2) the groundwater table is more than 25m below the testing site, and 3) preferably the soil is impermeable to reduce lining costs. After this site was identified, a mixing pit was created. The mixing pit in the San Fernando study was excavated to a depth of 1.5 meters and was 3 meters wide by 4 meters long. The soils at the landfill are high in clay particles, hence no liner was required.

FS was transported to the testing site by truck, and dumped into the pit in one off-loading. After filling the pit to the required level, lime was added and mixed. Lime was delivered at the site by the supplier in 50 kg bags. The lime was added manually one shovel-full at a time and by wearing dust masks the workers prevented breathing in the reactive powder. Then the lime was mixed into the FS. Although a small three horsepower mixing pump was available, manual mixing was performed: two to three operators kept the lime in suspension for 30 minutes with large wooden paddles. This time span proved to be sufficient to reach the desired level of mixing. It is suggested that an alternative way of mixing could be to add the lime directly into the tank of the truck. The progress of the treatment was monitored with a handheld pH monitor and a stop watch. To reach proper disinfection, the following goal had to be achieved: 30 minutes at pH 12; 60 minutes at pH 11.5; or 120 minutes at pH 11. The speed of the pH rise and the ultimate level of the pH will depend upon the amount of lime used per m^3 of FS, the quality of the hydrated lime, and the mixing. The San Fernando trials show that an application rate of 50 kg of lime per $5m^3$ of FS was adequate to achieve the required level of treatment (pH 11 for 2 hours). After that period, full pathogen destruction was observed as well as a proper solid liquid separation. Tests performed on other parameters such as heavy metal content revealed that the levels remained all under the limits set by the DOH of the Philippines.

After the process was completed, the pH decreased again toward neutral. After 24 hours, the clear liquid was siphoned off and discharged to a leachate pond and it could then be used to irrigate agricultural land, or for landscaping purposes. The solids may be applied as a soil amendment or dried and used for example as a cover for sanitary landfills.

For the San Fernando study, the cost of the hydrated lime was 455 Philippine Pesos (9 USD) per 50 kg bag, including delivery. Taking into account two employees, monitoring, excavation and miscellaneous costs, the total cost for treatment was approximately 200 Philippine Pesos (4 USD) per m^3 of FS treated. The results, costs and applicability make lime stabilisation a feasible technology for FS loads up to 15 m^3.

5.5 INNOVATIVE TECHNOLOGIES FOR FAECAL SLUDGE TREATMENT

There is currently a great deal of of research being conducted on innovative FS treatment technologies. Many of these innovations incorporate resource recovery, and this section therefore has a strong link with Chapter 10, Enduse of Treatment Products.

5.5.1 Vermicomposting

Earthworms are a member of the oligochaetes sub-class and they appear to be very effective in organic waste reduction. An example is the "vermi-filter", which treats diluted domestic wastewater sludge in a system inoculated with earthworms (Zhao *et al.*, 2010). Interestingly, the earthworms seemed to function in synergy with bacterial communities within the filter. Worms cannot survive in fresh faeces and need some kind of support in the form of layers of soil and vermi-compost. Vermicomposting is not a reliable method to ensure adequate pathogen removal. However, when carried out under proper conditions the technology of vermicomposting can lead to a complete removal of coliforms. Rodríguez-Canché *et al.* (2010) found helminth egg removal in experiments with vermicomposting on septic tank FS. Permissible levels for reuse in agriculture were achieved after 60 days, starting from the initial earthworm inoculation, faecal coliforms, Salmonella spp., and helminth ova were reduced to <1000 MPN/g, <3 MPN/g, and <1 viable ova/g on a dry weight basis, respectively.

Potential advantages and constraints of vermicomposting

In general, the advantages and constraints for vermicomposting are similar to the points for co-composting. However, vermicomposting cannot be carried out at the thermophilic temperatures of co-composting. Therefore if adequate pathogen reduction is not achieved during treatment, further treatment steps are required. Constraints are that the technology is still in development; the worms can be quite susceptible to toxic components (or higher concentrations in general), and the time span until matured compost is reached can be longer than for thermal composting. The production of worms can be beneficial provided there is a market for them.

5.5.2 Black Soldier flies

The Black Soldier fly (*Hermetia illucens*) originated in America, but is commonly found in temperate and warm climates. The fly larvae feed on decaying organic material, such as vegetables and fruits, or manure. The Black Soldier fly (BSF) larvae have been investigated for the degradation of organic wastes such as municipal solid wastes, animal manure, and FS (Diener *et al.*, 2009; Diener *et al.*, 2011; Qing *et al.*, 2011). This process relies on the natural growing cycle of BSF which need to feed only during the larval stage, then migrate for pupation, and do not feed anymore, even during the adult stage. Therefore, the risks of the BSF being a vector for disease transmission is very low, as it is not attracted by decaying organic matter when it can fly (Sheppard *et al.*, 1994). During their larval stage, BSF larvae achieve a rapid reduction of organic waste volumes of up to 75%, together with the removal of nutrients such as nitrogen and phosphorus (Diener *et al.*, 2009). This growth stage can vary from two weeks to four months depending on the availability of food, and thus allows for the treatment of wastes even when waste is not produced continuously.

BSF larvae have been shown to grow well solely on FS; however, Diener *et al.* (2009) observed that a mixture of FS and municipal solid waste can achieve higher and faster larvae mass production. This can be advantageous for the selling of the larvae as animal feeding to farmers (see Chapter 10). The FS residue remaining after the BSF larvae feed need to be further composted or anaerobically digested to produce a soil conditioner.

Figure 5.6 Black Soldier fly pre-pupae (photo: Stefan Diener).

Most of the work so far has been at a laboratory scale. Yet, the market seems to be developing. For example, the Biocycle company is working on upscaling a profitable business model by collecting human waste, treating it at their facilities, harvesting the larvae and preparing the product for sale (www.biocycle.com). Low operation costs of this technology and the market potential (crushed dried larvae as protein source) make it a promising technology. Some technical as well as entrepreneurial questions however are still to be answered.

Potential advantages and constraints of Black Soldier flies

The advantage of using BSF for treatment of FS is that it can be achieved with or without mixing with other organic wastes, and on a small scale. It allows revenue generation for small entrepreneurs with minimal investment. However, information on upscaling experience in low- and middle-income countries is not yet available, and therefore precise recommendations on the design and operation of this technology cannot be given for FS treatment (Diener *et al.*, 2011).

5.5.3 Ammonia treatment

Ammonia treatment can be applied for pathogen reduction. Pathogen inactivation by uncharged ammonia (NH_3) has been reported for several types of microorganisms, bacteria, viruses and parasites (Jenkins *et al.*, 1998; Pescon and Nelson, 2005). The principle of pathogen reduction with ammonia is based on the fact that ammonia (NH_3) enters cells, takes up intracellular protons for the formation of ammonium (NH_4^+) and as a charged ion disturbs the functioning of organisms (Park and Diez-Gonzalez, 2003). Ammonia addition to sludge has been applied for wastewater sludge, where it is commonly referred to as alkaline stabilisation (Allievi *et al.*, 1994; Mendez *et al.*, 2002).

More recently, investigations have been conducted on using the ammonia from excreta for pathogen reduction in FS. This can be done by collecting urine separately, and then mixing it with FS, as urine has a high ammonia concentration (Chapter 2). For sludge with low ammonia concentration, additional ammonia in the form of the synthetic urea can be added to enhance the treatment.

Potential advantages and constraints of ammonia treatment

In comparison to lime treatment, ammonia requires less stringent storage conditions. It seems particularly applicable in areas with urine diverting dehydrating toilets (UDDTs). In the cases where synthetic urea needs to be applied, the costs become higher, which may limit the economic feasibility and sustainability of the technology. Another constraint is the stability of nitrogen in treatment endproducts, and whether the full nutrient benefit can be achieved.

> **Case Study 5.4: Sanitisation by ammonia – die-off of Ascaris and the Safe Sludge project**
>
> To evaluate the pathogen reducing effect of ammonia, Fidjeland *et al.* (2013) monitored the viability of Ascaris eggs from FS during storage at laboratory scale at different ammonia concentrations and temperatures. At ammonia concentrations above 170 mM and 23 °C, 99.9% reduction of Ascaris egg viability could be achieved within 1.5 month - this corresponds to 2 L flush water per person and use. For flush water volumes of 6 L per person and use, the ammonia concentrations were lower (44 mM) and 6 months storage was required at 23 °C. At higher temperatures, the inactivation of Ascaris eggs was faster and the required ammonia concentration lower. By implementing toilet systems which use airtight storage and low flush water volumes, the intrinsic ammonia may be sufficient to sanitise the FS without additional treatment.

Another project called "Safe Sludge" was carried out from May 2011 until May 2013. The goal was to achieve the pathogen reduction of FS from urine diverting dry toilets (UDDTs) with the ammonia that is naturally present in urine. UDDT sludge is much drier with smaller volumes than most FS. This process appears to be more effective for UDDT sludge as it facilitates the contact time between ammonia and the pathogen containing faeces and no synthetic urea needs to be added. The urease enzyme is not active above a pH of around 9. As a result, the Safe Sludge disinfection process requires two stages: a 4-hour contact time for urine and faeces to hydrolyse urea, followed by the addition of an alkalinising agent (calcium hydroxide) to produce ammonia.

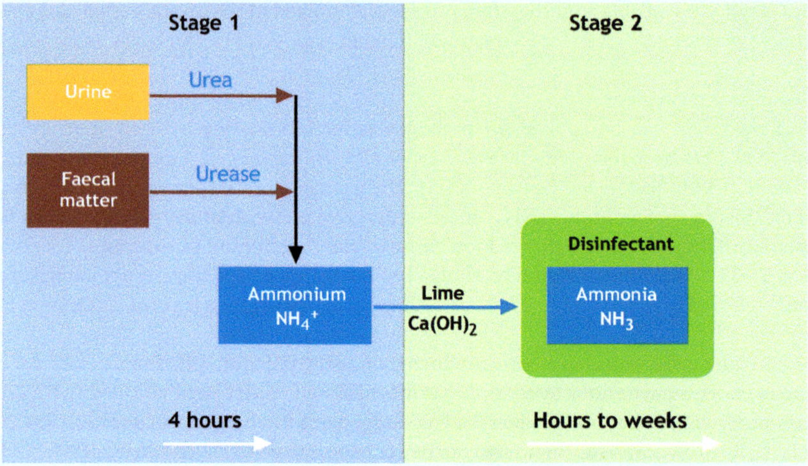

Figure 5.7 The Safe Sludge two stage disinfection process: hydrolysis of urea catalysed by urease (up to 4 hrs) and lime addition to increase pH and so achieve ammonium conversion into ammonia, a known disinfectant (estimated desinfection time: hours to weeks). Figure adapted from http://forum.susana.org (2013).

5.5.4 Thermal drying and pelletising

Thermal drying allows the removal of all types of liquids from FS (see Chapter 3). It has been applied in the management of wastewater sludge for many years, and the technology has been taken up and improved from its original application in other industries (e.g. paper industry). Several types of technologies exist, all based on the ability of evaporating water with heat. The endproducts are stable and in a granular form allowing easier storage or transport.

Direct or indirect thermal dryers are also referred to as convection or contact dryers, respectively (Lowe *et al.*, 2007). These systems require preliminary dewatering if used for sludge that is high in water content. In direct thermal driers, the hot air or gases are mixed with the dewatered sludge, as they pass through it, or are transported with it. In indirect thermal driers, a heat exchanger is used, which allows the heat convection to the sludge. In this case, the heat carrying media is often steam or oil, and does not come in direct contact with the sludge, which reduces the operational need to separate the sludge from the heat carrier. In both cases, the vapor produced by the evaporated water needs to be collected and transported out of the dryer. Gas treatment can be an issue depending on environmental requirements and the odors produced. Indirect thermal dryers produce less contaminated vapor.

Potential advantages and constraints of thermal drying

Thermal drying results in a significant reduction in volume as well as pathogen content. Dried sludge is easy to handle and to market, and can be used in agriculture (see Chapter 10). The main constraints are the expense, high energy requirements, the potential risks of fire or explosion due to the gas and dust in the system, and the high maintenance requirements.

Pelletising combines mechanical dewatering and thermal drying technologies. The dried pellets can then be used as an energy source or soil conditioner, and are relatively easy to transport and to market. Case studies 5.5 and 5.6 present examples with FS to produce a soil conditioner and fertiliser.

Case Study 5.5: Pellets as soil conditioner/fertiliser

(Adapted from Nikiema *et al.*, 2013)

In Ghana, experiments were carried out with dried FS to produce five different types of products, which were subsequently pelletised using locally constructed machinery (a version of a 380V pelletiser). As a source of dried FS, public latrine sludge was mixed with septic tank sludge in a 1:2 ratio dewatered and stabilised in an unplanted drying bed and co-composted (Nikiema *et al.*, 2013). A number of key operating parameters were identified: moisture content (10-55% in mass) and binder type and concentration (clay or starch - 0-10% in mass). Output parameters were the amount of high-quality pellets that were generated, and the length and stability of pellets.

From preliminary investigations cassava starch and clay were identified as possible binding materials, of which cassava starch proved to be the better candidate. Nikiema *et al.* (2013) recommend pre-treating the starch through the addition of hot water (85 ± 5°C) to dry the starch under manual stirring; a 3% mixture is recommended for addition to dried sludge to obtain the best pellets.

Case Study 5.6: Pelletising LaDePa machine from Durban, South Africa

Another example of combining drying and pelletising is the LaDePa (Latrine Dehydration and Pasteurisation) system developed by eThekwini Water and Sanitation (EWS, Durban, South Africa) in conjunction with their technology partner Particle Separation Systems. Here, a technology has been developed that can treat FS from pit latrines over a number of subsequent thermal and mechanical treatment steps. It separates detritus from dewatered sludge via screws which deposit pellets on a moving belt. These pellets are dried with air at 100°C (the so-called "Parsep dryer"), and pathogen reduction is achieved by means of medium wave infrared radiators. Although this is energy intensive, the energy consumption per person equivalent is approximately half of what a conventional activated sludge wastewater treatment plant consumes. After several modifications and redesigns the LaDePa is now available in modular form (in one container) to treat any sludge between 20 – 35 % solids and to pasteurise it to an 80 – 90 % solid product. The pellets can be sold and used as a fuel or as a soil amendment.

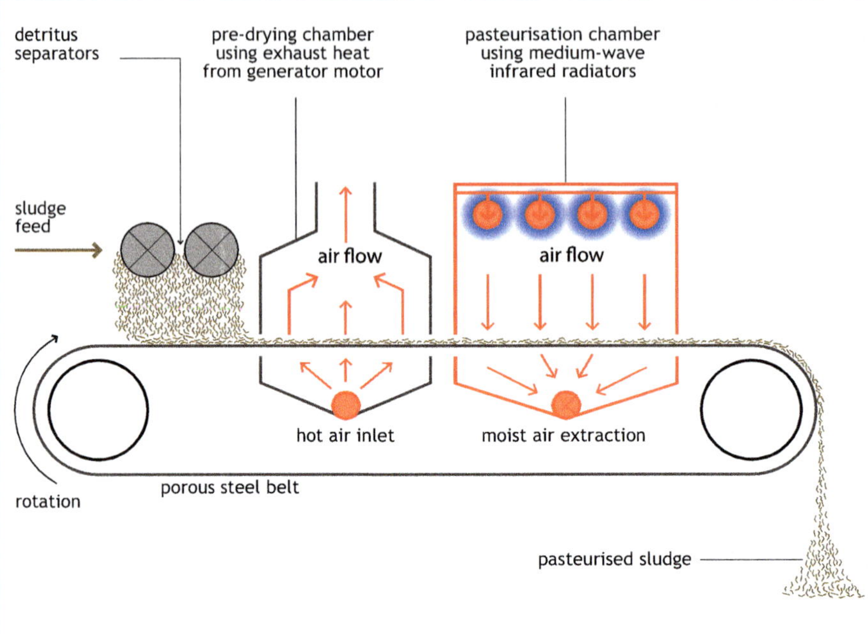

Figure 5.8 Schematic representation of the LaDePa machine.

A significant advantage of the LaDePa machine is that is designed to treat pit latrine sludge without any (manual) pre-sorting. Commonly, pit latrine sludge contains quite a high amount of non-organic municipal solid waste (see also Chapter 2, Section 2.9.7), which complicates treatment and the rendering of a useful product, but the LaDePa is able to remove the detritus.

Potential advantages and constraints of sludge drying and pelletising

The main advantages of these technologies are that they are compact, mobile and robust. Moreover, depending on which processes are used, the pellets are free from pathogens and therefore safe for agricultural use. Pellets can also be used as a dry fuel in industrial combustion, regardless of the pathogen content. However, in case of a breakdown of the system, costs and skilled knowledge requirements may be high. Moreover, the main constraint of mechanical drying and pelletisiing is the dependency on electrical power. The energy use, capital costs, and specialised knowledge required for maintenance are other drawbacks.

5.5.5 Solar drying

A special form of drying is applied in solar sludge driers. They also have been used on a large scale since the nineteenth century in Europe and USA for the treatment of wastewater sludge (Hill and Bux, 2011). This technology is generally constructed in greenhouse structures with transparent covers, concrete basins and walls. Sludge is disposed into these basins and processed for about 10 to 20 days. Options exist for batch or continuous operation, with devices to control the conditions in the greenhouses (e.g. ventilation, air mixing, temperature). The main factors influencing the evaporation efficiency in these systems are the solar variation, the air temperature and the ventilation rate, with initial dry solid content of the sludge and air mixing also influencing the process (Seginer and Bux, 2005). Short wavelengths light such as UV is blocked by the cover. Therefore, the pathogen reduction efficiency is slightly reduced, especially for faecal coliforms that are very sensitive to UV (Shanahan *et al.*, 2010). Final dried solid content of about 40% (after 12 days drying) up to about 90% (after 20 days drying) were found under different conditions by Shanahan *et al.* (2010) and Hill and Bux (2011) respectively.

Potential advantages and constraints of solar drying

The main advantages of this option are the low energy requirements, the limited complexity of the technology and low investment costs, and the high potential dewatering efficiency. The main constraints are the space requirements and the need for mechanical means to turn the sludge, as well as to ventilate the greenhouses. Although pilot tests are being carried out, for the moment no information is available on the use of this technology for the treatment of FS in low-income countries or on design and operating parameters that need to be considered for this purpose.

5.6 SELECTING TREATMENT TECHNOLOGIES

This chapter has provided an overview of treatment technologies. When selecting the optimal treatment and/or combinations of treatments, many factors have to be considered, which is complicated by the lack of long-term operating experience. The Technology Selection Scheme and FSM Planning from A to Z presented in Chapter 17 are designed to aid in this selection. In addition, Box 5.1 provides information on how to compare costs of treatment technology options.

Box 5.1: Cost comparison of treatment technology options

Linda Strande

Outlined in the Planning Section are all of the steps that need to be taken during the decision-making process, and Chapter 17 Planning Integrated FSM Systems summarises all of the steps in the Technology Selection Scheme (Figure 17.10) and FSM Planning from A to Z (Table 17.1). In addition, Chapter 13 explains different models of financial flows, and Chapter 12 explains different models of management. A cost comparison cannot be made outside of an integrated approach that includes Management and Planning, as the costs of Technology are directly interrelated and impacted by these other factors. Costs need to be compared on a life-cycle basis, i.e. over the planning horizon of the project, including all recurrent expenditures (e.g. transport, O&M, capacity-building and policy development). It is not only the cost that is important, but by whom and how it will be paid over the long-term. The best option is not always the cheapest, but that which ensures household satisfaction, broad coverage and cost recovery (see Section 17.5). Ultimately, the success of a FSM plan largely depends on the capacity of the stakeholders to enforce the financial mechanisms that have been planned, and the capacity of the stakeholders to operate and maintain the FSTP.

The difficulty of comparing the costs of FSM technologies presented in this chapter, are that they are mostly operational at the laboratory or pilot scale, and there is no long-term operating experience on which to base estimates on. Access to this cost data will improve overtime with increased installations, but in the meantime reasonable estimates have to be calculated. Another difficulty is the heterogeneity of costs from one context to the other.

There are two ways to calculate costs over the expected lifetime:

1 Net Present Value (NPV): this calculation entails determining the required capital investment, and all future cash flows over the expected lifetime. This amount is then converted to the present worth, i.e. the amount of money that would need to be invested today to cover all expenses during the lifetime. The lower the sum, the less expensive the option is. This value can be used to compare options with equivalent lifetimes.

2 Equivalent Annual Cost (EAC): this calculation entails determining the required capital investment, and all future cash flows over the expected lifetime. This amount is then converted to the amount of money that would be required on a yearly basis over the expected lifetime. The NPV can be converted into an EAC, and vice versa, but EAC can be used for comparisons among technologies with different lifetimes. Annual costs are calculated as the annualised capital cost over the expected lifetime with interest rate, in addition to annual operations and maintenance costs as shown in equation 5.3:

Equation 5.3:

$$AC_o = -C_o \left[\frac{(1+i)^{n_o} \cdot i}{(1+i)^{n_o} - 1} \right] - F_o$$

Where:

AC_o = annualised cost of sanitation system component (USD/capita/year)

C_o = capital cost of sanitation system component (USD/capita)

n_o = service lifetime of sanitation system (years)

i = real interest rate

F_o = annual operating cost of faecal sludge management system component (USD /capita/year)

Another possibility for comparisons among systems is to normalise the costs to tons of total solids treated. An example of estimating costs based on a pilot scale is provided by Steiner *et al.* (2002) based on planted drying bed research conducted in Thailand by Koottatep *et al.* (2001) and Surinkul (2002). Based on construction and operation during the pilot phase, annualised costs were estimated to be 1,500 USD/year (0.95 USD per capita or 186 USD per ton of total solids). Ongoing operations and maintenance (e.g. harvesting plants, sludge removal) were not included, but could be forecasted based on operation during the pilot.

This estimate only included the cost of the drying bed, and not the onsite containment system at the household level, collection and transport, or enduse of treated sludge and leachate. A comparison of complete FSM and sewer based systems, and how the costs are borne is provided by Dodane *et al.* (2012) and summarised in Table 5.2. The analysis was done based on existing side-by-side operational systems in Dakar Senegal. To compare FSM and sewer based systems, the entire service chain needs to be considered. When costs for the entire service chain are estimated, it is critical to evaluate by whom the costs will be borne (e.g. household, government, private sector).

Table 5.2 Comparison of FSM and sewer based systems on existing side-by-side operational systems in Dakar, Senegal (Dodane *et al.*, 2012)

ANNUALISED CAPITAL COSTS (PER CAPITA*YEAR)									
	SEWER BASED (SB)				FAECAL SLUDGE MANAGEMENT (FSM)				
	House	ONAS	Enduser	TOTAL	House	C&T	ONAS	Enduser	TOTAL
Household Connection[1]	0.00	-4.98	0.00		-2.74	0.00	0.00	0.00	
Collection Conveyance[2]	0.00	-30.20	0.00		0.00	-0.28	0.00	0.00	
Treatment Plant[3]	0.00	-7.49	0.00		0.00	0.00	-1.03	0.00	
TOTAL	0.00	-42.66	0.00	**-42.66**	-2.74	-0.28	-1.03	0.00	**-4.04**
ANNUAL OPERATING COSTS (PER CAPITA*YEAR)									
	SEWER BASED (SB)				FAECAL SLUDGE MANAGEMENT (FSM)				
	House	ONAS	Enduser	TOTAL	House	C&T	ONAS	Enduser	TOTAL
Collection Conveyance[4]	0.00	-6.64	0.00		-5.00	0.26	0.00	0.00	
Sanitation Tax[5]	-2.00	2.00	0.00		-2.00	0.00	0.00	0.00	
Treatment Plant[3]	0.00	-6.46	0.00		0.00	0.00	-0.84	0.00	
Valorisation Endproducts[6]	0.00	1.13	-0.01		0.00	0.00	0.01	-0.01	
TOTAL	-2.00	-9.97	-0.01	**-11.98**	-7.00	0.26	-0.83	-0.01	**-7.58**
CAPITAL AND ANNUAL OPERATING COSTS COMBINED (PER CAPITA*YEAR)									
TOTAL	-2.00	-52.63	-0.01	**-54.64**	-9.74	-0.02	-1.86	-0.01	**-11.63**

1 Household Connection (capital) = household sewer connection, septic tank
2 Collection Conveyance (capital) = sewer, pumping stations, vacuum trucks
3 Treatment Plant (capital and operating) = wastewater treatment plant, faecal sludge treatment plant
4 Collection Conveyance (operating) = sewer, pumping stations, onsite emptying fee, truck transport
5 Sanitation Tax (operating) = sanitation tax paid by every resident based on water consumption
6 Valorisation End-products (operating) = biogas, reclaimed water, biosolids

As will be discussed in Chapter 14.5.3, the comparison of treatment technologies is also complicated by factors such as the level of centralisation or decentralisation. FSM technologies tend to be more decentralised or semi-centralised than centralised sewer-based systems. In terms of meeting long-term urban growth requirements, decentralised technologies are more flexible as they can be built in a modular basis as needed (Maurer, 2009). On a management and capital cost basis, the economy of scale results in larger plants being more cost effective than smaller plants. However, when haulage of sludge is taken into account, it appears that smaller decentralised plants are more affordable as travel distances and time can be reduced. For this reason, it is important to consider the whole supply chain when making a decision. The correlation between scale and cost is not linear, and typically a breakeven point can be found. For example, in Japan decentralised wastewater treatment including reclamation is more affordable than conventional dual-pipe water delivery and sewer systems at flows greater than 100 m^3/day (Gaulke, 2006). All of these factors are very dependent on the local context and the specificities of each city (see Section 14.4.8).

5.7 CONCLUSIONS

As shown in Figure 5.1, this chapter provided an overview of established and emerging technologies for FS treatment. Important points to consider when selecting technologies include that there are different technologies for different treatment objectives, and they can be employed alone and/ or in combination. There are many factors to consider when selecting optimal treatment configurations, including final enduse, treatment objectives, potential advantages and constraints, and how to compare costs. Further material is covered in the management and planning sections of this book.

5.8 BIBILOGRAPHY

Allievi, L., Colombi, A., Calcaterra, E., Ferrari, A. (1994). Inactivation of faecal bacteria in sewage sludge by alkaline treatment. Bioresource Technology 49(1), p.25-30.

Arthur, R., Hammond, A.B. (2010). Potential Biogas Production from Sewage Sludge: A Case Study of the Sewage Treatment Plant at Kwame Nkarumah University of Science and Technology, Ghana. International Journal of Energy and Environment 1(6), p.1009-1016.

Biosolids Technology Factsheet. Facsheet on the use of composting for biosolids management produced by US EPA. Available from http://water.epa.gov.

Cofie, O., Koné, D. (2009). Co-composting of faecal sludge and organic solid waste Kumasi, Ghana – Case study of sustainable sanitation projects. Sustainable Sanitation Alliance (SuSanA).

Cofie, O.O., Agbottah, S., Strauss, M., Esseku, H., Montangero, A., Awuah, E., Koné, D. (2006). Solid–liquid separation of faecal sludge using drying beds in Ghana: Implications for nutrient recycling in urban agriculture. Water research, 40(1), p.75-82.

Cooperband, L. (2002). The art and science of composting – a resoiurce for farmers and compost producers. University of Wisconsin-Madison, Centre for Integrated Agricultural Systems.

Diener, S., Zurbrügg, C., Tockner, K. (2009). Conversion of organic material by black soldier fly larvae: establishing optimal feeding rates. Waste Management & Research, 27(6), p.603-610.

Diener, S., Zurbrügg, C., Gutierrez, F.R., Nguyen, D.H., Morel, A., Koottatep, T., Tockner, K. (2011). Black Soldier fly larvae for organic waste treatment - prospects and constraints. Proceedings of the 2nd International Conference on Solid Waste Management in the Developing Countries 13-15 February 2011, Khulna, Bangladesh.

Dodane, P.H., Mbéguéré, M., Ousmane, S., Strande, L. (2012). Capital and Operating Costs of Full-Scale Faecal Sludge Management and Wastewater Treatment Systems in Dakar, Senegal. Environmental Science & Technology 46(7), p.3705-3711.

EAWAG and IWMI (2003). Co-composting of Faecal Sludge and Municipal Organic Waste – A Literature and State-of-Knowledge Review. Available from http://www.eawag.ch/forschung/sandec/publikationen/ewm/dl/CCP_FS_orgWaste.pdf.

EAWAG, Stauffer B., Spuhler D. (2010). Imhoff Tank. In: Conradin, K., Kropac, M., Spuhler, D. (Eds.) (2010): The SSWM Toolbox. Basel: seecon international gmbh.

Fernández, R.G., Ingallinella, A.M., Sanguinetti, G.S., Ballan, G.E., Bortolotti, V., Montangero, A., Strauss, M. (2004). Septage Treatment Using Waste Stabilization Ponds. Proceedings, 9th International IWA Specialist Group Conference on Wetlands Systems for Water Pollution Control and 6th International IWA Specialist Group Conference.

Fidjeland, J., Magri, M.E., Jönsson, H., Albihn, A., Vinnerås, B. (2013). The potential for self-sanitisation of faecal sludge by intrinsic ammonia. Water Research 47(16), p.6014-6023.

Gaulke, L.S. (2006). Johkasou. On-site Wastewater Treatment and Reuses in Japan. Proceedings of the Institute of Civil Engineers. Water Management 159(2), p.103-109.

Hall, J. (1999). Ecological and economical balance for sludge management options. In: Langenkamp, H and Marmo L (eds) Workshop on Problems around Sludge, Stresa, Italy, 18-19 November 1999, p.155-172.

Heinss, U., Larmie, S.A., Strauss, M. (1998). Solids separation and pond systems for the treatment of faecal sludges in the tropics 05/98. Dubendorf, Switzerland: EAWAG/SANDEC.

Hill, P.E., Bux, M. (2011). Harnessing solar energy for biosolids management: a green approach to drying. Abstract, avaialable from http://info.ncsafewater.org.

Ingallinella, A.M., Sanguinetti, G.S, Fernández, R.G., Strauss, M., Montangero, A. (2002). Cotreatment of sewage and septage in waste stabilization ponds. Water Science and Technology 45(1), p.9-15.

Jenkins, M.B., Bowman, D.D., Ghiorse, W.C. (1998). Inactivation of Cryptosporidium parvum oocysts by ammonia. Applied Environmental Microbiology 64(2), p.784-788.

Klingel, F., Montangero, A., Koné, D., Strauss, M. (2002). Fecal Sludge Management in Developing Countries. A planning manual. EAWAG: Swiss Federal Institute for Environmental Science and Technology Sandec: Department for Water and Sanitation in Developing Countries.

Koné, D., Cofie, O., Zurbrügg, C., Gallizzi, K., Moser, D., Drescher, S., Strauss, M. (2007). Helminth eggs inactivation efficiency by faecal sludge dewatering and co-composting in tropical climates. Water Research 41 (19), p.4397-4402.

Koottatep, T., Polprasert, C., Oanh, N.T.K., Heinss, U., Montangero, A., Strauss, M. (2001). Septage dewatering in vertical-flow constructed wetlands located in the tropics.

Koottatep, T., Surinkul, N., Polprasert, C., Kamal A.S.M., Koné, D., Montangero, A., Heinss, U., Strauss, M. (2004). Treatment of septage in constructed wetlands in tropical climate–Lessons learnt after seven years of operation. Water Science and Technology 51 (9), p.119-126.

Kurup, B., Kurup, R., Mathew, K. and Ho, G. (2002) Co-treatment of septage in a municipal sewage treatment pond system. Water Science & Technology, 46 (9). p.315-321.

Lowe, P. (2007). Developments in the thermal drying of sewage sludge. Water and Environment Journal 9(3), p.306-316.

Mara, D.D. (2004). Domestic Wastewater Treatment in Developing Countries. London: Earthscan Publications.

Mara, D.D., Alabaster, G.P., Pearson, H.W., Mills, S.W. (1992). Waste Stabilization Ponds: A Design Manual for Eastern Africa. Lagoon Technology International. Leeds, England.

Maurer, M. 2009. Specific net present value: An improved method for assessing modularization costs in water services with growing demand. Water Research 43(8), p.2121-2130.

Mendez, J.M., Jimenez, B.E., Barrios, J.A. (2002). Improved alkaline stabilization of municipal wastewater sludge. Water Science and Technology 46, p.139-146.

Metcalf and Eddy (2003). Wastewater Engineering: Treatment and Reuse, 4th ed. McGraw Hill, New York.

Murray Muspratt, A., Nakato, T., Niwagaba, C., Dione, H., Baawuah, N., Kang, J., Stupin, L., Regulinski, J., Mbéguéré, M., Strande, L. (2014). Fuel potential of faecal sludge: calorific value results from Uganda, Ghana, and Senegal. Journal of Water Sanitation and Hygiene for Development, in press.

Nikiema, J., Cofie, O., Impraim, R., Adamtey, N. (2013). Processing of Fecal Sludge to Fertilizer Pellets Using a Low-Cost Technology in Ghana. Environment and Pollution 2 (4), p.70-87.

Papadopoulos F., Papadopoulos A., Parisopoulos G., Sdragas G., Metaxa I. (2007). The Treatment of Septage Using Stabilization Ponds. Fresenius Environmental Bulletin 16(4), p.385 - 392.

Park, G. W., and F. Diez-Gonzalez. 2003. Utilization of carbonate and ammonia-based treatments to eliminate Escherichia coli O157:H7 and Salmonella DT104 from cattle manure. Journal of Applied Microbiology 94(4), p.675-685.

Pescon, B.M., Nelson, K.L. (2005). Inactivation of Ascaris suum eggs by ammonia. Environmental Science and Technology 39, p.7909-7914.

Robbins, D. (2009). Septage Treatment – Lime Stabilisation. Published on http://sfluseptage.blogspot.nl/2010/05/septage-tretament-lime-stabilization.html.

Rodríguez-Canchéa, L.G., Cardoso Vigueros, L., Maldonado-Montiel, T., Martínez-Sanmiguel, M. (2010). Pathogen reduction in septic tank sludge through vermicomposting using Eisenia fetida. Bioresource Technology 101(10), p.3548-3553.

Rouse, J., Rothenberger, S., Zurbruegg, C. (2008). Marketing Compost. A Guide for Compost Producers in Low and Middle-Income Countries. Duebendorf: Water and Sanitation in Developing Countries (SANDEC), Swiss Federal Institute for Environmental Science (EAWAG).

Seginer, I., Bux, M. (2005). Prediction of Evaporation Rate in a Solar Dryer for Sewage Sludge. International Commission of Agricultural Engineering (CIGR, Commission Internationale du Genie Rural), E-Journal Volume 7.

Shanahan, E.F., Roiko, A., Tindale, N.W., Thomas, M.P., Walpole, R., Ipek Kurtböke, D. (2010). Evaluation of pathogen removal in a solar sludge drying facility using microbial indicators. International Journal of Environmental Research and Public Health 7(2), p.565–582.

Sheppard, D.C., Newton, G.L., Thompson, S.A., Savage, S.E. (1994). A value added manure management system using the Black Soldier fly. Bioresource Technology 50, p.275-279

Song, Z., Qin, J., Yang, G., Feng, Y., Ren, G. (2012). Effect of human excreta mixture on biogas production. Advanced Materials Research 347, p.2570-2575.

Steiner, M., Montangero, A., Koné, D., Strauss, M. (2002). Estimation of Collection, Haulage, Treatment and Disposal/ Reuse Cost. 1st Edition, October 2002. Sandec report, available from www.eawag.ch/sandec.

Still, D., Louton, B., Bakare, B., Taylor, C., Foxon, K., Lorentz, S. (2012). Investigating the potential of deep row entrenchment of pit latrine and waste water sludges for forestry and land rehabilitation purposes - WRC Project No. K5/1829. Water Reserach Commission (WRC), South Africa.

Strauss, M., Drescher, S., Zurbruegg, C., Montangero, A., Cofie, O., Drechsel, P. (2003). Co-composting of Faecal Sludge and Municipal Organic Waste - a Literature and State-of-Knowledge Review. Available from www.eawag.ch/forschung/sandec/publikationen/swm/dl/Strauss_2003.pdf.

Strauss, M., Larmie, S.A., Heinss, U., Montangero, A. (2000). Treating Faecal Sludges in Ponds. Available from www.eawag.ch/forschung/sandec/publikationen/ewm/dl/FS_WSP_overview2.pdf.

Tilley, E., Ulrich, L., Lüthi, C., Reymond, P., Zurbrügg, C. (2014). Compendium of Sanitation Systems and Technologies. Dübendorf: Swiss Federal Institute of Aquatic Science & Technology (EAWAG) 2nd revised edition.

Turovskiy, I. and Mathai, P.K. (2006). Wastewater Sludge Processing. John Wiley and Sons, New Jersey.

WSP (Editor) (2007): Philippines Sanitation Source Book and Decision Aid. Washington: Water and Sanitation Program.

Zhao, L., Wang, Y. Yang, J., Xing, M., Li, X., Yi, D., Deng, D. (2010). Earthworm-microorganism interactions: a strategy to stabilize domestic wastewater sludge. Water Research 44(8), p.2572-2582.

Chapter 6

Settling-Thickening Tanks

Pierre-Henri Dodane and Magalie Bassan

Learning Objectives

- Understand in what contexts settling-thickening tanks are an appropriate treatment technology.
- Understand the fundamental mechanisms of how settling-thickening tanks function.
- Have an overiew of potential advantages and disadvantages of operating a settling-thickening tank.
- Know the appropriate level of operations, maintenance and monitoring necessary to achieve solids-liquid separation in settling-thickening tanks.
- Be able to design a settling-thickening tank to achieve the desired treatment goal.

6.1 INTRODUCTION

Settling-thickening tanks are used to achieve separation of the liquid and solid fractions of faecal sludge (FS). They were first developed for primary wastewater treatment, and for clarification following secondary wastewater treatment, and it is the same mechanism for solids-liquid separation as that employed in septic tanks. Settling-thickening tanks for FS treatment are rectangular tanks, where FS is discharged into an inlet at the top of one side and the supernatant exits through an outlet situated at the opposite side, while settled solids are retained at the bottom of the tank, and scum floats on the surface (Figure 6.1). During the retention time, the heavier particles settle out and thicken at the bottom of the tank as a result of gravitational forces. Lighter particles, such as fats, oils and grease, float to the top of the tank. As solids are collected at the bottom of the tank, the liquid supernatant is discharged through the outlet. Quiescent hydraulic flows are required, as the designed rates of settling, thickening and flotation will not occur with turbulent flows. Baffles can be used to help avoid turbulence at the inflow, and to separate the scum and thickened sludge layers from the supernatant.

Following settling-thickening, the liquid and solid fractions of FS require further treatment depending on their final fate, as the liquid and solids streams are still high in pathogens, and the sludge has not yet been stabilised or fully dewatered (for combinations of technologies reader is referred to Chapter 5 and 17). Settling-thickening tanks can be used in any climate, but are especially beneficial when treating FS with a relatively low solids concentration, and/or in temperate or rainy climates. This is an important consideration in urban locations where space is limited, as it can reduce the required area of subsequent treatment steps. For example, achieving solids-liquid separation in settling-thickening tanks prior to dewatering with drying beds reduces the required treatment area (footprint) for drying beds.

When using settling-thickening tanks there should be at least two parallel streams to allow for an entire operational cycle of loading, maintenance and sludge removal. For increased sludge compaction and ease of operations and maintenance, tanks should not be loaded during compaction, if the sludge is left to thicken at the bottom of the tank, or during the desludging period, when the supernatant is drained and the scum and thickened sludge are removed. Tanks are usually operated with loading periods ranging from one week to one month, depending on the tank volume. When operated in parallel, each tank is only loaded 50% of the time.

In most existing implementations in low-income countries, the sludge removal is done with backhoes, pumps if the sludge is not too thick to pump, or strong vacuum trucks. On the other hand, in wastewater treatment plants clarifiers typically include mechanical devices to remove the settled sludge from the tank.

This chapter presents an overview of the fundamental mechanisms, design recommendations, operational conditions and performances of settling-thickening tanks for FS treatment. It is also possible to have larger scale settling ponds, which are similar to anaerobic ponds in wastewater treatment. The main differences between ponds and tanks are that more sludge can accumulate, the sludge is more difficult to remove, and longer retention times result in anaerobic digestion. Due to a lack of actual operational settling-thickening tanks for FS, the information in this chapter is based on theoretical knowledge, and on operating experiences in West Africa. In Kumasi, Ghana, there are 100 m^3 settling-thickening tanks that are employed prior to drying beds at a FS treatment plant (FSTP), as shown in Figure 6.2. The design guidelines presented in this chapter are readily adaptable to other contexts.

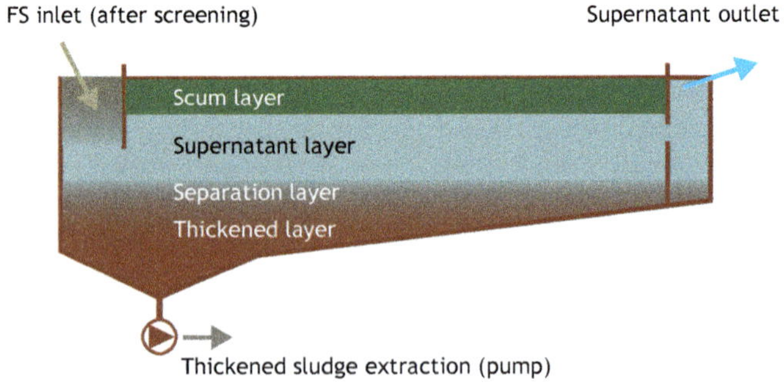

Figure 6.1 Schematic of the zones in a settling-thickening tank.

Figure 6.2 Left: the twin settling-thickening tanks of Rufisque faecal sludge management plant, in Dakar, Senegal. The operating cycle lasts two weeks (one week loading, one week desludging), and sludge is transferred to drying beds with a pump. Right: the settling-thickening pond in Achimota, Accra, Ghana, has an operational cycle of 8 weeks (4 weeks loading, 4 weeks desludging). As the sludge is thickened for four times as long, it is not possible to remove it with a pump, and manual extraction is carried out (photo: SANDEC).

6.2 FUNDAMENTAL MECHANISMS

Settling-thickening tanks rely on three main fundamental mechanisms: settling, thickening, and flotation, which are also described in more detail in Chapter 3. Anaerobic digestion also occurs in the tanks, although this is not a treatment goal of settling-thickening tanks, as anaerobic digestion results in gas production, and the resulting bubbles can hinder the solids-liquid separation through mixing and flotation of particles. A brief overview of these mechanisms is given in the following section.

6.2.1 Settling

In settling-thickening tanks the suspended solid (SS) particles that are heavier than water settle out in the bottom of the tank through gravitational sedimentation. The types of settling that occur are:

- discrete, where particles settle independently of each other;
- flocculant, where accelerated settling due to aggregation occurs; and
- hindered, where settling is reduced due to the high concentration of particles (Ramalho, 1977).

Discrete and flocculant settling happen rapidly in the tank. Hindered settling occurs above the layer of sludge that accumulates at the bottom of the tank, where the suspended solids concentration is higher. These combined processes result in a reduction of the solids concentration in the supernatant, and an accumulation of solids at the bottom of the tank.

Particles with a greater density settle faster than particles with lower densities. Based on the fundamentals of settling the distribution of types and shapes of particles in FS (and their respective settling velocities) could theoretically be used to design settling-thickening tanks. Although this theory is important in understanding the design of settling-thickening tanks, the reality is that when designing a settling tank, empirical values are determined and used for the design based on the characteristics of the FS in specific conditions.

The theoretical settling velocity of a particle is given by Equation 6.1. It is defined by the velocity attained by a particle settling in the tank as the gravitational strength overcomes the buoyancy and drag force that retain the particle in the top layer of the tank.

Equation 6.1: $$V_c = \left[\frac{4}{3} \cdot \frac{g \cdot (\rho_s - \rho) \cdot d}{C_d \cdot \rho} \right]^{1/2}$$

Where:

V_c = final settling velocity of the particle (m/h)
g = gravitational acceleration (m/s^2)
ρ_s = particle density (g/L)
ρ = fluid density (g/L)
d = particle diameter (m)
C_d = drag coefficient

The critical settling velocity, V_C, is selected based on the amount of solids that are to be removed. Theoretically, if the flow is laminar (i.e. not turbulent) and there is no shortcutting of the hydraulic flow in the tank, all the particles with a velocity greater than V_C will be removed. This allows the tank to be designed based on the percentage of desired particle removal in the settled sludge. As the flow in the tank is lengthwise, the length has to be designed to be long enough to ensure that particles with V_C have adequate time to settle out below the level of the outlet. Particles with $V_C < V_{CO}$ will not have time to settle out, and will remain suspended in the effluent (as shown in Figure 6.3). How V_C is selected for actual design purposes is discussed in Section 6.3.2.

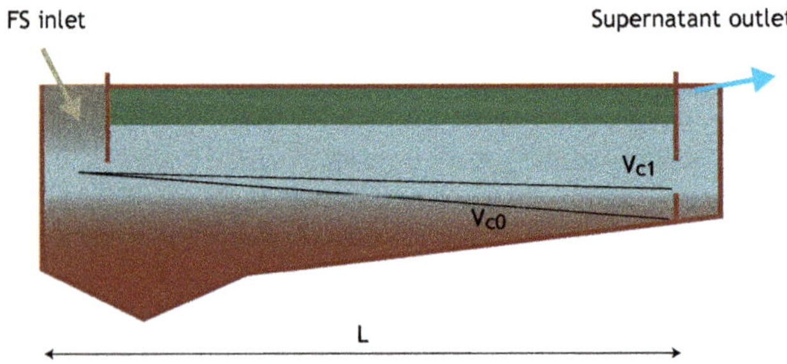

Figure 6.3 Schematic of the final settling velocity (V_C) needed for a particle to settle in a tank of length L.

6.2.2 Thickening

Particles that accumulate at the bottom of the tank are further compressed through the process of thickening. The settled particles are compressed due to the weight of other particles pressing down on them, and water is squeezed out, effectively increasing the concentration of the total solids. This happens as a result of gravity, when the concentration of SS is high and inter-particle strengths hinder the individual movement of particles. Allowing room in the tank for sludge storage as it settles and accumulates is an important consideration in the design of tanks, because as sludge accumulates, it effectively reduces the depth of the tank available for settling. This is also important in designing the ongoing operations and maintenance, and schedule for sludge removal.

6.2.3 Flotation

Similarly to the settling and thickening mechanisms, the influence of gravitational strength due to density differences explains flotation. Buoyancy is the upward force from the density of the fluid. For particles that float, the buoyancy is greater than the gravitational force on the particle. Hydrophobic particles such as fats, oils and greases, and particles with a lower density than water are raised to the top surface of the tank by flotation. Some particles are also raised to the surface by gas bubbles resulting from anaerobic digestion. This layer that accumulates at the top of the tank is referred to as the scum layer.

The scum layer is important to consider in the design process as it also effectively reduces the volume of the tank. The scum layer associated with FS settling can be significant, and cannot be overlooked. Significant scum layers can be seen on the surface of the settling-thickening tanks and ponds in Figure 6.2.

6.2.4 Anaerobic digestion

Anaerobic digestion also occurs in settling-thickening tanks, mainly in the thickened layer. The level of digestion depends on the degree of the initial stabilisation of FS, the temperature, and on the retention time inside the tank. This process degrades a part of the organic matter and generates gasses. Operational experience has shown that fresh FS that is not stabilised (e.g. from public toilets that are emptied frequently) does not settle well. This is because anaerobic digestion of fresh FS contributes to an increased upflow from gas bubbles, and FS that is not stabilised also contains more bound water. Thus, stabilised FS (e.g. from septic tanks) and/or FS that is a mixture of stabilised and fresh sludge are more appropriate for treatment in settling-thickening tanks (Heinss *et al.*, 1998; Vonwiller, 2007).

6.2.5 Solids-liquid zones

The interactions of these fundamental mechanisms result in the separation of the FS into four layers, as illustrated in Figure 6.1 (Heinss *et al.*, 1998; Metcalf and Eddy, 2003):

- A layer of thickened sludge at the bottom. The solid concentration is higher at the bottom than at the top of this layer.
- A separation layer between the thickened layer and the supernatant, as the transition between these is not immediate. Hindered settling occurs mainly in the separation layer, where the settled sludge is not completely thickened. Particles in the separation layer can be more easily washed out with the supernatant than particles in the thickened layer.
- A supernatant layer between the separation layer and the scum layer. This consists of the liquid fraction and the particles that do not settle out or float to the surface.
- A layer of scum at the top of the tank. This consists of the floating organic and non-organic matter, the fats, oils, and greases contained in FS, as well as particles that have been raised up by gas up-flow.

6.3 DESIGN OF SETTLING-THICKENING TANKS

This section provides recommendations for the design of settling-thickening tanks for the treatment of FS based on the current available knowledge. The tank design is based on the estimated volume of FS, and the resulting supernatant flow, and production of scum and thickened sludge layers. An adequate design needs to include regular and efficient removal of the scum and thickened sludge, which needs to be considered to optimise the solids-liquid separation. These design aspects are discussed below, and examples are provided in the case studies and the design example.

6.3.1 Laboratory tests and faecal sludge characteristics influencing the design

A good understanding of site specific FS characteristics is required in order to determine the tank surface and the volume of the scum, supernatant, separation, and thickened sludge layers. As discussed in Chapter 2, determining an accurate value for influent loading of FS can be challenging depending on the local infrastructure and existing management system. The design loading needs to take into account that FS quantities and characteristics can also vary seasonally. An empirical estimation of settling ability for the specific FS that the tank is being designed for needs to be determined for adequate design of the tank. Preliminary laboratory analyses should be conducted on the FS that is to be treated, especially in terms of settling ability, thickening ability, potential for scum accumulation and SS concentration (Strauss *et al.*, 2000). It is important to ensure that the FS used for these tests is that which will actually be treated. For example, if there is an existing network of collection and transport companies with vacuum trucks, sludge should be sampled from the trucks as this is what will be discharged at the treatment plant.

Figure 6.4 Imhoff cones being used in analyses of sludge volume index (photo: SANDEC).

The sludge volume index (SVI) is a laboratory method to empirically determine the settling ability of sludge based on the amount of suspended solids that settle out during a specified amount of time. To determine the SVI, first the suspended solids content of FS is determined, and then a graduated Imhoff cone is filled with the FS sample that is left to settle (see Figure 6.4). After 30-60 minutes, the volume occupied by the settled FS is recorded in mL/L. The SVI is then calculated by dividing the volume of settled FS by the SS concentration (in g/L), which gives the volume of settled sludge per gram of solids (see the example problem on the calcualtion of SVI below). The Imhoff tests do not provide exact estimates of the depth of the thickened layer, as they are batch tests and not continuous loading as in a settling-thickening tank. Imhoff cones with volumes greater than one litre provide a more representative result as the wall effect is reduced (Heinss *et al.*, 1999).

Based on experiences in the design of settling-thickening tanks for wastewater treatment plants, wastewater sludge with a SVI of less than 100 (mL/g SS) achieves good solids-liquid separation in settling-thickening tanks. Measurements with FS in Accra, Ghana and Dakar, Senegal showed that FS had a good settling ability and thickening ability with SVI of 30-80 mL/g (Heinss *et al.*,1998), and the personal experience of Dodane). SVI tests conducted in Dakar, Senegal showed that FS settled rapidly during the first 20 minutes, after which more thickening occurred and continued for 100 minutes (Badji *et al.*, 2011).

Example Problem: Calculation of sludge volume index (SVI)

A sample of FS from a septic tank in Burkina Faso has a SS concentration of 6.6 g/L.
The volume of the settled FS after 60 minutes is 198 mL/L.
The SVI = Volume of settled FS/SS concentration = 198/6.6 = 30 mL/g

This FS would be considered to be appropriate for treatment in a settling-thickening tank. With activated sludge, it is considered that ideal settling conditions are reached with SVI less than 100 mL/g SS (Pujol *et al.*,1990). For FS, the stability and origin also needs to be taken into account, but more studies are needed to assess the adequate limits.

6.3.2 Tank surface and length

The length of the tank needs to be sufficient and have adequate hydraulic distribution, to ensure that the entire tank surface area is used, and that particles have enough time to settle. The surface area of the settling-thickening tank can be calculated as shown in Equation 2, based on the upflow velocity (V_u) and the influent flow (Q_p) (Metcalf and Eddy, 2003).

Equation 6.2: $$S = \frac{Q_p}{V_u}$$

Where:
S = surface of the tank (m^2)
Q_p = influent peak flow (m^3/h)
V_u = upflow velocity (m/h)
$Q_p = Q \cdot C_p/h$,

Where:
Q = mean daily influent flow
C_p = peak coefficient
h = number of operating hours of the treatment plant (influent is only received during operating hours)

The upflow velocity (V_u) is defined as "the settling velocity of a particle that settles through a distance exactly equal to the effective depth of the tank during the theoretical detention period" (Ramalho, 1977). It is used to calculate the acceptable inflow that will allow for particles with the defined settling velocity to settle out. Particles with a settling velocity slower than V_u will be washed out with the supernatant. A value is selected for the desired percentage of suspended solids removal, and then the upflow velocity is selected to be equal to the final settling velocity of the lightest particles that will settle in the tank. For example, as shown in Figure 6.3, $V_u = V_{c0} > V_{c1}$. Thus, for a given FS influent, the upflow velocity in a tank surface corresponds to the removal of a given percentage of suspended solids. The peak coefficient is calculated by observation of when the greatest volumes of trucks are discharging at the FSTP. For example, in Dakar the peak period was observed to be 11:00 because trucks have their busiest emptying periods during the morning, and was calculated to be 1.6 times higher than the average.

V_u can be estimated based on SVI values. Despite the limits of the theoretical calculation for design purposes, methods and calculations to link SVI and V_u have been developed based on long-term experiences in activated sludge treatment (Pujol *et al.*, 1990). However, this type of empirical knowledge does not yet exist for FS. V_u = 0.5 m/h could be used for rectangular settling tanks treating FS that have a SVI less than 100 (personal experience, Pierre-Henri Dodane). Once the surface area has been calculated, the length: width ratio needs to be selected. For example (Heinss *et al.*, 1998) recommend a width to length ratio between 1:10 to 1:5. The lower the selected final settling velocity, the longer the tank needs to be, and the more particles that will settle out.

6.3.3 Tank volume

Once the surface area of the tank has been determined, the volume can be calculated, considering the depth of the four layers described in Figure 6.1. It is necessary to plan for the reduction in depth that will occur due to the accumulation of scum and thickened sludge, which will result in solids washed out with the supernatant if underestimated.

Based on field observations of settling-thickening tanks in Accra and Dakar (Heinss *et al.*, 1998), the following values are recommended for designing tanks for FS with similar characteristics:

- scum zone: 0.4 m (with 1 week loading, 1 week compaction and cleaning) to 0.8 m (with 4 weeks loading and 4 weeks compaction and cleaning);
- supernatant zone: 0.5 m; and
- separation zone: 0.5 m.

The depth of the thickened sludge zone needs to be calculated given the expected load inflow and the concentration of the thickened sludge (C_t). The design of a sufficient storage volume for the thickened sludge is crucial to avoid outflow of settled sludge during one operating cycle. Therefore, the expected operating cycle duration (i.e. loading, compaction and sludge removal) and methods for scum and thickened sludge removal need to be defined in the first place. The volume of the thickened sludge storage zone (V_t) can be calculated as shown in Equation 6.3 (Metcalf and Eddy, 2003).

Equation 6.3: $$V_t = \frac{Q \cdot C_i \cdot e \cdot N}{C_t}$$

Where:
V_t = volume of thickened sludge storage zone (m^3)
Q = mean FS daily inlet flow (m^3/day).
C_i = suspended solids mean concentration of FS load (g/L)
e = expected settling efficiency (= proportion of suspended solids separated, as %)
N = duration of the FS load for one cycle in days
C_t = suspended solids mean concentration of thickened sludge after the loading period (g/L)

The mean daily flow is used for the sludge accumulation estimate, but the peak flow is used for the tank surface and length design to ensure settling is achieved under all the expected operating conditions. The volume of the thickening zone is based on the expected settling of FS. It is not considered in the design, but longer storage times when the tanks are not loaded prior to sludge removal, result in increased thickening and compaction. In the field, average FS settling efficiencies of only about 60% have been observed, due to poor operation and maintenance and gas upflow (Heinss *et al.*, 1998). However, it is recommended to use 80% to estimate the maximum efficiency.

Care must be taken to ensure a relatively accurate estimate of C_t. An overestimation will lead to an insufficient storage volume and to a reduced settling efficiency, as solids may be washed out without being able to settle. An underestimation will lead to the design of an unnecessarily large storage volume and increase in construction costs. Table 6.1 presents examples of SS concentrations given the initial FS load and thickening duration.

Table 6.1 Concentration of sludge in the thickening zone of settling tanks in Accra and Dakar (Heinss *et al.*, 1998; Badji *et al.*, 2011)

Place of measurement	Concentration at inlet (g SS/L)	Thickening duration (day)	Concentration in thickened zone (g SS/L)
Dakar, FSTP	5	10	60-70
Accra , FSTP	15-20	9	60-85
Accra , FSTP	15-20	30	>100
Accra , FSTP	15-20	50	140
Accra , laboratory	40	7	100

The FS loading period needs to be defined given the FS characteristics, the expected total solid concentration of the thickened sludge, and the seasonal variations. The advantages of short loading and compaction periods are that the scum layer is maintained at a minimum depth, and the thickened sludge is easily removed by pumping, as it is not heavily compacted (Case Study 6.1).

Case Study 6.1 : Operation of settling-thickening tanks in Dakar, Senegal and Accra, Ghana
(Adapted from Heinss *et al.*, 1998; Badji *et al.*, 2011)

Settling-thickening tanks of different sizes have been in operation in Dakar (Senegal) since 2006 and in the Accra region (Ghana), since the late 1980s. Short loading periods of about one week were adopted for the FS treatment plants in Dakar, where the thickened sludge is mostly removed by pumps, and the most compacted sludge and scum is removed with vacuum trucks. The removal of scum requires powerful vacuum trucks, which are not always available. It is thus crucial to ensure the regular availability of mechanical means to remove the most compacted solid products to ensure the tank's efficiency and sustainability.

Settling-thickening tanks of the Cambérène treatment plant were designed with a nominal HRT of 8.6 hours. Due to initial underestimation of the FS volumes to treat, the settling-thickening tanks were overloaded and operated with an effective HRT of 1.7 h. Thus as discussed in Chapter 2, a preliminary study to assess the volumes and concentrations to be treated is required before designing tanks. Collection and transport activities should be assessed, including area served, the number of households, the frequency of collection from onsite systems, and the type of onsite systems.

Long loading periods of 4 weeks were adopted for the settling-thickening tanks in Accra, where the tanks have a larger volume to allow storage of greater quantities of FS. Due to the size, these types of tank are also referred to as settling-thickening ponds. The loading phase was operated over 4 weeks, with an additional compaction phase where they are not loaded over 3-4 weeks before the sludge removal. In this case, the scum layer is deeper, and the thickened sludge is more compact and therefore more difficult to remove. Front-end loaders have been used to remove both the thickened sludge and the scum, which have a high solids concentration. Large settling-thickening ponds can therefore be more difficult to operate.

6.3.4 Inlet and outlet configuration

Grit screening must be undertaken before the loading of FS into the settling-thickening tanks in order to facilitate maintenance (e.g. removal of coarse waste to avoid potential degradation of pumps). This is explained further in Chapter 5, Overview of Treatment Technologies.

The inlet zone should allow for the uniform and quiescent distribution of the flow in the whole tank and avoid short-circuiting. Therefore, baffles are recommended to help disperse the energy of the inflow, and to reduce the turbulence in the tanks. (Heinss *et al.*, 1998) recommend locating the inlet zone near the deep end of tanks to improve the solids settling. The pumps for the extraction of the thickened sludge must be adapted to remove concentrated sludge. Easy access points should also be included to allow the sampling of sludge in these zones, and to ensure that easy repair of pumps is possible.

The supernatant outlet zone should be located under the scum layer and above the thickened sludge storage layer. Baffles are useful to avoid washout of the scum with the supernatant. To ensure an optimal hydraulic flow, the outlet channel can be extended along the width of the wall (Heinss *et al.*, 1998). It must be at the opposite side of the inlet zone. Outlets that are positioned near to the shallower side of the tank reduce the carry-over of the settled solids from the thickening layer.

6.4 OPERATION AND MAINTENANCE OF SETTLING-THICKENING TANKS

At least two settling-thickening tanks should be operated alternately in parallel, in order to allow for sludge removal as tanks should not be loaded during this time. The loading of FS, and the compaction and removal of the thickened sludge and scum comprise the main phases of an operating cycle. These periods allow for the expected solids-liquid separation and thickening operations. While the tanks are not loaded, additional compaction occurs prior to the removal of thickened sludge and scum, due to the lack of hydraulic disturbance (Heinss *et al.*, 1998). During this time further solids-liquid separation occurs, and the SS concentration increases in the thickened sludge and scum.

6.4.1 Sludge and scum removal

The timing of the removal of sludge and scum as planned for in the design is essential to ensure that the settling-thickening tanks are functioning properly, and that there is adequate depth for the settling of particles, leading to a reduced solids-liquid separation.

Figure 6.5 shows an example of the volume reduction resulting from inadequate sludge removal practices. In this case, the scum layer was not removed during such a long period that as a consequence, weeds are seen growing on the surface. This should be avoided.

If it is observed that a higher volume of thickened sludge has accumulated than what was designed for, this means that the solid load is higher than expected, and operations should be appropriately altered. Sludge removal typically lasts a few hours to a day following the compaction period. Once in operation, detailed monitoring can be done to optimise compaction and sludge removal times based on actual operating conditions.

Figure 6.5 Example of inadequate operation and maintenance of a settling-thickening tank in West Africa. The scum was not removed during a long period, which allowed plants to grow on it. The volume of sludge and scum accumulated does not allow for proper operation of the tank, or for solids-liquid separation (photo: SANDEC).

Figure 6.6 Settling-thickening tank of Rufisque showing the scum layer (photo: SANDEC).

The first step in sludge and scum removal is typically removal of the scum layer. The scum layer generally has a high solids concentration that cannot be easily pumped and can remain after the thickened sludge is removed (Figure 6.6), in which case it needs to be manually removed. If possible, scum can be removed with shovels from both sides of the tank when the tank is narrow enough for access, or by mechanical means such as vacuum trucks with strong pumps. Scum can also be removed manually or sucked by a vacuum tanker after emptying the tank as it is done at the Cambérène treatment plant.

Next, the supernatant layer is frequently removed by pumping or by gravity (depending on the design). It can be pumped to the parallel settling-thickening tank or to the next step in the treatment chain. The thickened sludge can then be pumped or shoveled out of the tank after the supernatant has been removed. When a pump is used for extracting the thickened sludge, the supernatant layer does not need to be removed, as the supernatant layer can facilitate the pumping of thickened sludge as a pressure is maintained. As tanks are frequently over 2 m deep, adequate access for sludge removal (and for tank and pump cleaning) needs to be integrated into the design. The operator knows when it is time for sludge removal based on the loadings and times given in the design, and also by visual observation.

It is possible to design settling-thickening tanks with devices that continuously scrape and pump the thickened sludge out of the tanks, and remove the scum over the supernatant zone. These devices allow easier operation and increase the management flexibility, but increased operating and maintenance costs need to be taken into consideration (Chapter 11).

6.4.2 Start-up period and seasonal variations

As settling-thickening tanks rely mainly on physical processes, there is no special requirement for start-up periods. It is however useful to adjust the load time, assess the depths of the different zones and optimise the compaction time and sludge removal frequency. Seasonal variations of meteorological conditions and FS characteristics may influence the efficiency of the tanks. For example, loss of water through evaporation could increase the solids content of the scum. High temperatures may also increase the anaerobic digestion process, and therefore the height of the scum layer.

Case Study 6.2: Cambérène FST – settling tanks and sludge drying beds

(Adapted from Badji *et al.*, 2011; continued in Case Study 7.2)

Cambérène FSTP, the first treatment plant at scale serving Dakar city, was put in operation in 2006. It is composed of a combination of settling-thickening tanks (two tanks of 155 m^3 each) and unplanted drying beds (10 beds of 130 m^2 each). The thickened sludge is extracted from the settling-thickening tanks and transferred to drying beds by pumping. The effluent from the tank and the leachate from the drying beds are transferred to the wastewater treatment plant. Each week, one tank is used for receiving FS while the other tank is pumped out and cleared from the scum layer. The FS in Dakar is dilute with an average TS of 5 g/L. There is a high groundwater table in Dakar and the majority of sludge is from septic tanks.. The combination of settling/thickening tanks and drying beds was selected to thicken the dilute sludge before drying and in order to reduce the required area for drying beds.

From 2007 to 2009, daily measurements of the pollutant fluxes were conducted at the inlet and outlets of the two treatment stages. Continuous monitoring of sludge characteristics (concentration, dry matter content) in the settling-thickening tanks and drying beds was conducted, and reported by Badji *et al.* (2011), as summarised in Figure 6.7. Although the FSTP was designed for treating 100 m^3 of FS/day and 700 kg TS/day, the plant received 340 m^3 FS/day and 1,700 kg TS/day. The pollutant analysis in the plant in real operational condition is presented in the figure below.

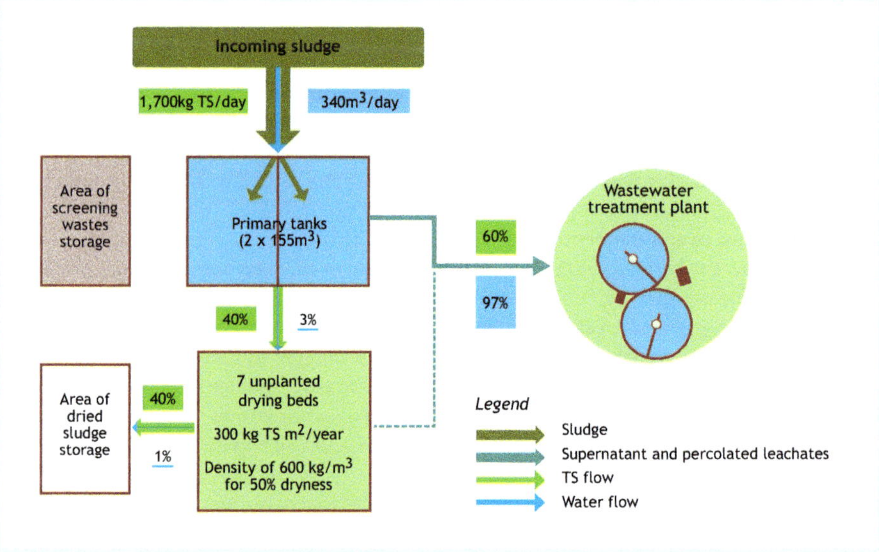

Figure 6.7 Efficiency and fluxes analysis of Cambérène faecal sludge treatment plant in real operational conditions (Badji *et al.*, 2011).

Experiences from Accra reveiled the possibility to obtain a TS concentration of 150 g/L with a similar settling tank. However, it was not known at the time that the influent in Ghana had a higher concentration to start with. Moreover, the Dakar operator left the sludge in the settling-thickening tank for one week only, while the plant in Ghana ran on longer HRTs. As a consequence, a thickened sludge concentration of 60-70 g TS/L was achieved after one week of thickening while 140 g TS/L was envisioned in the design. The achieved removal was far from the designer expectations. This shows the great importance of conducting preliminary studies in order to define sludge characteristics (e.g. concentration, inlet flow and thickening degree) based on local conditions, despite the difficulty – it can save significant amounts of money for the real operation of the plant. Operation and maintenance of the tanks were considered difficult by the operator, in particular entering the tank for scum cleaning. The pumping also required attention as the axle was often blocked by debris and needed to be cleaned. This led to delays in thickened sludge pumping, and as a consequence to an increased overloading in the parallel tank (see Figure 6.7).

6.5 PERFORMANCE OF SETTLING-THICKENING TANKS

The most important consideration in the performance of settling-thickening tanks is the separation of the liquid and solid fractions. The efficiency of the key mechanisms to achieve this are discussed here.

6.5.1 Solids-liquid separation

In the field, the mean settling efficiency of operating tanks and ponds is about 50-60% of SS in the settled volume. This efficiency can reach up to 80% where the tanks have been adequately designed and operated (Heinss *et al.*, 1999).

The concentration of the thickened sludge (C_t) achieved depends on the operating cycle duration and the initial FS characteristics (thickening ability), as presented in Table 6.1. Achieving 60 g SS/L in the thickened zone for a seven day load period seems a reasonable estimate. In Accra, with an operating cycle of about eight weeks, (Heinss *et al.*,1998) observed a total solid content of 150 g TS/L in the thickened layer.

The scum layer thickness and SS content depends mainly on the operating cycle duration, the FS characteristics and the evaporation process. (Heinss *et al.*, 1998) report a scum layer of 80 cm in settling-thickening tanks operated with cycles of 8 weeks. In the Dakar FSTP the observed scum layer had a depth of 10 to 20 cm after one week of loading.

6.5.2 Treatment performance

The main objective of settling-thickening tanks is solids-liquid separation, not stabilisation or pathogen reduction. Further treatment steps are required for both the thickened solids and supernatant. Dissolved organic matter, nutrients, and suspended particles will remain in the supernatant. Examples include 50% of influent COD in the settled sludge, and 50% in the supernatant (Badji *et al.*, 2011), and 10% influent BOD and 25% COD in the supernatant (Heinss, *et al.*, 1998). Total pathogen removal or inactivation is also negligible. Many larger pathogens such as Helminth eggs settle out, and the amounts that are partitioned in the solids will be correlated to SS removal efficiency. (Heinss *et al.*, 1998) observed that 50% of the total Helminth eggs were partitioned in the thickened sludge.

Table 6.2 Results of preliminary studies to determine design parameters

Initial raw FS concentration:	$C_{i(TS)}$= 7 g TS/L $C_{i(SS)}$= 5 g SS/L
FS origin:	Mainly septic tanks (stabilised FS)
Total volatile solids percentage	< 70%
Influent flow:	Q=140 m^3/day
FSTP opening time:	7 h/day 5 days/week 52 weeks/year
Daily peak flow coefficient:	C_p=1.6 (peak flow is often in the morning, after the first trucks rotation)
Concentration of thickened sludge (1 L Imhoff cones)	60 g SS/L
Settling ability (1 L Imhoff cones)	Good (SVI = 23 << 100)

6.6 ADVANTAGES AND CONSTRAINTS OF SETTLING-THICKENING TANKS

Settling-thickening tanks are efficient as a first treatment step as they rapidly achieve solids-liquid separation, they are relatively robust and resilient, and they reduce the volume of sludge for subsequent treatment steps.

Constraints of settling-thickening tanks include:

- lack of experience operating with FS, and lack of empirical data and results on which to base designs on;
- settled sludge still has relatively high water content and requires further dewatering;
- the liquid fraction remains highly concentrated in SS and organics; and
- pathogen removal is not significant, and the endproducts of settling tanks therefore cannot be discharged into water bodies or directly used in agriculture (for more details on appropriate enduse see Chapter 10).

6.7 DESIGN EXAMPLE FOR A SETTLING-THICKENING TANK

As mentioned in the previous sections, the design of settling-thickening tanks involves calculating the basin surface, the zone volumes and the hydraulic configurations.

6.7.1 Initial situation

In a real-life situation, sufficient preliminary studies are needed to allow for the specific design according to the local context characteristics. This example of a design calculation corresponds to a typical situation in which settling-thickening tanks can be implemented and is based on information obtained from preliminary studies as shown in Table 6.2.

6.7.2 Assumptions and design decisions

Based on these preliminary results, the following assumptions and design decisions can be made:

- a final settling velocity of V_c = 0.5 m/h based on SVI tests and experience.;
- the expected settling efficiency (e) is 80% of SS.;

- two parallel tanks are designed to allow the cleaning of one during the loading of the other;
- a loading period of one week (N = 5 = number of treatment plant opening days per week) to minimise anaerobic digestion and gas upflow. This means that each tank is loaded for one week out of every two weeks, while the extraction of thickened sludge and scum is carried out on the other tank;
- a short compaction period of 2-3 days. Hence, the removal of thickened sludge and scum occurs every 10 days by pumping, as the thickened sludge is still sufficiently liquid; and
- the operator has experience in wastewater treatment and therefore the thickened sludge pumping and tank cleaning is likely to be carried out correctly.

Technology

6.7.3 Design calculations

The tank surface (S) needed to allow for the selected final settling velocity (V_c) is estimated based on the influent peak flow (Q_p) as shown in the following equations.

Equation 6.4: $Q_p = Q \cdot C_p / 7 = 32\ m^3/h$

Where 7 = number of FSTP opening hours per day

Equation 6.5: $S = Q_p / V_c = 64\ m^2$

Thickening zone volume

The daily SS quantity of FS discharged (M) is calculated from the initial FS concentration (C_i):

Equation 6.6: $M = Q \cdot C_{i(SS)} = 700$ kg SS/day

The daily SS mass of thickened sludge (M_t) is then deduced from the SS settling efficiency (e):

Equation 6.7: $M_t = M \cdot e = 560$ kg SS/day

Where e = 80%. For a safe design, the value of e should be the maximum expected efficiency (not the mean).

The volume of the thickening sludge storage zone (V_t) is related to the mass of the particles trapped in the thickening zone (M_t) and the SS concentration achieved in the thickened sludge (C_t):

Equation 6.8: $V_t = M_t \cdot N / C_t = 47\ m^3$

Tank configuration

The surface of the tank should be long and narrow and facilitate the distribution of flow. The recommended width to length ratio ranges from 0.1 to 0.2. To reach a surface close to 64 m^2 (see Equation 6.5), the following configuration should be adopted:

Equation 6.9: $S(l \cdot L) = 3 \cdot 22 = 66\ m^2$

Zone depth

The following design characteristics are given for each zone:

- Scum zone: 0.4 m (value assumed to be safe for a 2 week cycle);
- Supernatant zone: 0.5 m (Heinss *et al.*, 1998);
- Separation zone: 0.5 m (Heinss *et al.*, 1998); and
- Thickening sludge zone: 0.75 m (based on 47 m^3 storage in a 66 m^2 tank).

A schematic diagram of the zone depths is shown in Figure 6.8.

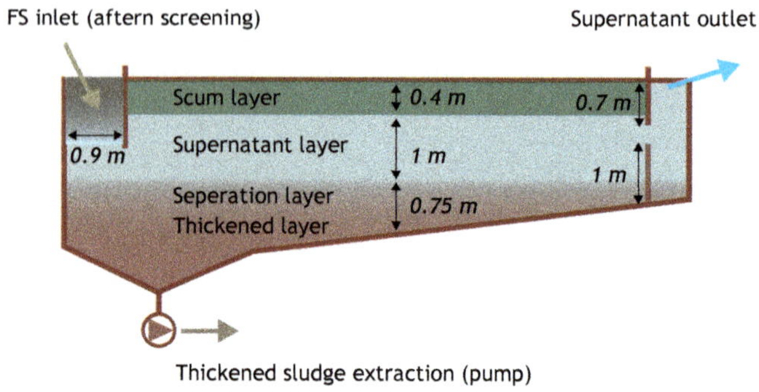

Figure 6.8 Schematic of tank configuration described in design example.

6.7.4 Mass flow analysis of faecal sludge treatment

In this example, the thickened zone was designed based on an 80% SS removal. In order to plan for options to further treat the supernatant, a more realistic settling efficiency (e) of 60% SS should be considered such that the supernatant contains 40% of $C_{i(SS)}$. These mass flows are shown in Figure 6.9, with the estimated SS flows to the treatment options for the supernatant and the thickened sludge.

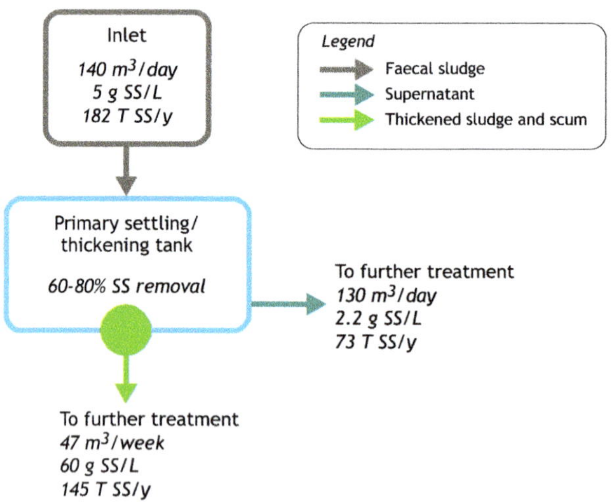

Figure 6.9 Schematic presentations of the treatment and mass flows for the theoretical example.

6.8 BIBLIOGRAPHY

Badji K., Dodane P.H., Mbéguéré, M., Koné, D. (2011), Traitement des boues de vidange: éléments affectant la performance des lits de séchage non plantés en taille réelle et les mécanismes de séchage. Actes du symposium international sur la Gestion des Boues de Vidange, Dakar, 30 juin – 1er juillet 2009, EAWAG/SANDEC.

Heinss, U., Larmie, S.A., Strauss, M. (1998). Solids Separation and Pond Systems for the Treatment of Faecal Sludges in the Tropics – Lessons Learnt and Recommendations for Preliminary Design, EAWAG/SANDEC, Report No. 05/98.

Heinss, U., Larmie, S.A., Strauss, M. (1999). Characteristics of faecal sludges and their solids-liquid separation. EAWAG/SANDEC: Janeiro.

Pujol, R., Vachon, A., Martin, G. (1990). Guide technique sur le foisonnement des boues activées, ed. FNDAE, Ministère de l'Agriculture et de la Foret.

Ramalho, R.S. (1977). Introduction to Wastewater Treatment Processes, Academic Press.

Strauss, M., Larmie, S.A., Heinss, U., Montangero, A. (2000). Treating faecal sludges in ponds. Water Science and Technology 42(10), p.283-290.

Metcalf and Eddy (2003). Wastewater Engineering: treatment, disposal, reuse. Tchobanoglous, G., Burton, F.L. eds. McGraw-Hill Book Company.

Vonwiller, L. (2007) Monitoring of the faecal sludge treatment plant Cambérène in Dakar. EAWAG, Dübendorf, Switzerland.

End of Chapter Study Questions

1. What are three fundamental mechanisms that explain how settling-thickening is achieved? Explain how they work.

2. List three advantages and three disadvantages associated with the operation of settling-thickening tanks.

3. What are the three factors that need to be calculated in order to design settling-thickening tanks?

4. In the design of settling tanks, why is it important to calculate the basin surface, the zone volumes, and the hydraulic configurations?

Technology

Chapter 7

Unplanted Drying Beds

Pierre-Henri Dodane and Mariska Ronteltap

Technology

Learning Objectives

- Have an understanding of an unplanted drying bed for sludge dewatering.
- Have an overview of the main components of unplanted drying beds, their characteristics and their effect on the performance of the beds.
- Know the appropriate level of operational and maintenance monitoring necessary for the operation of unplanted drying beds.
- Be able to design an unplanted drying bed to achieve the desired treatment objectives.

7.1 INTRODUCTION

Unplanted sludge drying beds are shallow filters filled with sand and gravel with an under-drain at the bottom to collect leachate. Sludge is discharged onto the surface for dewatering (Figure 7.1). The drying process in a drying bed is based on drainage of liquid through the sand and gravel to the bottom of the bed, and evaporation of water from the surface of the sludge to the air. The design as well as the operation of the drying bed is fairly straightforward, provided the sludge loading rate is well selected and the inlet points for depositing the sludge onto the bed are properly designed. Depending on the faecal sludge (FS) characteristics, a variable fraction of approximately 50-80% of the sludge volume drains off as a liquid (or leachate), which needs to be collected and treated prior to discharge (Tilley *et al.*, 2014). After reaching the desired dryness, the sludge is removed from the bed manually or mechanically. Further processing for stabilisation and pathogen reduction may be required depending on the intended enduse option. When considering the installation of a drying bed, the ease of operation and low cost needs to be considered against the relatively large footprint and odour potential.

7.2 TREATMENT PRINCIPLE

A FS treatment plant (FSTP) consists of several drying beds in one location. Sludge is deposited on each of these drying beds where it remains until the desired moisture content is achieved. It is subsequently mechanically or manually removed for disposal or further treatment and reuse.

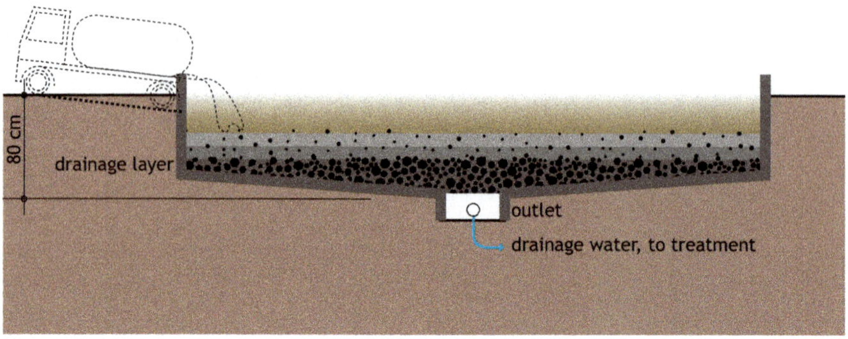

Figure 7.1 Schematic overview of an unplanted sludge drying bed (Tilley *et al.*, 2014). A splash plate is not visible but is an essential element (Section 7.5).

The drying process is based on two principles. The first principle is percolation of the leachate through sand and gravel. This process is significant with sludge that contains large volumes of free water (Section 3.2), and is relatively fast, ranging from hours to days (Heinss *et al.*, 1998). The second process, evaporation, removes the bound water fraction and this process typically takes place over a period of days to weeks. Heinss *et al.* (1998) reported removal of 50 to 80% by volume due to drainage, and 20 to 50% due to evaporation in drying beds with FS. This range is typical for sludge with a significant amount of free water, but there is more evaporation and less percolation with sludge that has more bound water. For example, no leachate was observed in a study with preliminary thickened sludge (Badji, 2011). In planted sludge drying beds evapotranspiration also contributes to water loss, which is explained in Chapter 8.

7.3 UNPLANTED SLUDGE DRYING BED DESIGN PARAMETERS

When designing a drying bed, there are several influencing factors that need to be taken into consideration. These aspects vary from location to location, and can be grouped under climate factors and the type of sludge to be treated. Other key parameters that have an impact on the sludge drying process include the sludge loading rate, the thickness of the sludge layer, and the total bed surface. All these aspects are discussed in the following sections.

7.3.1 Climate factors

Climate factors affecting the operation of unplanted drying beds include the following:

- Humidity: high humidity reduces the contribution of evaporation to the drying process;
- Temperature: higher temperatures, also in combination with relatively low humidity and high wind, will enhance the total amount of water removed via evaporation;
- Rainfall: in locations where rainfall is frequent and occurs for long periods of time intense, a drying bed may not be feasible. Pronounced rainy seasons can be accommodated for by not using the beds in that period, or by covering them with a roof. Rainfall will may rewet the sludge, the intensity of which depends on the phase of drying.

Figure 7.2 Freshly loaded and partially dewatered faecal sludge on unplanted drying beds at Niayes faecal sludge treatment plant, Dakar, Senegal (photo: Linda Strande).

7.3.2 Type of faecal sludge

The origin of the sludge is important when using drying beds . Septic tank sludge has less bound water and is hence more readily dewatered than fresh FS. In other words, it is considered to contain a lower specific sludge resistance for dewatering. It therefore can be applied in a thicker sludge layer or at a higher total solids loading rate or at a higher sludge loading rate. Sludge from public toilets is typically not digested: particles have not settled (see also Chapter 2). Because it has a higher specific sludge resistance for dewatering less water will be removed, a longer sludge drying time may be required, or it may not be appropriate for drying beds.

Pescod (1971) carried out experiments with fresh pit latrine sludge on drying beds and obtained a wide variation in drying results – some comparable to more stable sludge. Generally a proper solid liquid separation is difficult to obtain with fresh public toilet sludge. An alternative is to mix this type of sludge with older, more stabilised sludge (e.g. septic tank sludge) to enhance the dewaterability (Koné *et al.*, 2007; Cofie *et al.*, 2006).

Case Study 7.1: Designing a sludge drying bed in Kumasi, Ghana
(Adapted from Cofie and Koné, 2009).

In order to pre-dry sludge for a co-composting pilot plant (Case Study 5.1) a small sludge drying bed was designed for Kumasi, Ghana. The climate is sub-equatorially wet with two rainy seasons, a major one from late February to early July and a minor one from mid-September to early November. FS is collected from onsite sanitation systems (septic tanks, pit latrines and unsewered public-toilets) by vacuum trucks within the city of Kumasi and transported to the project site. Of the 500 m^3 /d of FS produced, 1.5 m^3/day is treated in the pilot plant. Two unplanted drying beds were built with a surface area of 25 m^2 each (to hold 15 m^3 excreta with a depth of 30 cm). They consisted of different layers of a gravel-sand filter material of different thickness and particle sizes. The technical details and characteristics that were taken into account for the design are listed in Table 7.1.

Table 7.1 Technical details and characteristics recommended for faecal sludge dewatering in drying beds

Sizing of the beds:	Production of filter layers:
25 – 30 cm sludge layer on beds	Reduce pressure flow via splitting chamber, inlet channel, and splash plates
100-200 kg TS/m²/year (TS stands for total solids)	Drying bed removal efficiency:
0.08 m²/cap	97% SS (suspended solids), 90% COD (chemical oxygen demand), 100% HE (helminth eggs)
Untreated sludge characteristics:	**Dried sludge production:**
Partly stabilised (septage or mixture of septage and public toilet)	0.1 m³ per m³ fresh FS
Sludge with ≤ 30 % share of public toilet sludge	Hygienisation necessary prior to use in agriculture as biosolids
Sand characteristics:	**Leachate:**
Sand particles do not crumble	Quality fairly comparable to tropical wastewater
Sand easily available locally	Salinity too high for irrigation
Sand thoroughly washed prior to application onto the gravel base	Leachate treatment

Based on the technical details presented in Table 7.1 the following design for the drying beds was determined:

3 FS truck loads/cycle (1 truck carries ~5m³)	Hydraulic load on drying beds: 30 cm/cycle
3 dewatering cycles/month	Surface of sludge drying beds: 50 m²
Volume of FS treated: 15 m³/cycle = 45 m³/month = 1.5 m³/d	FS volume reduction through dewatering assumed: 90%
Ratio of public toilet sludge to septage sludge = 1:2	Dried sludge produced: 1.5 m³/cycle = 4.5 m³/month

The leachate from the drying beds is collected in a leachate storage tank and discharged into the facultative stabilisation pond of the Buobai FSTP before final discharge into a nearby stream. The dried FS is removed from the drying beds once it can be removed by spade (after 10 days) and stored prior to co-composting (Case Study 5.1).

7.3.3 Sludge loading rate

The sludge loading rate (SLR) is expressed in kg TS/ m^2/year. It represents the mass of solids dried on one m^2 of bed in one year. Pescod (1971) states that any general number linking the total amount of sludge to be dried to a sludge loading rate, bed surface area and loading depth can only be an estimate, as the local conditions vary greatly. However, it is possible to indicate a range of sludge loading rates which typically vary between 100 and 200 kg TS/m^2/year in tropical climates, with 100 for poorer conditions and 200 for optimal conditions, while approximately 50 kg SS/m^2/year is commonly used in temperate climates in Europe (Duchêne, 1990). Poor conditions entail high humidity, low temperature, long periods of rainfall, and/or a large proportion of fresh FS. Optimal conditions comprise a low humidity, high temperature, a low amount of precipitation, and stabilised sludge. It may be possible in some cases to achieve an even higher sludge loading rate. Cofie *et al.* (2006) for example applied sludge at a loading rate of up to 300 kg TS/m^2/year. Badji (2011) also found a SLR of 300 kg TS/m^2/year to be effective for dewatering thickened FS with 60 g TS/L, while about 150 kg TS/m^2/year was estimated to be an effective rate for a FS with 5 g TS/L in the same climatic conditions. Optimal local operating conditions need to be determined through pilot-scale experiments.

7.3.4 Thickness of the sludge layer

A review of the literature shows that sludge is typically applied in a layer of 20 to 30 cm in depth, with a preference for 20 cm. It may seem a better option to apply a thicker sludge layer as more sludge can be applied to one bed; however, this will result in an increased drying time, and a reduction in the number of times the bed can be used per year. For any particular sludge dried under the same weather conditions, Pescod (1971) found that an increase in the sludge layer of only 10 cm prolonged the necessary drying time by 50 to 100%.

It is also important that the sidewalls of the drying beds are high enough to accommodate different loadings. For example, if a layer of 20 cm is applied with a water content of 90%, the initial height before the water is drained-off will be much greater than 20 cm. If the beds receive sludge discharged from a truck as opposed to settling tanks, the walls need to be higher than the planned 20 to 30 cm of sludge layer to allow for the increased volume of liquid.

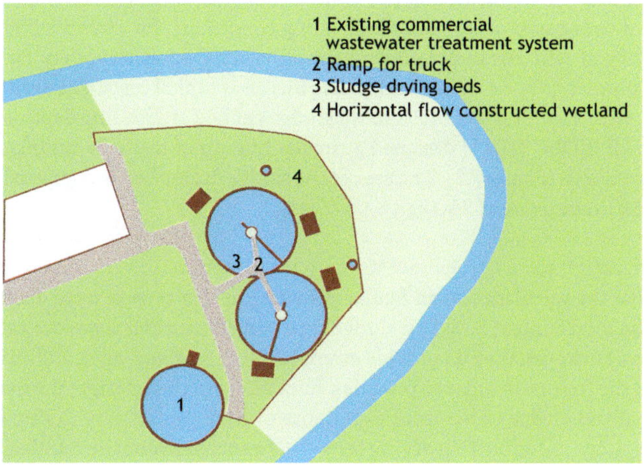

Figure 7.3 Proposed lay-out for a faecal sludge treatment plant with unplanted sludge drying beds. The beds are laid out in a circular design with one inlet. The leachate is to be treated in horizontal flow constructed wetlands (HPCIDBC, 2011).

Figure 7.4 Loading of the beds at Niayes faecal sludge treatment plant, Dakar, Senegal (photo: Linda Strande).

7.3.5 Number of beds

The number of beds required depends on the amount of sludge arriving at the plant per unit of time, the sludge layer thickness and the allowable sludge loading rate. For instance, for two weeks of drying duration and FS arriving 5 days per week, a minimum of 10 beds is required. The number of beds can then be increased or decreased considering the optimal sludge layer thickness. It is also important to adapt the number of beds based on the actual operating conditions, for example frequency of sludge removal, or frequency of rain. An increased number of beds increases the safety factor for adequate treatment with variable FS, or poor operation, but also increases capital costs. Cofie *et al.* (2006) utilised two beds of 25 m^2, with a loading rate of 7.5 m^3 of sludge per bed at a loading depth of 30 cm. For the Kathmandu valley, HPCIDBC (2011) designed a circular line-up of the beds, arranged in two circles, with one inlet per two beds (Figure 7.3). The area of the sludge drying bed is calculated as 43 m^2 with a total of 28 beds and a loading rate of 250 kg $TS/m^2/year$.

7.3.6 Summary of design parameters

It must be noted that the calculations and figures provided in this section were determined through local research for the local context based on sludge type and climate and therefore cannot be taken as applicable to all cases. However, they do provide examples of acceptable ranges, and an indication of the interdependency of the factors. In order to provide a suitable drying bed design, the designing engineer needs to obtain local knowledge either from experience or from preliminary drying tests under local conditions. The first stage in conducting drying tests will be to determine the number of days required in order to obtain a desired total solids content of the sludge, or at least to obtain a sludge that can be readily removed. If for example the results from these drying tests indicate a two week drying period, including one day for loading and two days for removal, one bed can be filled 26 times per year. Further example calculations are given in Section 7.7.

7.4 CONSTRUCTION OF AN UNPLANTED SLUDGE DRYING BED

A drying bed treatment facility consists of the beds with an inlet and an outlet, a leachate collection and drainage system, a designated area outside of the beds for storage and continued drying of the sludge, and potentially settling-thickening tanks. Sludge can be loaded directly from trucks onto the beds. In this case, various configurations exist such as creating one inlet for two beds, with a splitter to divide the sludge between the beds (Cofie *et al.*, 2006), by designing the bed with a ramp for the inlet of the sludge. Alternatively, a holding or settling tank can be installed into which the sludge is first discharged before being pumped into the drying beds. A splash plate must be used to prevent erosion of the sand layer and to allow even distribution of the sludge (Tilley *et al.*, 2008). This is crucial, as without a splash plate, the sand layer would be destroyed during the very first loading operation. Bar screens at the inlet are essential to keep rubble and trash present in the sludge from entering the bed. This is important to allow for proper use or disposal of the sludge after drying. The drying bed is typically a rectangular shape excavated from the soil, with a sealed bottom. As was shown in Figure 7.1, the bottom of the bed slopes downwards towards where the drainage system is is installed such that the leachate can drain to the discharging point or further treatment. As the leachate is high in suspended solids, organic material, and nutrients, it needs to be treated before it can be discharged to the environment, according to the quality required for reclamation or for receiving water bodies (see Chapter 10 for further details).

7.4.1 Gravel and sand

Layers of gravel and sand are applied on top of the drainage system. When constructing drying beds, it is essential to use washed sand and gravel in order to prevent clogging of the bed from fine particles. This is important both for the initial construction, and for further supplemental additions of sand. The gravel layers function as a support and there are typically two or three layers with two different diameters of gravel (Figure 7.1). The distribution of diameter size in the layers is based on avoiding clogging from small particles washing into the drain. The lower layer usually contains coarser gravel with a diameter of around 20-40 mm and the intermediate layer contains finer gravel with a diameter between the coarse gravel and the upper sand layer, for example 5-15 mm. Locally available materials will also have an influence on the design. For example, Cofie *et al.* (2006) made use of gravel with a diameter of 19 mm applied in a 15 cm supporting layer underneath 10 cm of gravel with a 10 mm diameter. To avoid the migration of particles from the sand layer into the gravel layers, a third layer of small gravel can also be used according to what is locally available, for example 2-6 mm.

A sand layer is placed on top of the gravel. The sand layer enhances drainage and prevents clogging, as it keeps the sludge from lodging in the pore spaces of the gravel. The diameter of the sand is crucial as sand with a larger diameter (1.0-1.5 mm) can result in the relatively fast accumulation of organic matter, thereby increasing the risk of clogging, This risk is reduced if sand with a smaller diameter (0.1-0.5 mm) is used (Kuffour *et al.*, 2009).

When selecting sand for the bed, it is important to note that the sand will need to be replaced occasionally, as a certain amount of the sand is bound to the sludge and will therefore be removed when the sludge is removed. It is therefore recommended that the sand that is chosen is easily obtained. Duchène (1990) reported a loss of a few centimetres of sand for each 5-10 drying sequences, whereas at the Cambérène FSTP in Dakar 5 cm is lost after 25 drying sequences (Badji, 2008).

The sand also needs to be replaced when there is a build-up of organic matter and the bed starts to clog. Kuffour *et al.* (2009) observed a link between the rate of clogging and the rate of organic matter build-up on the sand. As organic matter builds up faster on sand with larger particles, a bed filled with larger diameter sand is more likely to clog. Cofie *et al.* (2006) had to replace the sand twice in a series of 8 dewatering cycles over 10 months due to clogging in a pilot scale implementation. For a full scale application, HPCIDBC (2011) estimated a sand exchange period of three years at a sludge loading rate of 250 kg TS/m^2/year, a sludge filling height of 20 cm and a one week drying period (applicable to Nepali conditions).

7.4.2 Sludge removal

In order for the sludge to be removed properly, it needs to be dry enough that it can be shovelled. Pescod (1971) carried out experiments with different types of sludge and treatment technologies, including lagoons and drying beds, and found sludge with a TS content of at least 25% fit for removal. The drying time of a specific sludge type depends on a number of factors, one of which is the sludge dewatering resistance. The higher the sludge dewatering resistance, the lower the drainage rate which leads to a prolonged drainage time. Sludge is removed mechanically or manually, with shovels and wheel barrows being the most common manual method (Figure 7.5).

In order to remove the sludge, a ramp must be provided to allow wheel barrows or other equipment to access the bed. If a drier sludge is required, this can be achieved by evaporation after it is removed from the drying bed. The dried sludge is frequently stored in heaps for periods of up to one year, during which time pathogen reduction can occur. It is however, recommended that a more controlled treatment is employed in order to produce reliable and consistent endproducts.

Rewetting of the sludge is considered problematic if rainfall occurs before the free water of the sludge is completely drained. In this case, the moisture content of the sludge increases again and the drying period is prolonged. When the sludge is already dry enough to expose the sand layer through the cracks in the sludge, rain water can pass straight through the sludge and drains through the drying bed.

Figure 7.5 Removing sludge from unplanted drying beds at Cambérène treatment plant, Dakar, Senegal (photo: Linda Strande).

Case Study 7.2: Cambérène faecal sludge treatment plant (continued from Case Study 6.2)

As presented in Case Study 6.2, the Cambérène FSTP is a combination of settling/thickening tanks and unplanted drying beds. The drying beds were designed based on a 200 kg TS/m^2/year loading and a 20 cm deep sludge layer. The operator considers the sludge sufficiently dried when it can be easily removed with a spade, i.e. when the sludge is not sticking to the sand layer anymore. In the climatic conditions of Dakar, this corresponds to 30-35 days drying, even during the rainy season. The dry matter content reaches about 50%, which is an average, with a drier layer on top and 20-30% dry matter in the deeper layer of the sludge. As the operator takes one week more for organising the dried sludge removal, each of the 10 beds of 130 m^2 takes 40 day cycles. This leads to an effective loading rate of 340 kg TS/m^2.year. As a consequence, the operator usually uses only 6-7 beds instead of the 10 beds.

Figure 7.6 Unplanted drying beds, sludge removal and accumulation at Cambérène treatment plant, Dakar, Senegal (photos: Pierre-Henri Dodane).

The leachate is still highly concentrated (2,500 mg TS/L, 1,900 mg SS/L, 3,600 mg COD/L). The dried sludge is removed manually by shovel. One worker needs about two days for removing the 7 cm deep dried sludge layer from a 130 m^2 bed. The dried sludge density is about 300 kg/m^3. Cambérène FSTP produces about 600 m^3/year of dried sludge. The dried sludge is first stored behind the drying beds and later collected by public works companies for soil enrichment.

7.5 QUALITY OF DRIED SLUDGE AND LEACHATE

The main purpose of a drying bed is to achieve dewatering; i.e. a physical separation between liquid and solids. Drying beds are therefore not designed with stabilisation or pathogen removal in mind, although some biodegradation may occur. Therefore, any pollutants present in the FS are not removed and either remain in the sludge or are present in the leachate.

Table 7.2 Analyses of leachate from sludge drying beds in Kumasi, Ghana (from Koné *et al.*, 2007)

	First day	Last day	Difference
pH	8.2	7.9	-0.3
EC (μS/cm)	21,900	11,400	-10,500
SS (mg/L)	600	290	-310
COD (mg/L)	5,600	3,600	-2,000
BOD (mg/L)	1,350	870	-480
NH_3-N (mg/L)	520	260	-260
TKN (mg/L)	590	370	-220
NO_3^--N (mg/L)	50	170	120

Koné *et al.* (2007) carried out experiments with mixtures of septic tank and public toilet sludge, and analysed the leachate on the first and the last day of filtration for a variety of parameters (Table 7.2). Although the measured concentrations were lower on the last day, the leachate was still far from environmentally safe for disposal with for example a BOD concentration of 870 mg/L. Hence, according to the final use or standards for receiving water bodies, the leachate should be collected and treated as a concentrated liquid waste stream, for example in ponds (see Chapter 5 - Montangero and Strauss, 2002), or recovered for an appropriate enduse as described in Chapter 10.

Koné *et al.* (2007) also analysed FS from druing beds for Ascaris and Trichuris eggs. The results of public toilet sludge and septic tank sludge are presented in Table 7.3. Sludge was applied in different ratios to unplanted sludge drying beds at a loading rate between 196 and 321 kg TS /m^2/year, and left to dry until the TS content was at least 20%. Dewatering on the drying beds alone was not sufficient to inactivate all helminth eggs, and a total count of up to 38 *Ascaris* and *Trichuris* eggs was recovered after dewatering, of which 25–50% were viable (Koné *et al.*, 2007). This illustrates the need for additional storage time or other treatment options for increased pathogen reduction.

Table 7.3 Prevalence of Ascaris and Trichuris eggs in Kumasi's public toilet and faecal sludge (Koné *et al.*, 2007)

	Ascaris		Trichuris		Total	
Public toilet sludge						
Sample 1	13[a]	(38%)[b]	2	(13%)	16	(34%)
Sample 2			9	(52%)	9	(52%)
Septic tank sludge						
Sample 3	3	(23%)	2	(0%)	5	(13%)
Sample 4	94	(53%)	24	(58%)	118	(54%)
Sample 5	29	(37%)	15	(25%)	44	(32%)

a Number of eggs/g TS

b Percentage (%) of viable eggs in parentheses

7.6 DESIGN EXAMPLE

This section provides two examples on design requirements for unplanted drying beds.

7.6.1 Example 1: Known drying time (two weeks per bed at a loading depth of 20 cm)

An example of the calculations made in Section 7.4 is provided in this section. A plant is to receive 500 kg of total solids per day, at a density of 50 kg TS/m^3. Based on preliminary tests it was found that 15 cm of this type of sludge takes 11 days to reach the desired final total solid content. Including one day for filling and two days for excavation, one bed receiving this type of sludge needs two weeks for a full drying cycle and can therefore be used 26 times per year. At a loading rate of 500 kg TS/day or 10 m^3/day, one bed of 67 m^2 is filled each day. Assuming that the trucks arrive only on week days, 10 beds will be filled in two weeks. After two weeks, the first bed can be used again. Based on these considerations, a minimum of 10 beds is required for this plant to receive and treat the incoming sludge. Adding a few extra beds is not only recommended for increased flexibility in case of changes in quality and quantity of the FS; but is also essential to enable necessary maintenance to the plant, such as sand replacement. The number of extra beds that can be added depends on the investment potential and anticipated changes in sludge quantity and quality.

7.6.2 Example 2: Design for settled sludge under good climate conditions

In this example, a plant is being designed for sludge with a concentration of 30 g TS/L arriving at the plant at a load of 50 m^3/day in a setting with good climate conditions (see section 7.3.1 for a divison and definition of climate conditions). The plant receives sludge only on weekdays, for 52 weeks of the year. The annual mass of sludge received can be calculated from equation 7.1:

Equation 7.1: $M = c_i \cdot Q_i \cdot t$

In which M is the sludge load in kg TS per year, c_i is the average total solids concentration in the sludge arriving at the plant in g TS/L, Q_i is the flow in m^3 per delivery day, and t is the number of delivery days per year. For the described situation, this comes to:

Equation 7.2: $M = 30 \cdot 50 \cdot 5 \cdot 52 = 390{,}000$ kg TS/year.

Since the plant will be built in a region with desirable climate conditions (Section 7.4.1), a sludge loading rate of 200 kg TS/m^2/year can be applied. Therefore, taking the yearly sludge load into account, a drying bed with an area of 390,000 (kg TS/year) / 200 kg (TS/m^2/year) = 1,950 m^2 is required. For a sludge loading height of 0.20 m and a loading rate of 50 m^3/day, a capacity of 250 m^2/day needs to be available. Assuming that one bed can accommodate 250 m^2/day, a minimum of 8 drying beds are required to treat 1,950 m^2.

With these beds, the drying duration will be one week, with one day left for the operator to remove the sludge. To make the operation and maintenance easier and more robust, it could be recommended that the drying duration is two weeks. Hence, 10 beds are needed. The drying beds total surface will thus be 2,500 m^2, and the effective sludge loading rate is 160 kg TS / m^2/year. Sludge is applied once a day to consecutive beds with a 20 cm layer.

7.7 INNOVATIONS AND ADAPTATIONS IN SLUDGE DRYING BEDS

Drying beds could potentially be modified in order to increase drying rates and reduce sand loss. Aspects that have been investigated include the installation of piping systems, drying in greenhouses, the use of wedge wire, mixing and coagulants. These are discussed in the following sections.

7.7.1 Piping systems

Radaidah and Al-Zboon (2011) investigated the modification of a wastewater sludge drying bed whereby solar heating was used to heat up water prior to circulating it through the sludge drying bed in order to enhance the drying process. It was found that wastewater sludge treated on a standard bed dried from 96% to 33% moisture over a period of 18 days, but when dried on this modified bed using water heated to 70°C, the same result could be achieved after only 10 days of drying. After an 18 day period the sludge was further dried to achieve 8% moisture. This modified system would be most suitable for areas with limited space and where there is sufficient sun light. This type of system would be more expensive, but it does offer an interesting modification of the standard drying bed. This could also be achieved with recovery of industrial waste heat (Diener *et al.*, 2012)

7.7.2 Greenhouses

Bux *et al.* (2002) experimented with covering beds with glass panels in order to enhance the drying of sludge from the pharmaceutical industry. A reduction in the drying time of 25-35% was reported. It is important to note that any system involving covering of the beds needs to be well ventilated, either actively or passively, in order to facilitate the transport of the water saturated air away from the bed. Drying in greenhouses is also a technology actively applied for wastewater sludge in the US, often combined with an active mixing device and blowers to enhance the drying process (Huber Technology, 2013). Various researchers are currently working on adapting lower cost options for FS, for example the FaME project (www.sandec.ch/fame; Figure 7.6).

7.7.3 Wedge wire

A further option is to use stainless steel wedge wire as a surface to enhance sludge drying and drainage, or to reduce the amount of sand that partitions with the sludge upon removal (Tchobanoglous *et al.*, 2002). Whilst this is effective for the drying of wastewater sludge its effectiveness for FS drying has not yet been reported.

Figure 7.6 Pilot scale research facility at Cambérène treatment plant, Dakar, Senegal. Evaluating rates of dewatering with passive and active ventilation with greenhouses (photo: Linda Strande).

7.7.4 Additives to the sludge to enhance drying

Pescod (1971) makes reference to a study conducted by Luong in Bangkok, Thailand where alum (potassium aluminium sulphate) was added to the FS in order to increase the rate of drying. This study found that conditioning with alum should only be carried out during the wet season, as there was no significant advantage in conditioning the sludge during the dry season. Research on coagulants for FS treatment is also being conducted as part of the FaME project.

7.8 CONCLUSIONS

Based on the information provided in this chapter, it can be concluded that while some knowledge on the use of unplanted sludge drying beds for FS treatment exists, more detailed research is required in order to provide clear guidelines on their design and operation, and to assist in understanding and overcoming problems.

7.9 REFERENCES

Badji, K., 2008. Traitement des boues de vidange : éléments affectant la performance des lits de séchage non plantés en taille réelle et les mécanismes de séchage (Faecal sludge treatment: parameters affecting the unplanted drying beds efficiency and drying mechanism at real scale). Engineer degree, Génie des Procédés, Ecole Supérieure Polytechnique, Dakar, Senegal.

Badji K., Dodane P.H., Mbéguéré, M., Kone D., (2011), Traitement des boues de vidange : éléments affectant la performance des lits de séchage non plantés en taille réelle et les mécanismes de séchage. Actes du symposium international sur la Gestion des Boues de Vidange, Dakar, 30 juin – 1er juillet 2009, EAWAG/SANDEC.

Bux, M., Baumann, R., Quadt, S., Pinnekamp, J., Mühlbauer, W. (2002). Volume reduction and biological stabilization of sludge in small sewage plants by solar drying. Drying Technology 20 (4-5), p. 829-837.

Cofie, O., Agbottah, S., Strauss, M., Esseku, H., Montangero, A. (2006). Solid- liquid separation of faecal sludge using drying beds in Ghana: implications for nutrient recycling in urban agriculture. Water Research 40(1), p. 75-82.

Cofie, O., Koné, D. (2009). Case study of sustainable sanitation projects: Co-composting of faecal sludge & organic solid waste. Kumasi, Ghana. Available from www.susana.org. Accessed September 2013.

Diener, S., Reiser, J.C., Murray, A., Mbéguéré, M., Strande, L. (2012). Recovery of industrial waste heat for faecal sludge drying. SANDEC News no. 13, p.16.

Duchène, P. (1990). Les systèmes de traitement des boues des stations d'épuration des petites collectivités. Documentation technique FNDAE 09, Ministère de l'Agriculture et de la Forêt, France, in French. Available from www.fndae.fr/archive/PDF/fndae09-a.pdf.

HPCIDBC (2011). Status and Strategy for Faecal Sludge Management in the Kathmandu Valley, High Powered Committee for Integrated Development of the Bagmati Civilization, Kathmandu, Nepal.

Heinss U., Larmie S.A., Strauss, M. (1998). Solids Separation and Pond Systems for the Treatment of Faecal Sludges in the Tropics. SANDEC Report No.5/98 Second Edition, EAWAG/SANDEC, Duebendorf Switzerland.

Huber Technology (2013). Huber Solar Active Dryer SRT (PDF). Website www.huber-technology.com, accessed May 21, 2013.

Koné, D., Cofie, O., Zurbrügg, C., Gallizzi, K., Moser, D., Drescher, S., Strauss, M. (2007). Helminth eggs inactivation efficiency by faecal sludge dewatering and co-composting in tropical climates. Water Research 41(19), p. 4397-4402.

Kuffour, A. R., Awuah, E., Anyemedu, F.O.K., Strauss, M., Koné, D., Cofie, E. (2009). Effect of using different particle sizes of sand as filter media for dewatering faecal sludge. Desalination 248, p. 308-314.

Montangero, A., Strauss, M. (2002). Faecal Sludge Management, SANDEC/EAWAG Lecture Notes.

Pescod, M. B. (1971). Sludge handling and disposal in tropical developing countries. Journal of Water Pollution and Control Federation 43(4), p. 555-570.

Radaidah, J. A., Al-Zboon, K. K. (2011). Increase the efficiency of conventional sand drying beds by using intensive solar energy: a case study from Jordan. Presented at the 2011 2nd International conference on environmental science and technology. IPCBEE vol. 6, IACSIT Press, Singapore.

Strauss, M., Montangero, A. (2002). FS Management – Review of Practices, Problems and Initiatives. DFID Engineering Knowledge and Research Project - R8056. Consultancy report to GHK, the United Kingdom, 73 p.

Tchobanoglous, G., Burton, F.L., Stensel, H.D. (2002). Wastewater Engineering: Treatment and Reuse, 4th edition / revised. McGraw-Hill Engineering and Computer Science Books.

Tilley, E., Lüthi, C., Morel, A., Zurbrügg, C., Schertenleib, R. (2014). Compendium of Sanitation Systems and Technologies. Swiss Federal Institute of Aquatic Science and Technology (EAWAG) and WSSCC. Dübendorf, Switzerland. 2nd revised edition. Available from www. sandec.ch.

End of Chapter Study Questions

1. Describe the main components of unplanted drying beds, and the basic fundamentals of their operation.
2. Name two key mechanisms for the dewatering of sludge with unplanted drying beds uPDBs.
3. List four critical factors that need to be taken into consideration when designing unplanted drying beds.
4. Describe what types of treatment objectives can be met with unplanted drying beds.

Technology

Chapter 8

Planted Drying Beds

Ives Magloire Kengne and Elizabeth Tilley

Learning Objectives

- Know what a planted drying bed for sludge dewatering is.
- Have an overview of the vegetation types that may be used, the role they play in sludge dewatering and criteria for their selection.
- Know the appropriate level of operational and maintenance monitoring necessary for the operation of planted drying beds.
- Be able to design a planted drying bed to achieve the desired treatment objectives.

8.1 INTRODUCTION

Planted drying beds (PDBs), also sometimes referred to as planted dewatering beds, vertical-flow constructed wetlands and sludge drying reed beds, are beds of porous media (e.g. sand and gravel) that are planted with emergent macrophytes. PDBs are loaded with layers of sludge that are subsequently dewatered and stabilised through multiple physical and biological mechanisms (Kadlec and Knight, 1996).

PDBs were initially developed to stabilise and dewater sludge from small activated sludge treatment plants in Europe and US (Kadlec and Knight, 1996; Lienard and Payrastre, 1996; Nielsen, 2003). This technology has also been successfully adapted in other parts of the world for use with various types of sludge, including faecal sludge (FS) from onsite sanitation technologies. In northern climates, PDBs have been shown to achieve higher rates of water removal, solids reduction and increased oxidation in the summer compared to winter (Edwards *et al.*, 2001), thus making them ideal for tropical countries where there is less climactic variation and more constant solar radiation. Since 1996, SANDEC/EAWAG (The Department of Water and Sanitation in Developing Countries at The Swiss Federal Institute of Aquatic Science and Technology) and their research partners have jointly undertaken field experiments to determine treatment efficiencies and to establish the design and operational guidelines of pilot beds treating FS from onsite sanitation technologies. A pilot-scale facility has been operating successfully in Thailand for nearly a decade (Koottatep *et al.*, 2005) while in Africa, trials at yard-scale have been conducted at the University of Yaoundé I (Cameroon, Figure 8.1) and at full scale at the Cambérène Treatment Plant in Dakar, Senegal (EAWAG/SANDEC, 2009; Kengne *et al.*, 2008).

Although there are limited examples of full-scale PDBs treating FS in operation, current research has produced promising results and it is expected that PDBs will be adopted more extensively around the world, especially in tropical regions of low-income countries.

FS is repeatedly loaded onto PDBs, with up to 20 cm of FS per loading (Kadlec and Wallace, 2009), where it accumulates for several years depending on the loading rate, the capacity of the system and mineralisation rates (Nielsen, 2003). Due to the limited number of existing operational beds and the variety of operational conditions, it is currently difficult, if not impossible to give a single value or precise range for the duration of sludge accumulation. Long-term bed permeability is maintained by the dynamic system of percolation canals created by the continuously growing root system of the plants. The volume of sludge on the PDB reduces continuously (through moisture loss and degradation), and the plants maintain porosity in the sludge layer thereby significantly reducing the need for sludge removal compared to unplanted drying beds (which require sludge removal every two to three weeks) (Strauss *et al.*, 1997). Emergent macrophytes are therefore essential to the improved performance for stabilisation, pathogen reduction and clogging of PDBs over unplanted drying beds.

The dewatering, organic stabilisation and mineralisation performance of the PDB depends on a variety of factors such as the media type and size, the type of plants, the maturity of the beds, climatic factors, and the sludge characteristics, as well as operational factors such as the hydraulic loading rate (HLR), the solids loading rate (SLR), and the loading frequency (Breen, 1997; Prochaska *et al.*, 2007; Van Cuyk *et al.*, 2001). As the bed matures, microbial communities become more established and stable. The following sections discuss, in detail, the operational conditions and design parameters which currently define the best practice for PDBs used for FS .

Figure 8.1 Pilot-scale planted drying beds for faecal sludge treatment at the University of Yaoundé I (photo: Linda Strande).

8.2 MACROPHYTES

Macrophytes are plants found in wetlands, marshes and swamps, and are distinguished by their ability to grow when partially or fully submerged in water. There are four types of macrophytes: freely floating, submerged, floating leaved and emergent. Freely floating macrophytes have leaves that float on the surface of the water, usually with submerged roots. Submerged macrophytes are usually rooted in the bottom soil with the vegetative parts predominantly submerged. Floating leaved macrophytes are those that are rooted but have floating leaves, while emergent macrophytes are rooted in soil below shallow water with the leaves and stems emerging above the surface of the water.

While macrophytes produce many seeds, reproduction by germination does not generally take place due to the hindering aquatic conditions (Hutchinson and Dalziel, 1972). Successful reproduction usually occurs by way of cuttings, stolons or rhizomes. Rhizomes are dense, underground stems from which vertical shoots grow upward, and from which roots grow out and down. On the stem, new buds form at nodes and the space between the nodes is termed the internode. The rhizomes are important as they provide a large surface area for bacteria to grow on; and in PDBs bacteria are responsible for the degradation of organics and mineralisation of sludge. The rhizosphere is the area surrounding the rhizomes which has a higher concentration of oxygen due to release by the roots (Section 8.3.2). Figure 8.2 shows schematic diagrams of the rhizomes and the node and internode structure that are typical of macrophytes.

Emergent macrophytes are generally the best suited for PDBs because they are the most productive of all aquatic macrophytes. In other words, the rate of multiplication and generation of new biomass is very high. They can establish and extend their roots and rhizomes through the sludge layers while the stem continues to grow up through the accumulating sludge layer. Leaves growing above the sludge layer are able to make use of solar radiation for photosynthesis and transpiration.

Figure 8.2 Left: rhizome structure of *E. pyramidalis* showing roots and two newly formed buds; Right: application of faecal sludge on drying beds with young shoots that have nodes and internodes (photos: Ives Kengne).

Although a variety of macrophytes exist in nature, there are a limited number of emergent macrophytes that grow well under PDB conditions, which vary between aerobic, anoxic and anaerobic (depending on the loading), with the added burden of variable pH, salinity, and high nutrients (De Maeseneer, 1997; Uggetti *et al.*, 2012). The high and highly variable nutrient levels in sludge means that any plant used in a drying bed must be able to tolerate a wide range of growing conditions and must be able to withstand the shocks associated with sludge loading and drying. Sludge from public toilets, for example, is high in salts (conductivity up to 15 mS/cm) and ammonia (2 to 5,000 mg/L), which are toxic to most plants (Clarke and Baldwin, 2002). To offset these potentially lethal conditions, public toilet sludge must be diluted with sludge that has lower concentrations of salt and ammonia (e.g. from septic tanks) in order to provide conditions that are suitable for the specific type of macrophytes growing in the drying beds.

In European applications, reeds (*Phragmites sp.*) and cattails (*Typha sp.*) are the most common types of emergent macrophytes used in PDBs (Kadlec and Knight, 1996; Kim and Smith, 1997; Koottatep *et al.*, 2005). *Phragmites australis* is an invasive species and its use is restricted in the US and New Zealand (Uggetti *et al.*, 2012). Other indigenous options are currently being explored such as antelope grass (*Echinochloa sp.*) and papyrus (*Cyperus papyrus*). Based on preliminary testing, both of these show promising results for use in PDBs.

A macrophyte that is to be used in PDBs should have the following characteristics (De Maeseneer, 1997):

- fast growing under diverse conditions;
- high transpiration capacity;
- tolerance to different water levels and drought conditions;
- tolerance to extremes of pH and salinity;
- deep growing rhizome and root system;
- ability to build new roots on the nodes when they become encased in sludge; and
- readily available, indigenous and non-invasive

Although *Phragmites australis* (reeds) is the most frequently used species in PDBs (De Maeseneer, 1997; Hardej and Ozimek, 2002), there are potentially many other local, untested species with these characteristics that are capable of achieving similar, if not improved levels of treatment. A summary of the most commonly used macrophytes for FS treatment is presented in Table 8.1.

Table 8.1 Macrophytes commonly reported for the treatment of faecal sludge (Kengne *et al.*, 2008; Nielsen, 2005; Koottatep *et al.*, 2005)

Plant species	Common name	Water type	Habitat	Water regime
Phragmites sp.	Reeds	Fresh to brackish	Marshes; swamps	Seasonal to permanent inundation, up to 60 cm
Typha sp.	Cattail	Fresh to marshes	Pond margins	Seasonal to permanent inundation, up to 30 cm
Cyperus papyrus	Papyrus	Fresh to marshes	Pond margins, lakes	Seasonal to permanent inundation, up to 30 cm
Echinochloa sp.	Antelope grass	Fresh to brackish	Marshes; swamps	Seasonal to permanent inundation, up to 40 cm

8.3 TREATMENT MECHANISMS

The treatment of sludge in PDBs is achieved through a combination of physical and biochemical processes. In wet, rainy climates, macrophytes play an essential role in almost all processes, and are responsible for the higher levels of treatment in terms of stabilisation and pathogen removal in PDBs compared to unplanted drying beds (Brix, 1997; Kadlec and Knight, 1996). Macrophytes therefore play an essential role in the following:

- stabilising the beds to prevent media erosion and clogging, and improving the drainage;
- increasing moisture loss (through evapotranspiration, in contrast to only evaporation in unplanted drying beds);
- providing a surface area for microbial growth within the sludge layer;
- transferring oxygen to the sludge layer (i.e. within the rhizosphere); and
- absorbing heavy metals and nutrients.

However, while PDBs, and their ability to wick away moisture via transpiration, make them an applicable technology in humid or rainy climates, the macrophytes of a PDB could wilt and die off in a climate that is too hot and dry, especially if the sludge does not provide sufficient moisture. However, if the PDB can be operated to induce 'ponding', i.e. keeping a certain amount of water in the beds by turning off the drainage outlet valve of the ponds or by adjusting the level of the outlet valve, then PDBs can be operated efficiently, even in a very dry climate.

The following sections explain the key treatment mechanisms that occur in PDBs and the ways in which macrophytes assist in these processes. This information should be weighed against the other technology options presented in this book, and the appropriate technology solution should be selected based on site-specific conditions (Chapter 17).

8.3.1 Infiltration (percolation)

When sludge is applied on the beds, solids are retained on the surface of the filtering matrix (either the porous substrate or the existing sludge layer), while the liquid drains vertically through the media where it is collected for further treatment (Kadlec and Knight, 1996). One of the main operational concerns with unplanted drying beds is the formation of erosion channels which can lead to short-circuiting and uneven treatment. However, in PDB, the dense macrophyte root system impedes erosion and helps to stabilise the sludge layers. The natural movement of the plants due to wind, and the growth of roots encourage water to drain around the stems and the tubular spaces created around them, thereby leading to improved drainage. As macrophytes grow, they break up and loosen the accumulating sludge, which also maintains good conditions for filtration (Brix, 1994). When macrophytes die, the decaying roots and rhizomes leave behind small pores and channels which allow infiltration and circulation of air, and encourage aerobic conditions (Brix, 1994).

8.3.2 Evapotranspiration

Transpiration is the process by which water is lost into the atmosphere from the leaf and stem surfaces of the plant. The presence of macrophytes therefore aid in sludge drying by absorbing and then releasing moisture via transpiration. In temperate climates, evapotranspiration (the sum of evaporation and transpiration) rates of up to 2.5 cm/day can be achieved in reed stands on very hot days. (De Maeseneer, 1997). The rate could be even higher in tropical regions depending on climatic conditions such as wind speed and relative humidity. Evapotranspiration from the macrophytes results in increased moisture loss and volume reduction, compared to unplanted drying beds. In one study a comparison was made between PDB and unplanted drying beds using sludge from a biological wastewater treatment plant. The PDB achieved over 95% volume reduction over a year (6 months loading, 6 months resting) with a total sludge loading depth of up to 493 cm, while the unplanted drying beds; achieved less than 90% reduction in volume. Dry matter content for the same beds, reached up to 69% for the PDBs while the unplanted beds achieved only 31%. This increased performance in PDBs is thought to be due to evapotranspiration and percolation of the macrophytes (Stefanakis and Tsihrintzis, 2012a). While sludge from wastewater treatment plants is not necessarily directly comparable to unstabilised FS, the majority of the PDB research has been conducted on wastewater sludge and these results are therefore used to provide examples throughout this chapter.

8.3.3 Stabilisation/mineralisation

Stabilisation (also referred to as humification) is the conversion of organic matter into more stable, organic components. Mineralisation is the process by which biologically available inorganic nutrients are released during the degradation of organic material (e.g. the degradation of amino acids results in the release of ammonia). The process of stabilisation and mineralisation leads to the release of inorganic nutrients, which are essential plant and microbial nutrients, thereby contributing to improved fertility of the macrophytes. Even FS that has undergone bacterial decomposition for years (e.g. in a septic tank), may require further stabilisation if it still has a high BOD. Stabilisation also reduces the odour of the sludge and destroys the pathogenic organisms. For instance, storage time has been found to contribute to weakening the external membranes of helminth eggs, which may be degraded by bacteria and fungi present in the sludge layer (Sanguinetti *et al.*, 2005).

The surface of rhizomes provide attachment areas for bacteria and other microorganisms, and the resulting microbial density and activity can lead to improved sludge mineralisation as well as improved water and nutrient uptake (Bialowiec *et al.*, 2007; Brix, 1997; Chen *et al.*, 2007; Gagnon *et al.*, 2007).

Metrics of mineralisation are not universally agreed upon. However indicators of the degree to which sludge on a PDB has been mineralised are the reduction in total volatile solids (TVS) and the ratio of TVS to total solids (TS) content, which indicate the change in readily degradable material. Mineralisation primarily takes place during resting periods between sludge loadings as it occurs more rapidly in aerobic conditions. When sludge is applied to the beds, oxygen is less available due to the water saturated conditions and high concentrations of biodegradable organic matter. One study using sludge from a biological wastewater treatment plant showed that the VS content of the sludge was reduced from 0.74 VS/TS to 0.59 VS/TS (Stefanakis and Tsihrintzis, 2012a) while another showed a final value of 0.52 VS/TS (Uggetti *et al.*, 2012). The VS content has been found to be much lower in the lower layers of the sludge bed compared to the upper layer, due to the increased retention time in the bottom layers leading to increased oxidation (Stefanakis *et al.*, 2009).

8.3.4 Oxygen transfer

Untreated FS contains little, if any, dissolved oxygen and is therefore generally anoxic or anaerobic. However, oxygen can be transferred into the sludge through various physical and biological mechanisms and create anoxic and aerobic zones. These varying concentrations of oxygen allow for complex processes (e.g. nitrification and denitrification) to occur in the PDB leading to improved levels of treatment compared to unplanted beds.

Rooted macrophytes have adapted to growing in water-saturated soil, where the pore spaces are filled with water and the conditions are anaerobic. Macrophyte roots obtain oxygen through an internal transfer system that moves oxygen from the leaves and stems down into the roots and rhizomes. The internal lacunar (circulatory) system may occupy up to 60% of the total tissue volume, depending on the species (Brix, 1994). Some of the oxygen that arrives at the root is leaked into the rhizosphere. This leaked oxygen then creates aerobic conditions in the immediately surrounding area, which supports a variety of aerobic bacteria, and helps promote aerobic degradation and nitrification. Leakage occurs primarily at the root-tip and the rate of oxygen release depends on the permeability of the root-walls, and the internal oxygen concentration among other factors. This rate is difficult to quantify, but oxygen release rates from roots of *Phragmites* have been calculated between 0.02 and 12g/ m^2/day (Brix, 1994).

As the top layer of sludge dries, it cracks, thereby creating gaps through which oxygen can further penetrate the sludge layer (Figure 8.3). These cracks are more pronounced in hot, arid climates, and are predominant in areas of the beds containing few rhizomes as the rhizomes hold the sludge together (Stefanakis and Tsihrintzis, 2012a). Hot, dry conditions are therefore beneficial for crack-induced oxygen transfer, yet extreme conditions could cause plants to wilt and die. This emphasises that the technology choice of planted or unplanted drying beds needs to be carefully chosen given the local conditions.

8.4 PERFORMANCE INDICATORS

The performance efficiency of a PDB is usually judged based on water content, the quantity and form of nutrients, and the degree of stabilisation and pathogen removal of the treated sludge. The following sections explain in more detail how these performance indicators are measured and assessed, and two case studies (Thailand and Cameroon) are presented to illustrate realistic performance data that has been achieved in the field.

8.4.1 Dewatering

Sludge dewatering refers to the removal of water from sludge for improved handling and reuse. This is generally assessed by measuring the TS concentration. Total solids (or dry matter) are one of the key design parameters for FS treatment plants (FSTPs). In tropical regions, it is possible to achieve dry matter (DM) percentages of at least 30% (Kengne *et al.*, 2009a) by treating sludge on PDBs.

Figure 8.3 An example of the formation of cracks in sludge dryings beds planted with *E. pyramidalis* (photo: Ives Kengne).

8.4.2 Nutrient removal

The fate of nutrients in FS treatment is very important as it will determine the enduse opportunities of the sludge and the treatment required for the effluent. Nitrogen (N) and phosphorus (P) can be recovered for beneficial enduses, but also have potentially damaging effects on surface and ground water if discharged to the environment. One study where sludge from a biological wastewater treatment plant was treated on a PDB showed that with an SLR between 30 and 75 kg dm/m^2/year, total Kjeldahl nitrogen (TKN) was reduced by 35 to 42% from an initial level of 55 mg TKN/g DM, compared to a 24% reduction in an unplanted drying bed, thereby indicating the role of macrophytes in nutrient cycling. It is thought that the main processes of nitrogen transformation in PDBs include uptake and assimilation by plants and microbiota, volatilisation, and denitrification in anaerobic zones (Kadlec, 2009). The macrophytes also play other roles in denitrification, for example as a carbon source and attachment sites for denitrifying microorganisms. Phosphorus removal, on the other hand, is found to be fairly similar between planted and unplanted beds; the primary removal mechanism for phosphorus appears to be sorption onto the porous media and plant roots (Stefanakis and Tsihrintzis, 2012a).

Plant uptake of nitrogen and phosphorus from PDBs was shown to range between 0.2% and 5% of the total nutrient load depending on climactic conditions, loading rates, and other factors (Stefanakis and Tsihrintzis, 2012a). Although no specific data for PDBs exists, nitrate reduction in constructed wetlands can account for 60 to 70% of nitrogen loss (Cooke, 1994) and similar removal rates could be expected in PDBs depending on the level of water saturation.

Nutrient recovery is achieved through plant uptake and subsequent harvesting. If the plants are allowed to die and decompose on the bed surface, the nutrients will be recycled back into the sludge. Harvesting the plants and using them for fodder or other beneficial uses is one of the main benefits of PDBs. A study conducted in Cameroon in PDBs vegetated with papyrus (*C. papyrus*) and antelope grass (*E. pyramidalis*) showed that the annual papyrus harvest generated 20 to 30 dry t/ha above-ground biomass, whereas the below-ground biomass varied between 80 and 150 dry t/ha (rhizomes are left in place for continued growth). A full harvest of *E. pyramidalis* shoots three times a year generated an annual biomass of at least 100 to 150 dry t/ha, as opposed to a below-ground biomass of 30 to 70 dry t/ha. By harvesting the plants, between 236 and 383 g N/m^2/year and 60 to 92 g P/m^2/year is removed from the PDBs in the case of papyrus, and between 216 and 330 g N/m^2/year and 55 to 84 g P/m^2/year for the aerial plant parts of the antelope grass. The removal of roots and rhizomes during desludging will generate an additional 55 to 124 g N/m^2/year and 33 to 36 g P/m^2/year (Kengne *et al*., 2008).

Phosphorus that is not in the leachate is mainly present as particulate forms in the sludge layer, or absorbed onto the media and root system. Nitrogen is mainly removed through nitrification and denitrification processes, both of which are increased in the presence of plants, which explains the increased treatment performance of leachate in PDB.

Technology

Case Study 8.1 : Loading frequency experiments in Thailand
(Adapted from Koottatep *et al.*, 2005)

Drying beds planted with cattail (*Typha*) were used in Bangkok, Thailand as a treatment for FS with average concentrations of 15.4 g/L TS, 18.7g/L COD, 1.1 g/L TKN and 0.4 g/L NH_3-N.

With a sludge loading rate that varied between 80 and 250 kg TS/m^2/year, removal efficiencies were found to range from 66 to 88% TS, 78 to 99% TCOD, 82 to 99% TKN and 40 to 98% NH_3-N. Approximately 65% of the liquid passed through the under drain to produce a leachate with concentration ranges of 1.9 to 6.01 g/L TS, 0.1 to 2.2 g/L TCOD, 0.006 to 0.25 g/L TKN, and 0.005 to 0.2 g/L NH_3-N. The remaining 35% of liquid was lost through evapotranspiration or retained in the accumulated sludge layer.

Varying the loading frequency between one and three times a week did not significantly impact the treatment performance, but loading twice a week assisted in supporting the growth of cattails without having to retain the leachate in the beds (which is achieved by closing the outlet valve to prevent drainage). However, in order to minimise the workload associated with the feeding, application was undertaken once a week and leachate retention was introduced as a permanent measure. This provided adequate moisture for the cattails to prevent wilting during dry weather.

Case Study 8.2: Loading rate experiments in Cameroon
(Adapted from Kengne *et al.*, 2011)

In Cameroon, experiments using FS from various onsite sanitation facilities (i.e. septic tank, public latrines and traditional pit latrines) were conducted using loading rates of 100 kg/m^2/year and 200 kg/m^2/year for *C. papyrus* and *E. pyramidalis*. The results showed that when the beds were loaded once a week the loading rate had no significant impact on the dewatering performance. On average, the concentrations of TS and TSS decreased from 3.7% and 27.6 g/L in the raw sludge to less than 0.5% and 2.1 g/L in the leachate, respectively. COD concentrations were reduced from 31 g/L in the raw sludge to less than 0.8 g/L in the leachate, while NH_4^+ was reduced from 0.6 g/L to less than 0.09 g/L. The TKN concentration in the leachate averaged between 0.1 and 0.2 g/L. Good nitrification was achieved with an average concentration of between 0.2 to 0.5 g/L, probably due to an increase of the oxygen concentration when passing through the filtering media.

Beds loaded with 100 kg TS/m^2/year rarely clogged and had an average dry matter content of more than 30%. Approximately 50% of the applied sludge volume was collected as leachate. Loading rates greater than 200 kg TS/m^2/year, resulted in a greater occurrence of clogging in the papyrus beds than those of *E. pyramidalis*, leading to decreased drainage of leachate from the beds.

8.4.3 Fate of heavy metals

In general, FS should not have high concentrations of heavy metals, unless the sludge receives discharge from industrial sources (Kroiss, 2004; Molla *et al*, 2002; Towers and Horne, 1997). Low concentrations of metals like chromium (Cr), cadmium (Cd), lead (Pb), copper (Cu), nickel (Ni), manganese (Mn), zinc (Zn), and iron (Fe) can be present in FS from additives added to onsite technologies, chemicals or batteries disposed of into the systems, exposure to contaminants from trucks that also transport industrial sludge, or because they are consumed and excreted by humans.

Research conducted using *Phragmites australis* and treated sludge from an activated sludge wastewater treatment plant found that the metals Cr, Cd, Pb, Cu, Ni, Mn, Zn, and Fe partitioned unevenly throughout the PDB, and did not accumulate in the macrophytes at significant levels. The metals were taken up in unequal quantities by the plants in decreasing order of Cr > Fe > Zn > Mn > Cu > Pb > Ni > Cd. The reeds were found to be fairly tolerant to metal concentrations and did not show any signs of toxicity, despite absorbing slightly increasing amounts of metals each year. Analysis showed that the metals were most concentrated in the roots, followed by leaves and stems. The quantities absorbed by the plants were however not significant and accounted for less than 3% of the total metal concentrations in the sludge (Stefanakis and Tsihrintzis, 2012b).

During treatment the concentration of metals in sludge typically increases as the organic matter is reduced through decomposition. However, in one study, the filter bed media was found to be the biggest sink of metals, accounting for 47% of the influent content. Sedimentation, adsorption and precipitation (as metal oxides, carbonates, and sulphides) are the primary mechanisms through which the gravel and sand layers trap and retain metals as they pass through the bed. It was found that only 16% of the influent metals were present in the leachate (Stefanakis and Tsihrintzis, 2012b).

8.4.4 Pathogen removal

When identifying the quality of sludge that can be used for a particular enduse, a multi-barrier approach for pathogen removal is followed rather than the application of strict limits. For example, sludge that is to be used as a fuel for combustion or for growing animal forage, does not require the same degree of pathogen reduction as sludge that has the potential to come into contact with crops for human consumption. Chapter 10 (Enduse of Treatment Products) addresses this issue in further detail.

The primary concern for sludge that is to be used in agriculture is pathogen content. Predation, dehydration and retention time are the main mechanisms in PDBs that result in pathogen reduction in FS, an increase in pathogen reduction comes with an increase in retention time. Helminth eggs are very resistant to environmental stress (e.g. dehydration and heat) and are an important indicator of sludge quality. Ingallinella *et al*., (2002) summarise various reports and show that treatment of FS in a PDB reduced the concentration of Helminth eggs from between 600-6000 helminth eggs/ L of FS to 170 eggs /g TS, with an egg viability of between 0.2 and 3.1% (Ingallinella *et al*, 2002). Other research showed that PDBs were able to achieve a complete elimination of helminth eggs in the leachate, but not in the solids, where 79 helminth eggs/g TS were measured (Kengne *et al*., 2009b).

8.4.5 Other considerations

Apart from this direct role in FS treatment, macrophytes are aesthetically pleasing and may provide a habitat for a range of wildlife like birds and reptiles (Brix, 1994). However, the presence of insects and other disease vectors (e.g. rodents, mosquitoes) could pose potential health hazards if not properly managed. Communities surrounding PDBs are generally more accepting of a treatment technology that appears to be 'natural' and in many cases, may not even be aware that the PDB is in fact artificial and used for FS treatment (De Maeseneer, 1997). Therefore, although there are no direct measurements for appearance, an additional advantage of PDBs is the aesthetic one which should be taken into account when selecting a treatment technology.

Table 8.2 summarises the performance indicators of PDBs observed under a variety of experimental conditions.

Table 8.2 Summary of performance indicators of planted sludge drying beds around the world

Country	SLR (kg TS/ m^2/year)	% Solids and moisture reduction	% Nutrients and organics	Other metrics	Plant used	References
France[1]	≈ 70	85% (TS)	70% (COD) 79% (TKN) 66% (NH_4-N)		*Phragmites australis*	Lienard and Payrastre, 1996
USA[1]	9.8-65	99% (TSS)	95% (COD) 90% (TKN) 42% (NH_4-N)		*Phragmites australis*	Burgoon *et al.*, 1996
USA[1]	16-106	46-49% (TVS) 15-47% (TS)			*Phragmites australis*	Kim and Smith, 1997
Poland[1]	-	94.6% (volume reduction), 43-65% (moisture content)			*Phragmites australis*	Obarska-Pempkowiak *et al.*, 2003
Thailand[2]	250	74-86% (TS) 96-99% (SS) 20-25% (DM content of dewatered sludge after 4 years)	78-99% (COD) 70-99% (TKN) 50-99% (NH_3)	< 6 viable helminth eggs/g of TS	*Typha augustifolia*	Koottatep *et al.*, 2005
Cameroon[2]	200	70.6-99.9% (TS) 78.5-99.9% (SS) 30% (DM content of dewatered sludge)	73.4-99.9% (COD) 69.2-99.3% (TKN) 50-99% (NH_3)	100% (helminth eggs)	*Echinochloa pyramidalis*	Kengne *et al.*, 2009
Senegal[2]	200	97% (TS) 99% (SS) 99% (DCO)	91% (NH_4^+) 97% (PO_4^{3-})		*Echinochloa pyramidalis*	Tetede, 2009

[1] Wastewater sludge [2] Faecal sludge

8.5 DESIGN AND CONSTRUCTION

Despite early successes with PDBs in Europe, and recent experiments with FS in low-income countries, PDBs for the treatment of FS are still in the early stages of development. Little research has been conducted with full-scale, operational plants. Few systems have been adequately monitored or have not been operating long enough to provide sufficient data that also allow for definitive design and construction guidelines. A number of design and operational uncertainties cannot be clarified until further research is conducted, and operating experience is compiled. However, it is currently accepted that the design should attempt to mimic PDBs used for treatment of wastewater sludge.

Construction costs are generally lower than for conventional sludge treatment technologies, and PDBs require less space than waste stabilisation ponds. Though mechanically simple (there are few moving parts) the technology requires careful design, construction and acclimatisation in order to achieve adequate results. Table 8.3 lists the general design considerations that should be taken into account for the construction of PDBs based on the results of existing plants. An example of a PDB design is provided in Case Study 3.

Table 8.3 General design considerations for the construction of planted drying beds (adapted from Davis,1995)

Factor	Parameters to consider	Remarks
Site selection	Land use and access	• Located centrally to reduce transport distances • Located away from dwellings to avoid odours or insects from spreading • Located with adequate truck access and away from residential areas to reduce noise
	Land availability	• Site should be large enough to accommodate present requirements as well as any future expansion
	Site topography	• Select (whenever possible) a site that will allow for gravitational flow to reduce energy and pumping costs
Structure	Cells	• Excavate basins or build up earth embankments around cells to create depth • A freeboard should be high enough to allow accumulation of sludge over a period of at least 3 to 4 years. A freeboard of 1.5 to 2 m is generally recommended • Multiple cells (in parallel) are recommended so that cells can be loaded sequentially and allow for a resting phase • Dykes can be used to separates cells and to avoid short-circuiting • The bottom should be sloped slightly (1-3%) • Allow for space between cells for machinery and maintenance activities (e.g. plant harvesting, sludge removal, etc.)
	Liners	• Must be sealed to avoid possible contamination of groundwater or intrusion of water into the beds. Synthetic liners are preferable, but compacted clay can also be used
Flow structures	Inlet	• Flow control structures should be simple and easy to adjust. Channels or gated pipes are generally used
	Outlets	• A weir, spillway or adjustable riser pipe should be installed to allow for the adjustment of water levels if necessary (i.e. to retain water in the cell to avoid plant wilting)
System life		• The operational life of the beds is determined by the loading rate, stabilisation rate and the number of beds. The number of beds should be determined based on the expected amount of sludge to be treated
Climate and weather		• Retaining water in the cells may be necessary to avoid the side effects of drought and high temperature (see "Outlets" above) • Increase the resting period (time between two consecutive loadings) when there is excessive rainfall
Filter matrix (substrate)		• Substrates can include sand, gravel (medium to coarse rock) or other coarse media • The upper substrate layer should have a coefficient of uniformity higher than 3.5 to avoid rapid clogging (this can be achieved after sieving or washing to remove fine particles) • A small amount of soil or organic material may be required to allow the growth of plants during the early stages • The bed must be kept moist, but not flooded until seeds have germinated or rhizome fragments produce new shoots
Vegetation		• Choose indigenous, non-invasive macrophytes that have been proven to grow on sludge • Select shoots or fragments with no visible signs of nematode attack • Plant or harvest in the rainy season to assist with growth or regrowth
Ventilation		• Increased air flow as well as better hydraulic flow conditions of the liquid can be achieved using hollow blocks or ventilation pipes*
Feeding system		• Uniform sludge distribution (preferably in the middle of the beds) avoids dead zones and uneven plant growth • Feeding should occur one to three times a week, depending on the season

* Comparative studies that examined the effect of installing aeration pipes (perforated PVC columns to convey air through the bed layers) found that they did not directly improve the dewatering process, although they did assist with plant growth, which improves evapotranspiration (Stefanakis and Tsihrintzis, 2012a). PVC columns may therefore be included in a PBD design, although they are not necessary.

Case Study 8.3: Design and construction of a planted drying bed in Thailand

In 1996, the Asian Institute of Technology (AIT) in collaboration with SANDEC/EAWAG, constructed a pilot-scale PDB to treat FS produced in Bangkok. This FS treatment scheme was comprised of the following units; i) screening for pre-treatment (retention of coarse material); ii) balancing and mixing tank (to achieve a certain degree of homogenisation of sludge from various sources); and iii) three PDBs attached to a waste stabilisation pond and a vertical-flow constructed wetland bed for leachate polishing. Each of the PDBs measured 5 m × 5 m at the surface of the filter bed (and 6.2 m x 6.2 m at the rim of the freeboard) and was lined with ferro-cement.

The depth of the filter media was designed to be 65 cm to prevent protrusion of the cattail roots and rhizomes through the bottom of the media (root length is between 30 and 40 cm). A 10 cm layer of fine sand, 15 cm layer of small gravel, and 40 cm layer of large gravel (from top to bottom) were used to create the filter matrix in each PDB unit. A freeboard height of one meter was allowed for the accumulation of the dewatered sludge. Narrow-leaf cattails (*Typha augustifolia*), collected from a nearby natural wetland, were planted on top of the sand layer in each bed unit at an initial density of 8 shoots/m^2. An underdrain and ventilation system consisting of hollow concrete blocks, each with a dimension of 20 cm x 40 cm x 16 cm and perforated PVC pipes with a diameter of 20 cm were installed at the bottom of the bed, under the filter media. Ventilation pipes of the same diameter were mounted on the drainage system and extended approximately one meter over top edge of the units to take advantage of natural draught ventilation to provide increased oxygen into the sludge layer and help reduce anaerobic conditions. The leachate of each PDB unit was collected in a 3 m^3 concrete tank for sampling and analysis.

Figure 8.4 A currently out of use pilot-scale drying bed at the Asian Institute of Technology (AIT), showing ventilation pipes (photo: Linda Strande).

Table 8.4 Summary of design elements used for faecal sludge planted drying beds in Thailand

Component	Details
Bed slope	1-3%
Side slope	50-100%
Drainage system	Coarse gravel, hollow concrete blocks or perforated pipes
Ventilation	Ventilation pipes connected to the drainage system
Filter Material	From bottom to top Large gravel (dia. = 5 cm) at a depth of 45 cm Medium gravel (dia. = 2 cm) at a depth of 15 cm Sand (dia. = 0.1 cm) at a depth of 10 cm
Vegetation	Cattails *(Typha augustifolia)*
Freeboard	1.0 m
Feeding system	Uniform distribution (in the middle of bed units)
Pre-treatment	Coarse bar screen

8.6 OPERATION AND MAINTENANCE

As with any treatment technology, proper operation and regular maintenance are essential for optimum performance and improved life span. An operating cycle generally consists of a start-up phase with reduced loading to acclimatise the plants, followed by loading at the design rate with intermittent plant harvesting and desludging. These aspects are discussed in the following sections.

8.6.1 Commissioning/ start-up

PDBs are technically simple, but biologically complex and must therefore be carefully operated during start-up to ensure that the macrophytes have a chance to acclimatise to growing under conditions of high-strength FS. During the start-up phase, the beds should be irrigated with untreated wastewater or diluted FS. One study found that during commissioning of a PDB with agricultural (pig) slurries, macrophytes were loaded with 25 mm of sludge twice in one month, 8-months after being planted. This time frame for acclimisation and low sludge loading rate (3 kg TS/m^2/year) was found to be sufficient to prepare the macrophytes for the full loading regime (Edwards *et al.*, 2001). Planting macrophytes during the rainy or wet season is also recommended to help the macrophytes endure the commissioning phase. Depending on the climate and operational conditions, a start-up phase lasting from months to an entire year can be necessary before loading the bed at the design loading rates. On average, a 6 month start-up is recommended (Kengne *et al.*, 2011). Cattails have been found to be more sensitive than reeds during the commissioning phase and may need extra time before they can withstand full loading. However, as shown in Case Study 8.3, two to three months has been adequate for acclimisation (Stefanakis and Tsihrintzis, 2012a). Plant density is another important factor and planting rates can vary from 4 plants/m^2 to 12 plants/m^2 (Edwards *et al.*, 2001). Only vigorous and young shoots, free of parasites, should be selected for the PDBs to ensure that the macrophytes survive and thrive. As the plants develop and increase in density, so too will the evapotranspiration rates (Stefanakis and Tsihrintzis, 2012a). Case Study 8.4 presents two examples of PDB commissioning conditions in West Africa.

Case Study 8.4: Commissioning planted drying beds in West Africa
(Adapted from SANDEC/EAWAG, 2009)

In Cameroon, young shoots, or fragments of *E. pyramidalis* shoots, having at least one internode, and old fragments of rhizomes of *C. papyrus* weighing 300 to 350 g (fresh weight) were allowed to grow for 6 weeks in the media saturated with raw domestic wastewater prior to sludge application. FS was applied over the next 6 months in increasing concentrations before reaching the full loading rate of 100 to 200 kg TS/m^2/yr (Kengne *et al.*, 2011). The plant density before sludge application was 11 shoots per m^2 for *E. pyramidalis* and 9 rhizomes (with 1 to 4 shoots/rhizomes) per m^2 for *C. papyrus*.

In Senegal, the starting phase for a full scale PDB with *E. pyramidalis* took four months during which time the beds were loaded with the supernatant from a FS settling-thickening tank. After this time, the PDBs were loaded with FS with a concentration ranging from 13 to 235 kg/m^2/year. Plant densities at the start up ranged from 9 to 12 shoots/m^2.

8.6.2 Loading rates and sludge accumulation

Before loading the beds, vacuum trucks should discharge the sludge into a holding-mixing tank that is fitted with a bar screen to retain coarse material and garbage and prevent it from clogging the bed. Furthermore, the tank has the benefit of acting as a buffering unit to regulate the flow of sludge onto the bed; some type of holding-mixing unit should always be installed before the bed is loaded.

Figure 8.5 A holding -mixing tank with a bar screen is used in Senegal to prevent garbage from clogging the bed (photo: Linda Strande).

Technology

Data on PDBs operating at nominal loading rates vary according to area, and indicate the importance of climate on the operating parameters. In general, hot, dry conditions that allow for increased rates of evapotranspiration allow for increased sludge loading rates. In Europe, loading rates with wastewater sludge have generally been low (not more than 80 kg/m^2/year), while results from FS treatment in tropical countries have revealed that PDBs can be loaded with almost three times this amount. For example, a series of experiments conducted at AIT with FS, showed that a cattail-based PDB was operated with up to 250 kg/m^2/year (Koottatep *et al.*, 2005). Similarly, in Dakar, trials of PDBs vegetated with *E. pyramidalis* performed well when loaded with FS at concentrations of up to 235 kg/m^2/year. In Cameroon, treatment of FS at yard scale show that a PDB planted with *C. papyrus* could be operated efficiently at 100 kg/m^2/year while a bed planted with *E. pyramidalis* could be loaded with 200 kg kg/m^2/year. However, attempts to increase the loading to 300 kg/m^2/year generally resulted in severe clogging of beds (Kengne *et al.*, 2011). Between 1996 and 2003, experimental drying beds were operated at the Asian Institute of Technology (AIT) in Bangkok, Thailand and the solids (kg TS/m^2) were monitored in the dried sludge and the effluent. The results of the mass balance are presented in Table 8.5.

It is interesting to note that on average, about 47% of the solids were retained in the dried layer of sludge, about 12% passed through the bed and were captured in the leachate (see below for a discussion on leachate) and 42% were 'unaccounted' for. The 'unaccountable' solids were lost due to a combination of mineralisation and/or sorbtion onto/integrated into the filter media. These results illustrate why media regeneration is necessary, and the importance of further treatment for the leachate treatment due to the high solids concentrations.

8.6.3 Feeding frequency and resting phase

Loading of PDBs is always intermittent and the frequency varies from site to site. Typically, loading occurs one to three times a week by means of valves and siphons or pumping devices installed in a buffer tank, which is preferable to loading directly from a truck. Once loaded with a layer of sludge, the bed is allowed to drain completely, during which time the pores of the filter matrix are emptied of leachate, and refilled with air. The next application of sludge effectively seals off these small pockets of air. Once this occurs, oxygen, which is instrumental in the nitrification process is rapidly depleted (Kadlec and Wallace, 2009). Therefore the resting time between loading periods is very important as it prevents biological clogging and allows pores to refill with oxygen (Stefanakis and Tsihrintzis, 2012a).

However, if the resting times between FS loading is increased, more PDBs would be required to treat the same volume of sludge. Using a semi-empirical equation, researchers determined that in order to maximise water-loss and minimise costs, 11 days was the optimum number of days between loadings (Giraldi and Iannelli, 2009). This is in keeping with other reported practices of between one and three weeks (Stefanakis and Tsihrintzis, 2012a).

Table 8.5 Total Solids (TS) mass balance of faecal sludge from septic tanks on planted drying beds after 300 days of operation (adapted from Koottatep and Surinkul *et al.*, 2004)

	Unit #1		Unit #2		Unit #3		Average
	(kg TS/m^2)	(%)	(kg TS/m^2)	(%)	(kg TS/m^2)	(%)	(%)
Faecal sludge	187		115		112		-
Dried sludge	93	50	60	52	43	38	47
Percolate	20	11	14	12	13	12	12
Unaccounted	74	39	41	36	56	50	42

8.6.4 Plant harvesting and regrowth

As mentioned in Section 8.5.2 a benefit of PDBs is that the macrophytes can be harvested for beneficial enduse (covered in more detail in Chapter 10). Macrophytes grown in PDBs generate two to three times the biomass that is produced in naturally occurring wetlands, due to the availability of nutrients, especially nitrogen and phosphorus (Warman and Termeer, 2005). Harvesting generally occurs on a regular basis (e.g. during desludging), but could also be dictated by other considerations such as the need to sell the plants for enduse purposes (e.g. fodder) or to mitigate insect attacks (Altieri and Nicholls, 2003; Pimental and Warneke, 1989). It has been found that insects can have a great impact on larger plants, especially in dense monocultures, which may require the removal of older plants to allow new and vigorous shoots to take over. *E. pyramidalis*, which is highly sought after as fodder in some regions, can be harvested up to three times a year (Kengne *et al.*, 2008).

Currently, harvesting is carried out manually since most of the PDBs have been operated at experimental or pilot scale. Mechanical methods will probably be introduced when PDBs are operated at full scale. Harvesting is done by cutting plants at the surface, not by pulling out the whole plant. This prevents damage to the filter, and if the rhizome is left intact, the leaves and stalks can readily regrow.

8.6.5 Bed emptying

Finding the optimum loading rate is important for the operation and maintenance of PDBs to ensure that the sludge layer does not accumulate and become too thick and require desludging before it is fully drained. On an experimental scale, it has been found that a loading rate of 100 kg TS/m^2/year, results in the accumulation of approximately 30 to 40 cm/year of sludge, compared to 50-70 cm/year if loading rate of 200 kg TS/m^2/year is used. For PDBs with a freeboard of 1.5 m to 2 m these loading rates would result in a 3-5 year operation life before desludging is required (Kengne *et al.*, 2011). Prior to removal, sludge can be left for several months without additional loading which results in greater pathogen and moisture reduction. For example, a significant increase of 25-43% in dry matter content was achieved when pilot-scale beds were left for one month prior to desludging in Cameroon, and the helminth (Ascaris) egg concentration was reduced to less than 4 viable eggs/g TS from 79 eggs/g TS and a viability 67% (Kengne *et al.*, 2009b).

Sludge removal is currently carried out manually, although mechanical desludging machines may be employed in the future. Depending on how carefully the bed was desludged, it may be necessary to reconstitute the substrate of the bed, either by adding to or replacing the upper layer (sand or fine gravel), or by replacing the entire bed.

8.6.6 Leachate

Leachate is the liquid that filters down through the sludge layer and the porous media. It should be collected and treated with a subsequent treatment technology prior to discharge to the environment. However, the leachate can also be used for irrigation or aquaculture (covered in more detail in Chapter 10). If the PDBs are located at a wastewater treatment plant, the leachate can be treated with the wastewater. Other possibilities include dedicated onsite technologies such as waste stabilisation ponds (Chapter 5; Strauss *et al.*, 1997). Measurement of the leachate characteristics over time shows that most parameters have a peak concentration following sludge loading (COD, PO_4^{3-}, TSS, VSS) followed by a rapid decline, indicating a flushing phenomenon and/or the dynamic treatment mechanisms at work in the bed. A study conducted with sludge from a biological wastewater treatment plant, showed an 80% reduction in COD (initially 2,500 mg/L) during the first 10 minutes after loading, and over 92% COD reduction after two days. Additionally, initial ammonium concentrations of more than 350 mg/L decreased rapidly and were reduced by 90% within the first 10 minutes after loading. This decrease in ammonia was accompanied by an increase in nitrate concentration, thereby reflecting the rapid nitrification process (Stefanakis and Tsihrintzis, 2012a). Research at AIT illustrated that approximately 12% of the total solids remain in the leachate (Table 8.5). The same research on parallel beds also

Technology

showed that 45% of the total liquid in the loaded sludge ended up as leachate (while 5% remained in the sludge layer, and 50% was lost to evapotranspiration). Furthermore, the leachate was only found to contain about 5% of the total nitrogen applied, with the majority (82%) being taken up and a small percentage (13%) remaining in the sludge layer (Koottatep and Surinkul, 2004). In general, leachate stops draining from the bed one to two days after loading. Leachate production is highly variable; high shock loads and intermittent flows must be taken into consideration in the design of any subsequent treatment process.

8.6.7 Factors affecting performance

The main causes of poor operational performance include poorly constructed filters; inadequate capillary connections; an inadequate number of beds, insufficient bed area; or overloading during commissioning and subsequent operation (Nielson, 2005). Other factors such as the settling of particulate matter, fast-growing biofilm, chemical precipitation and salt formation, and dense root development have also been mentioned as further reasons for clogging (Molle *et al.*, 2006). Operational problems can be overcome by proper dimensioning of the PDBs which takes the dewatering potential of the sludge into account and does not rely solely on calculations of the sludge volume production. The loading program should be designed to prevent the sludge layer from accumulating too quickly as this can inhibit the growth of such that the macrophytes. Table 8.6 summarises suggested operational parameters for PDBs and the operational aspects that need to be taken into consideration.

8.7 COSTS AND BENEFITS

One of the most attractive features of PDBs compared to other sludge treatment technologies is the fact that they have low capital, operating, maintenance, supervision and energy costs (Stefanakis and Tsihrintzis, 2012a). PDBs do not require chemical flocculants, centrifuges or belt presses (Edwards *et al.*, 2001). However, PDBs can be more expensive than unplanted drying beds, both in terms of the capital costs (e.g. purchasing macrophytes), and operational costs (e.g. plant harvesting, weeding and vector control), but they have the advantage of requiring less desludging (e.g. once every few years versus every two to three weeks).

Table 8.6 Operational parameters for a planted drying bed

Treatment component	Details	Remarks
Loading	60-250 kg TS/m^2/year	Depending on the sludge source and conditions
Feeding frequency	1-3 times a week	Depending on the weather conditions, the dry matter content of the sludge and the plant species
Resting	2 days to several weeks	Depending on the weather conditions, the dry matter content of the sludge and the plant species
Plant acclimatisation	Start-up with plant density of 4-12 shoots/m^2 Apply domestic wastewater and gradually add FS until the plants achieve a height of 1 m	Start-up during a rainy or wet season is recommended
Plant harvesting	Up to 3 times/year, following a few years of operations or during desludging	Depending on plant type, the growth status and valorisation option. Valid especially for *Echinochloa pyramidalis*

A study in Italy, attempted to quantify the costs associated with building and operating a PDB for the treatment of wastewater sludge. Although the values obtained are not representative of costs worldwide, they do provide some useful insights. Construction costs, including the plants, other materials and labour were estimated to be in the region of 350 USD/m^2 while the operating costs, including plant harvesting, sludge transport and disposal were calculated to be 180 USD/m^2 (Giraldi and Iannelli, 2009). Considering a sludge production rate (from primary wastewater treatment) of 16kg TS/capita/year, and assuming loading rates between 30 and 75 kg TS/m^2/year these PDBs could treat the sludge of between 1.7 and 4 capita/m^2 (Stefanakis and Tsihrintzis, 2012a). Since a large portion of the operating costs are associated with transport (e.g. transport to the disposal site and transport of an endproduct from the site), local transport costs can significantly impact on the total. Furthermore, construction costs will vary depending on the availability and cost of local labour and materials (Giraldi and Iannelli, 2009).

8.8 EXAMPLE PROBLEM

In order to demonstrate the calculations that are required in designing and constructing a PDB, an exercise is presented below as a practical example. Table 8.7 provides information that can be used to assist with the required calculations.

8.8.1 Practice question

After conducting a preliminary study, a municipality would like to design a PDB to dewater FS having the following characteristics:

Estimated annual FS emptied: 5,000 m^3/year

Average TS content of raw FS: 30,000 mg/L (or 30 kg TS/m^3)

Using this information:

Determine the total solids of faecal sludge per year:

5,000 m^3/year x 30 kg TS/m^3: 150,000 kg TS/year

Determine the specific area required for the planted-sludge drying bed

Choose the TS loading rate: 200 kg TS/m^2/year

Area required: $150{,}000 \text{ kg TS/year} \times \dfrac{1}{200 \text{ kg TS/m}^2\text{/year}} = 750 \text{ m}^2$

This specific area can be divided into several beds according to the topography of the site. Assuming that the topography of the area is uniform, the specific area can be split into 5 beds of 150 m^2 each Additional areas for bar screen, mixing tanks, leachate tanks and vacuum trucks need to be taken into consideration. The minimum area is about 20% of specific area.

Table 8.7 Suggested design parameters of planted sludge drying beds for faecal sludge dewatering

Design parameter	Suggested ranges	Unit
FS production rate	1.5	L/capita/day
TS content	30	mg/L
Solid loading rate	200	kg TS/m^2/year
FS application frequency	1 to 2	times/week

8.9 CONCLUSIONS AND RECOMMENDATIONS

PDBs are a relatively new technology for treating FS from septic tanks and other onsite sanitation technologies in low- and middle-income countries. Extensive experience in Europe and the US has produced robust results, but the data are not entirely applicable to FS due to the sludge type and strength, and the climatic conditions. Currently, many experimental and pilot scale beds are being investigated in various parts of the world, especially tropical climates where solar radiation and evapotranspiration is high. PDBs have long been known as a reliable technology for sludge treatment, but have become increasingly attractive for FSM in rapidly growing cities in low- to middle-income countries as they are less costly to build than conventional wastewater sludge treatment technologies, can be built using local materials and labour, and require little maintenance, few to no chemicals and minimal energy to operate successfully. Although the macrophytes require some time to acclimatise to the nutrient-rich sludge, the PDB can then operate for up to 10 years without desludging and the macrophytes can be harvested for beneficial use. The stabilised sludge layer can also be used as a soil amendment and organic fertiliser.

However, PDBs require a significant amount of space (0.25 to one m^2/capita) and therefore, the technology is not well-suited to dense, urban areas. Furthermore, the bed must be accessible by trucks that transport sludge, and should therefore be built on or near roads that are easily traversed by large vehicles. Although resilient, macrophytes may be prone to insect attacks and parasitism. Therefore, although maintenance is not constant, it must be diligent. In recent years, much research has been carried out in order to determine optimum parameters for the design and operation of the most robust PDBs as possible. There are, however still questions that remain unanswered, such as:

- the effects of feeding frequency on bed performance;
- the vulnerability and resilience of macrophytes to insect attacks;
- the effects of high conductivity and ammonia;
- the most effective treatment methods for leachate;
- the long-term (10+ year) performance of the beds; and
- the cost-benefit analysis of the system.

Each of these aspects should be researched under different loading rates, with different types of FS and under different climactic conditions. Although research remains important, priority should be given to upscaling and promoting PDBs whenever possible and appropriate. Time must not be wasted on perfecting this technology, but rather building on current knowledge and disseminating evidence as it is gathered.

8.10 BIBLIOGRAPHY

Altieri, M. A., Nicholls, C. I. (2003). Soil fertility management and insect pests: harmonizing soil and plant health in agroecosystems. Soil and Tillage Research 72(2), p.203-211.

Bialowiec, A., Wojnowska-Baryla, I., Agopsowicz, M. (2007). The efficiency of evapotranspiration of landfill leachate in the soil-plant system with willow Salix amygdalina L. Ecological Engineering 30(4), p.356-361.

Breen, P. F. (1997). The performance of vertical flow experimental wetland under a range of operational formats and environmental conditions. Water Science and Technology 35(5), p.167-174.

Brix, H. (1994). Functions of macrophytes in constructed wetlands. Water Science and Technology 29(4), p.71-78.

Brix, H. (1997). Do macrophytes play a role in constructed treatment wetlands? Water Science and Technology 35(5), p.11-17.

Chen, W., Chen, Z., He, Q., Wang, X., Wang, C., Chen, D. (2007). Root growth of wetland plants with different root types. Acta Ecologica Sinica 27(2), p.450-457.

Clarke, E., Baldwin, A. H. (2002). Responses of wetland plants to ammonia and water level. Ecological Engineering 18(3), p.257-264.

Davis, L. (1995). A handbook of constructed wetlands: A guide to creating wetlands for--agricultural wastewater, domestic wastewater, coal mine drainage, stormwater in the Mid-Atlantic Region. Vol 1. General considerations (Vol. 1). Washington, DC: USDA-NRCS, EPA Region III.

De Maeseneer, J. L. (1997). Constructed wetlands for sludge dewatering. Water Science and Technology, 35(5), 279-285.

EAWAG/SANDEC. (2009). Recueil des résultats de recherche sur la gestion des boues de vidange du projet de collaboration ONAS-EAWAG/SANDEC- Phase I, 2006-2009. Dakar: EAWAG/SANDEC.

Edwards, J. K., Gray, K. R., Cooper, D. J., Biddlestone, A. J., Willoughby, N. (2001). Reed bed dewatering of agricultural sludges and slurries. Water, Science and Technology 44(10-11), p.551-558.

Gagnon, V., Chazarenc, F., Comeau, Y., Brisson, J. (2007). Influence of macrophytes species on microbial density and activity in constructed wetlands. Water Science and Technology, 56(3), 249-254.

Giraldi, D., Iannelli, R. (2009). Short-term water content analysis for the optimization of sludge dewatering in dedicated constructed wetlands (reed bed systems). Desalination 246(1-3), p.92-99.

Hardej, M., Ozimek, T. (2002). The effect of sewage sludge flooding on growth and morphometric parameters of Phragmites australis (Cav.) Trin. ex Steudel. Ecological Engineering 18(3), p.343-350.

Hutchinson, J., Dalziel, J. M. (1972). Flora of west Tropical Africa (Vol. Vol. III). London: Crown Agents for Overseas governments and administrations.

Ingallinella, A. M., Sanguinetti, G., Koottatep, T., Montangero, A., Strauss, M. (2002). The challenge of faecal sludge management in urban areas – strategies, regulations and treatment options. Water, Science and Technology 46(10), p.285-294.

Kadlec, R. H., Knight, R. L. (1996). Treatment wetlands. Boca Raton, FL.: Lewis Publishers.

Kadlec, R. H., Wallace, S. (2009). Treatment wetlands (2nd edition ed.). Boca Raton, FL: CRC Press.

Kengne, I. M., Akoa, A., Soh, E. K., Tsama, V., Ngoutane, M. M., Dodane, P. H. (2008). Effects of faecal sludge application on growth characteristics and chemical composition of Echinochloa pyramidalis (Lam.) Hitch. and Chase and Cyperus papyrus L. Ecological Engineering 34(3), p.233-242.

Kengne, I.M., Dodane, P.-H., Amougou Akoa, Koné, D., 2009a. Vertical flow constructed wetlands as sustainable sanitation approach for faecal sludge dewatering in developing countries. Desalination, (248) p291-297.

Kengne, I. M., Amougou Akoa, Koné, D. (2009b). Recovery of biosolids from constructed wetlands used for faecal sludge dewatering in tropical regions. Environmental Science and Technology 43, p.6816-6821.

Kengne, I. M., Soh Kengne, E., Akoa, A., Bemmo, N., Dodane, P.-H., & Koné, D. (2011). Vertical-flow constructed wetlands as an emerging solution for faecal sludge dewatering in developing countries. Journal of Water, Sanitation and Hygiene for Development 01(1), 13-19.

Kim, B. J., Smith, D. (1997). Evaluation of sludge dewatering reed beds: A niche for small systems. Water Science and Technology 35(6), p.21-28.

Koottatep, T., Surinkul, N., Polprasert, C., Kamal, A. S. M., Koné, D., Montangero, A. (2005). Treatment of septage in constructed wetlands in tropical climate: lessons learnt from seven years of operation. Water Science and Technology 51(9), p.119-126.

Kroiss, H. (2004). What is the potential for utilizing the resources in sludge? Water Science and Technology 49(10), p.1-10.

Lienard, A. Payrastre, F. (1996). Treatment of sludge from septic tanks in reed beds filters pilot plants. In: IWA (Ed), 5th Int. Conf. on Wetlands Systems for Water Pollution Control, Vol. I, IWA, Vienna, p.1-9.

Molla, A. H., Fakhru'l-Razi, A., Abd-Aziz, S., Hanafi, M. M., Roychoudhury, P. K., Alam, M. Z. (2002). A potential resource for bioconversion of domestic wastewater sludge. Bioresource Technology 85(3), p.263-272.

Molle, P., Lienard, A., Grasmick, A., Iwema, A. (2006). Effect of reeds and feeding operations on hydraulic behaviour of vertical flow constructed wetlands under hydraulic overloads. Water Research 40(3), p.606-612.

Nielsen, S. (2003). Sludge drying reed beds. Water Science and Technology 48(5), p.101-109.

Nielsen, S. (2005). Sludge reed beds facilities – Operation and problems. Water Science and Technology 51 (9), p.99–107.

Pimental, D., Warneke, A. (1989). Ecological effects of manure, sewage sludge and other organic wastes on arthropod populations. Agricultural Zoology Reviews 3, p.1-29.

Prochaska, C. A., Zouboulis, A. I., Eskridge, K. M. (2007). Performance of pilot-scale vertical-flow constructed wetlands, as affected by season, substrate, hydraulic load and frequency of application of simulated urban sewage. Ecological Engineering 31(1), 57-66.

Strauss, M., Larmie, S.A., Heinss, U. (1997). Treatment of sludges from on-site sanitation - Low-cost options. Water Science and Technology 35 (6), p.129-136

Stefanakis, A. I., Akratos, C. S., Melidis, P., & Tsihrintzis, V. A. (2009). Surplus activated sludge dewatering in pilot-scale sludge drying reed beds. Journal of Hazardous Materials, 172(2-3), p.1122-1130.

Stefanakis, A. I., Tsihrintzis, V. A. (2012a). Effect of various design and operation parameters on performance of pilot-scale Sludge Drying Reed Beds. Ecological Engineering 38(1), p.65-78.

Stefanakis, A. I., Tsihrintzis, V. A. (2012b). Heavy metal fate in pilot-scale sludge drying reed beds under various design and operation conditions. Journal of Hazardous Materials (213-214), p.393-405.

Towers, W., Horne, P. (1997). Sewage sludge recycling to agricultural land: the environmental scientist's perspective. Journal of the Commission for International Water and Environmental Management 11, p.162-132.

Uggetti, E., Ferrer, I., Carretero, J., Garcia, J. (2012). Performance of sludge treatment wetlands using different plant species and porous media. Journal of Hazardous Materials, 217-218, 263-270.

Van Cuyk, S., Siegrist, R., Logan, A., Masson, S., Fischer, E., Figueroa, L. (2001). Hydraulic and purification behaviors and their interactions during wastewater treatment in soil infiltration systems. Water Research 35(4), p.953-964.

Warman, P. R., Termeer, W. C. (2005). Evaluation of sewage sludge, septic waste and sludge compost applications to corn and forage: yields and N, P and K content of crops and soils. Bioresource Technology 96(8), p.955-961.

End of Chapter Study Questions

1. Describe the main components of PDBs, and the basic fundamentals of their operation.
2. Explain what macrophytes are, and list four essential roles they play in FSM.
3. Identify four performance indicators that are important for monitoring the performance of PDBs to ensure they are meeting treatment objectives.
4. Finding the optimum loading rate is important for the operation and maintenance of PDBs, explain why this is important.
5. What are challenges and benefits of using the PDB technology for FSM in dense urban areas?

Chapter 9

Co-treatment of Faecal Sludge in Municipal Wastewater Treatment Plants

Carlos M. Lopez-Vazquez, Bipin Dangol, Christine M. Hooijmans and Damir Brdjanovic

Technology

Learning Objectives

- Understand the biodegradability and fractionation of organic matter and nitrogen compounds in faecal sludge.
- Understand the principles, key considerations and potential impacts of co-treatment of faecal sludge in sewer-based wastewater treatment systems.
- Determine volumes of faecal sludge that can be effectively co-treated in wastewater treatment plants.
- Understand the potential negative ramifications of co-treating faecal sludge in sewer-based wastewater treatment systems.

9.1 INTRODUCTION

The use of onsite sanitation technologies can be a sustainable solution to meet sanitation goals in a faecal sludge management (FSM) service chain, as long as the faecal sludge (FS) from these systems is collected, transported, treated, and then used for resource recovery or safely disposed of. One possibility for FS treatment is co-treatment with sewer-based wastewater treatment technologies. However, it is common knowledge that the majority of wastewater treatment plants (WWTPs) in low-income countries have failed, and improper co-treatment with FS has even been the cause of some failures. Hence, the objective of this chapter is to illustrate through modelling of WWTPs how these failures occurred, and the extreme difficulties with co-treatment that must be addressed to avoid failures. First, the chapter addresses activated sludge processes, and then anaerobic technologies including upflow anaerobic sludge blanket (UASB) reactors, digesters, and ponds. Co-treatment in ponds is also discussed in Chapter 5.

Based on the results of this chapter, co-treatment of FS with wastewater is not recommended for the vast majority of cases in low-income countries. If a co-management option is desired, a better option would potentially be co-management of FS with the sludge produced during wastewater treatment (i.e. biosolids). Many of the enduse and resource recovery options presented in Chapter 10 are appropriate for this, and could provide increased revenue from resource recovery. The tools in this chapter are relevant to evaluate existing, operational WWTPs, and for evaluating future WWTP designs.

In addition, the uncontrolled dumping of FS into sewers needs to be carefully regulated and prevented. The considerably higher solids content of FS (Chapter 2) may lead to severe operational problems such as solids deposition and clogging of sewer pipes. This is mostly because the diameter and slope of sewers are designed for the transport of municipal wastewater typically containing 250 to 600 mgTSS/L (Henze *et al.*, 2008) rather than the 12,000 to 52,500 mgTSS/L present in FS (Table 2.3). Hence, the first step in designing a co-treatment system includes determining how the FS will be transported to the treatment facility and discharged into the influent stream.

WWTPs are typically not designed for FS loadings, and process disruptions and failures are frequently observed. Common problems with co-treatment of FS in WWTPs range from the deterioration of the treated effluent quality to overloading tanks and inadequate aeration (Andreadakis, 1992; Al-sa'ed and Hithnawi, 2006; Heinss and Strauss, 1999; Strauss *et al.*, 2000; Chaggu, 2004; Harrison and Moffe, 2003; Lopez-Vazquez, 2008; Lake, 2010; Lake *et al.*, 2011; Wilson and Harrison, 2012; Still and Foxon, 2012).

Despite the potential operational problems, certain guidelines indicate that low volumes of FS could be co-treated in municipal WWTPs (ATV, 1985; USEPA, 1984, 1994). The USEPA states that that up to 3.6% of the maximum plant design capacity load can be FS (i.e. from septic tanks) (EPA, 1994). However, these recommendations are mostly based on biochemical oxygen demand (BOD_5) which does not account for the total organic and inorganic content present in FS or provide enough relevant information on the different biodegradable fractions (Henze and Comeau, 2008). Instead, chemical oxygen demand (COD) measurements are recommended to be used since total COD can be subdivided into useful organic fractions to assess the design and evaluate the performance of biological wastewater treatment processes. This chapter presents the impact of FS co-treatment in municipal WWTPs, based on expected average FS strength and COD and total nitrogen (TN) fractionations. This approach is recommended to evaluate whether co-treatment may be feasible without causing any process disruption or deterioration.

9.2 FAECAL SLUDGE BIODEGRADABILITY AND FRACTIONATION

9.2.1 Characterisation ratios

When evaluating FS characteristics to determine the potential for co-treatment, in addition to classic parameters such as COD, BOD and TSS, the ratios between these parameters also provide useful information. Ratios of parameters for public toilets and septic tanks are presented in Table 9.1.

The ranges of values in Table 9.1 are quite large and therefore only provide a rough estimation of the potential biodegradability. The ratios must also be used with caution. As compared to common values observed with wastewater, they suggest that FS is not readily biodegradable. The low VSS to TSS ratios indicate 23-50% inorganic content. The COD:BOD_5 ratio of 5.0 for public toilets indicates that, if degradable, the organics biodegrade slowly. In contrast, the COD:BOD_5 of 1.43 - 3.0 for septic tanks indicates the sludge is biodegradable, which probably is not the case, as septic tank sludge typically has a much longer storage time with significant stabilisation (e.g. years as opposed to days). This illustrates the need for a more reliable and informative method to determine the biodegradability of FS.

Table 9.1 Characterisation ratios for public toilet and septic tank faecal sludge to evaluate biodegradability for treatment purposes (calculated based on Table 2.3 and adapted from Henze *et al.*, 2008)

Ratios (g/g)	Public toilets	Septic tanks	Medium strength municipal wastewater
VSS:TSS	0.65-0.68	0.50-0.73	0.60-0.80
$COD:BOD_5$	5.0	1.43-3.0	2.0-2.5
COD:TKN	0.10	1.2-7.8	8-12
BOD_5:TKN	2.2	0.84-2.6	4-6
COD:TP	109	8.0-52	35-45
BOD_5:TP	17	5.6-17.3	15-20

The organic content to nitrogen ratios also indicate that organic concentrations are not sufficient for nitrogen removal by denitrification, as they are far below the lowest reported for nitrogen removal (Henze and Comeau, 2008). FS should only be considered for co-treatment in processes that include nitrogen removal if the influent wastewater has a high COD:TKN or BOD_5:TKN ratio (i.e. 12-16 and 6-8, respectively). In contrast, the COD:TP and BOD_5:TP ratios are relatively high, which suggests that there could be sufficient organic matter for biological phosphorus removal.

9.2.2 Biodegradability and fractionation

Fractionation is the breakdown of organic matter into groups based on biodegradability and physico-chemical properties. Frequently, (bio)degradability is measured by BOD_5. However, this method has limitations such as the incomplete determination of all the organics since the unbiodegradable fractions cannot be determined by this analytical technique, as underlined by Roeleveld and van Loosdrecht (2002) and Henze and Comeau (2008). Thus, the use of COD is preferred to assess the organic matter for design, control, monitoring and mathematical modelling of wastewater treatment processes. Advantages of COD over BOD_5 include: (i) a rapid analysis (e.g. hours as opposed to 5 days), (ii) more detailed and useful information including all degradable and undegradable organics, and (iii) the potential for the organics balance to be closed (on a COD basis). Of the two COD analytical determination methods, the dichromate method is preferred, as the permanganate method does not fully oxidise all organic compounds (Henze and Comeau, 2008).

The biodegradable fraction can be divided into readily and slowly biodegradable compounds. Readily biodegradable organics are assumed to be relatively small molecules that can dissolve in water and be rapidly consumed (e.g. volatile fatty acids and low molecular weight carbohydrates). Slowly biodegradable organics are considered to be more complex, and require extracellular breakdown prior to uptake and utilisation by microorganisms (Dold *et al.*, 1980). They are assumed to be colloidal and particulate compounds that can also be removed by physical-chemical means (e.g. coagulation-flocculation and settling).

The unbiodegradable fractions (often also referred to as inert) are not degraded, or degraded so slowly that they are not transformed during their transport in the sewer or residence time in WWTPs. They are also further divided into soluble and particulate organic groups. It is assumed that particulates can be removed by physical separation (e.g. settling), but the soluble unbiodegradable organics cannot be removed by biological or physical-chemical methods. Thus, when soluble unbiodegradable organics reach the sewage treatment plants, they pass through the system in the liquid phase with the same influent and effluent concentrations (Ekama, 2008). In wastewater treatment systems, the soluble unbiodegradable organics have a profound impact on effluent quality and the particulate unbiodegradable organics on sludge production and solids accumulation.

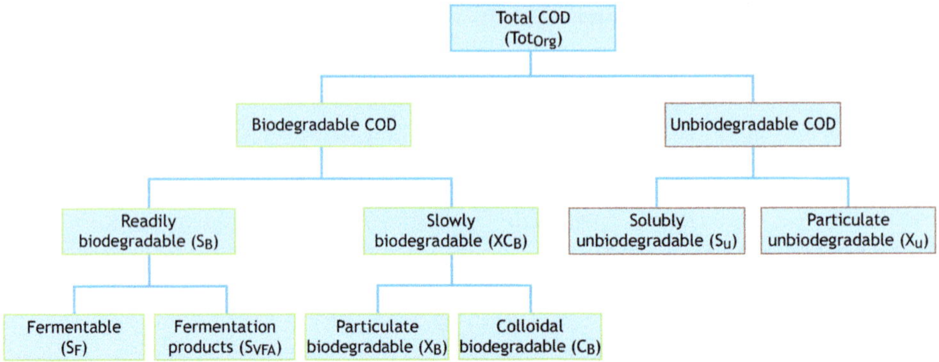

Figure 9.1 Organic matter (COD) fractionation diagram (adapted based on Melcer, 2003 and Corominas *et al.*, 2010).

It is important to underline that organic compounds contain different carbon, nitrogen and phosphorus components. It is preferable to determine and express carbon components in terms of COD (in view of the advantages of this analytical technique over others). Figure 9.1 illustrates the different COD fractions of the organic compounds as well as the common abbreviations for the different fractions (Corominas *et al.*, 2010):

X = particulate
S = soluble
C = colloidal
B = biodegradable
U = unbiodegradable
F = fermentable
VFA = products of fermentation

Thus, the total organic matter concentrations present in wastewater given as the sum of the different biodegradable and unbiodegradable COD fractions as shown in Equation 9.1:

Equation 9.1: $$Tot_{Org} = S_F + S_{VFA} + X_B + C_B + X_U + S_U \text{ (mgCOD/L)}$$

Recognising that organic nitrogen is the nitrogen content of the different organic compound groups, and adding the other inorganic nitrogen compounds (such as ammonia, nitrite and nitrate), the nitrogenous compounds can also be fractionated as (Figure 9.2):

Tot_N = total Kjeldahl nitrogen (TKN)
$Tot_{Ig,N}$ = total inorganic nitrogen
$Tot_{Org,N}$ = total organic nitrogen
NH_X = total free and saline ammonia
NO_X = total nitrite plus nitrate
$Tot_{Org,B,N}$ = total organic biodegradable nitrogen
$Tot_{Org,U,N}$ = total organic unbiodegradable nitrogen

Organic nitrogen can be divided into similar fractions such as COD because nitrogen is another component of the same organic groups. Thus, organic biodegradable nitrogen compounds are divided into particulate biodegradable ($XC_{B,N}$), which (bio)degrades more slowly, and soluble biodegradable ($S_{B,N}$), that is easily biodegradable.

The unbiodegradable organically bound nitrogen comprises particulate unbiodegradable and soluble unbiodegradable fractions (X_U,N and $S_{U,N}$ respectively). Because these organic groups are not degraded and remain unaffected by the biological processes, they remain intact, keeping their nitrogen (and COD and phosphorus) composition and characteristics. Therefore, in a treatment plant $X_{U,N}$ accumulates in the system and is added to the sludge mass, whereas $S_{U,N}$ leaves the plant through the effluent because it does not settle out and is not biologically removed. So, the unbiodegradable COD and organic nitrogen is simply the COD and nitrogen content of the unbiodegradable organics.

Therefore, Tot_N can be expressed as shown in Equation 9.2:

Equation 9.2: $$Tot_N = S_{NHx} + S_{NOx} + XC_{B,N} + S_{B,N} + X_{U,N} + S_{U,N} \quad (mgN/L)$$

In addition to the organic and nitrogenous compounds, wastewater also contains inorganic suspended solids (ISS) as part of the total suspended solids (Table 2.3). Bacteria are able to utilise small concentrations of ISS as trace elements or micronutrients for cell growth (e.g. magnesium, potassium and calcium compounds). However, they are not considered biodegradable. Consequently, the ISS tend to accumulate in wastewater treatment proportionally to the solids retention time (SRT) (Ekama, 2008).

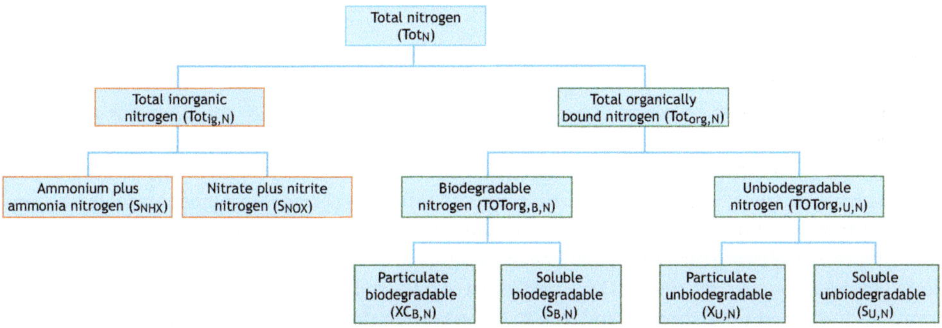

Figure 9.2 Nitrogen fractionation diagram (adapted based on Melcer, 2003 and Corominas *et al.*, 2010).

Technology

Table 9.2 Defined COD, TN and TSS concentrations for fresh and digested faecal sludge and high, medium, and low strength (Dangol, 2013; Hooijmans *et al.*, 2013)

Sludge type	Strength	COD (mg/L)	Total N (mg/L)	TSS (mg/L)
Fresh	High	250,000	5,000	100,000
	Medium	65,000	3,400	53,000
	Low	10,000	2,000	7,000
Digested	High	90,000	1,500	45,000
	Medium	45,000	400	25,000
	Low	3,000	200	1,500

9.2.3 Faecal sludge strength

FS can be classified as digested and fresh, and as high, medium and low strength, based on the COD and total nitrogen (TN) concentrations (Dangol, 2013; Hooijmans *et al.*, 2013). The values in Table 9.2 were defined by Dangol (2013) and Hooijmans *et al.* (2013) for modelling purposes based on values reported in the literature (Koné and Strauss, 2004; Heinss *et al.*, 1998; Elmitwalli *et al.*, 2006; Luostarinen *et al.*, 2007; Henze and Comeau, 2008; Halalsheh *et al.*, 2011; Ingallinella *et al.*, 2002).

Fractionations of WWTP influents have been carried out since the beginning of mathematical modeling of activated sludge systems, and examples can be readily found in the literature (Ekama *et al.*, 1986; Henze *et al.*, 1987). In contrast, literature reporting the fractionation of FS is not readily available. Examples found in literature are reported in Table 9.3. Interestingly, two different groups can be identified regardless of the strength, FS with higher fractions of biodegradable organics (up to 81% of the total COD), and FS with lower fractions of biodegradable organics (of around 43%). Consequently, the latter is more digested containing about 57% unbiodegradable organics.

Overall, the biodegradable organics in fresh FS can reach up to 82% of total COD (Table 9.3). The differences in biodegradable organics can be explained by the retention time of FS in the onsite sanitation system. Short retention times (e.g. days in public toilets) do not allow for significant stabilisation, whereas longer retention times (e.g. years in septic tanks) do. Elmitwalli *et al.* (2011), through mathematical simulations, estimated that after 90 days of accumulation in onsite systems the biodegradable fractions in fresh FS decreased from 0.81 to 0.25, whereas the unbiodegradable fractions increased from 0.19 to 0.75. This suggests the importance of matching treatment technologies to sludge types, e.g. biogas generation would be more suitable with sludge that is emptied frequently, or treated in situ. Interestingly, the COD fractionations of fresh and digested FS do not show considerable variations in spite of their strength and origin. Nevertheless, data is still limited and more studies are needed to be conclusive.

One study has reported N-fractionations of FS, as summarised in Table 9.4 (Dangol, 2013). N-fractionation of digested and fresh FS was estimated following a similar approach to Ekama (2008) for influent wastewater, and Lake (2010) for septic tank sludge. Based on the assumption that onsite systems partly function as anaerobic digesters (Montangero and Belevi, 2007), the biodegradation of organics leads to the production of fermentable organics and fermentation products (S_F and S_{VFA}, respectively) and to the release of inorganic nitrogen compounds (mostly NH_4^+ since a 6-8 pH range is usually observed) from the hydrolysis of organic nitrogen (Sötemann *et al.*, 2005). Thus, the biodegradable organic nitrogen fractions in FS can be included and therefore lumped on the free and saline ammonia (FSA) because they are eventually (and rapidly) hydrolysed. This assumption was based on the long retention times, and high solids and biomass concentrations found in onsite systems (Dangol, 2013).

Table 9.3 Faecal sludge COD fractionation

Origin	Total COD (mg/L)	XCb (slowly biodegradable)		Xu (particulate unbiodegradable)		Xa (acidogenic bacteria)		Sf (fermentable organic matter)		Svfa (volatile fatty acids)		Su (soluble unbiodegradable)		Sum of bio-degradable fractions	Sum of nonbio-degradable fractions
		(mgCOD/L)	Fraction	(mgCOD/L)	Fraction	(mgCOD/L)	Fraction	(mgCOD/L)	Fraction	(mgCOD/L)	Fraction	(mgCOD/L)	Fraction		
Fresh faecal sludge															
Vacuum toilet for black water (VBW)[1]	10,000	6,940	**0.69**	1,110	**0.11**	480	0.05	240	0.02	940	0.09	290	0.03	0.81	0.19
Vacuum toilet for faeces separation (VF)[1]	65,000	42,380	**0.65**	7,215	**0.11**	3,120	0.05	2,145	0.03	8,580	0.13	1560	0.02	0.82	0.18
Dry toilet (DT)[1]	45,000	31,230	**0.69**	4,990	**0.11**	2,160	0.05	1,080	0.02	4,230	0.09	1,310	0.03	0.81	0.19
Dry toilet for faeces with urine separation (DT)[1]	20,0000	130,400	0.65	22,200	0.11	9,600	0.05	6,600	0.03	26,400	0.13	4,800	0.02	0.82	0.18
Filter bag (FB)[1]	250,000	163,000	**0.65**	27,750	**0.11**	12,000	0.05	8,250	0.03	33,000	0.13	6,000	0.02	0.82	0.18
Bio-toilet mixed with saw dust[2]			**0.80**		**0.20**		-		-		-		-	0.80	0.20
Average fractions			**0.69**		**0.13**	**-**	**0.05**	**-**	**0.03**	**-**	**0.12**	**-**	**0.03**	**0.81 ± 0.01**	**0.19 ± 0.01**
Digested faecal sludge[3]															
High strength septic sludge[4]	90,000	34,118	**0.38**	53,882	**0.60**			1,176	0.01			824	0.01	0.39	0.61
Low strength septic sludge[4]	6,000	2,235	**0.37**	3,565	**0.59**			118	0.02			82	0.01	0.39	0.61
Septic sludge[5]	2,186	568	**0.26**	1,218	**0.56**			262	0.12			138	0.06	0.38	0.62
Septic tank sludge Jordan winter (18.4 °C)[6]	2,969	1,318	**0.44**	814	**0.27**			484	0.16			353	0.12	0.61	0.39
Septic tank sludge Jordan summer (21.9 °C)[6]	6,425	615	**0.10**	2,254	**0.35**			1,949	0.30			1,607	0.25	0.40	0.60
Average fractions			**0.31**		**0.47**				**0.13**				**0.09**	**0.43 ± 0.10**	**0.57 ± 0.10**

[1] Gaillard (2002); Elmitwalli *et al.* (2006); Luostarinen *et al.* (2007) [2] Lopez-Zavala *et al.* (2004)
[3] Biodegradable COD fractions estimated based on the STOWA protocol (Roeleveld and van Loosdrecht, 2002) [4] Henze *et al.* (2002) [5] Lake (2010) [6] Halalsheh *et al.* (2011)

Table 9.4 Nitrogen fractionation for digested (septic tank) and fresh faecal sludge (Dangol, 2013)

Fraction	Notation	Value	
		Digested faecal sludge	Fresh faecal sludge
Free and saline ammonia (FSA)	S_{NHx}	0.20	0.46
Soluble biodegradable	$S_{B,N}$	-	-
Particulate biodegradable	$XC_{B,N}$	-	-
Organic unbiodegradable particulate nitrogen	$X_{U,N}$	0.05	0.01
Organic unbiodegradable soluble nitrogen	$S_{U,N}$	0.75	0.53
Total nitrogen	Tot_N	1.00	1.00

9.3 CO-TREATMENT IN ACTIVATED SLUDGE WASTEWATER TREATMENT SYSTEMS

9.3.1 Influence on removal efficiencies and effluent quality

When co-treating FS in activated sludge WWTPs, the COD and TN concentrations in the reactor and effluent will increase proportionally to the FS strength and influent volumes. In addition, concentrations of soluble unbiodegradable COD and TN will reduce the treated effluent quality because they cannot be removed by either physico-chemical or biological processes. Thus, influent volumes of high- and medium-strength FS will need to be limited to comply with effluent standards. As shown in Figures 9.3 and 9.4, this is confirmed through mathematical modelling of an activated sludge treatment plant with an installed capacity of 100,000 p.e. (20,000 m^3/d) treating medium strength municipal wastewater and performing biological nitrogen removal (Henze *et al.*, 2008; Dangol, 2013). As observed, the influent COD and TN concentrations increase proportionally to the volumes of FS in the influent, reaching the highest concentrations with high strength fresh FS (Figure 9.3).

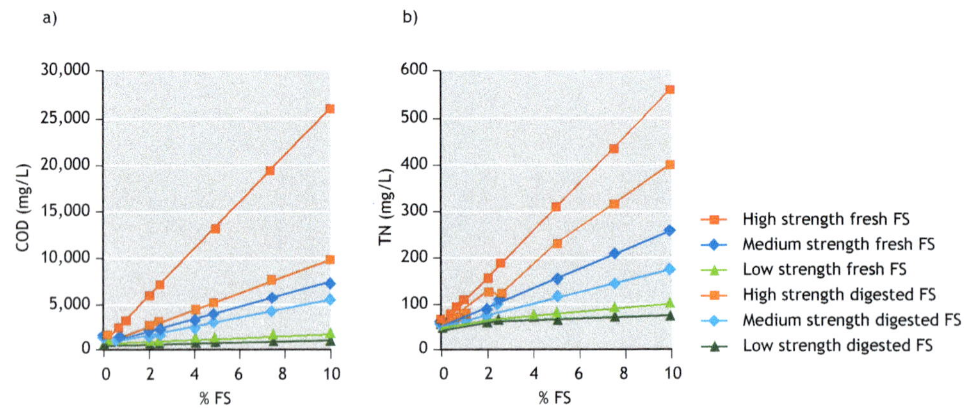

Figure 9.3 Effects of the combined discharge of municipal wastewater and faecal sludge (expressed as a percentage of the total influent discharged to the plant) on: (a) influent COD and (b) influent TN concentrations of an activated sludge wastewater treatment plant (Dangol, 2013).

It was also confirmed that the higher concentrations of soluble unbiodegradable fractions leads to higher effluent COD and TN concentrations (Figure 9.4). Thus, the soluble unbiodegradable COD and TN concentrations will set the first limit for the allowable FS volumes based on the compliance of certain effluent standards. For example, to meet the effluent requirements of <100-120 mgCOD/L and <10 mgTN/L, only 1.75-2.0% or 0.75-1.0% of the total influent could be comprised of medium- or high-strength FS, respectively. However, if plants do not have enough spare capacity (e.g. aeration, tank volumes, settling tanks and sludge handling facilities) the actual allowable volumes will probably be much lower due to the considerably higher loads discharged to the plants. For instance, 1% addition (equivalent to 200 m^3/d or 40 tankers of 5 m^3) of low strength digested FS (containing 3,000 mg COD/L as shown in Table 9.2) leads to a COD load increase of 600 kgCOD/d. This corresponds to an increase of 6,667 p.e. (assuming 1 p.e. = 90 gCOD/d), which may have a marginal effect on the 100,000 p.e. plant capacity. However, using the same approach, 1% discharge of medium- or high-strength fresh FS can equal the contribution of about 144,500 p.e. and 555,500 p.e., respectively, although this is at the upper limit of what would still allow for adequate plant operation.

Low-strength FS (e.g. from pit latrines with long residence times or infrequent emptying) does not have the same pronounced effects because of the lower concentrations of unbiodegradable COD and TN. However, assuming that there is enough spare capacity (e.g. aeration, tank volumes, settling tanks and sludge handling facilities), it will not meet the effluent requirements when it approaches 10% of the influent volume (corresponding to an increase of 66,667 p.e. and up to 222,220 p.e. for digested and fresh FS, respectively). This is similar to the recommendation of Still and Foxon (2012) of keeping the FS-to-influent wastewater ratio at no more than 1-10 to avoid a process failure at the plant.

9.3.2 Effects on oxygen demand

Aerobic treatment systems have limited aeration capacities. Co-treatment with FS can result in a severe increase in the oxygen demand due to the high concentrations of biodegradable COD and TN of FS. As observed in Figure 9.5, the effects of influent FS are so high that they can increase the relative oxygen demand (ΔFO_{TOT}) by 200%, with only 1% high-strength FS by volume in the influent, or 2% with medium-strength fresh FS. Prior to co-treatment with FS, the oxygen demand of the FS needs to be determined to evaluate whether the plant has enough aeration capacity to avoid process disruption.

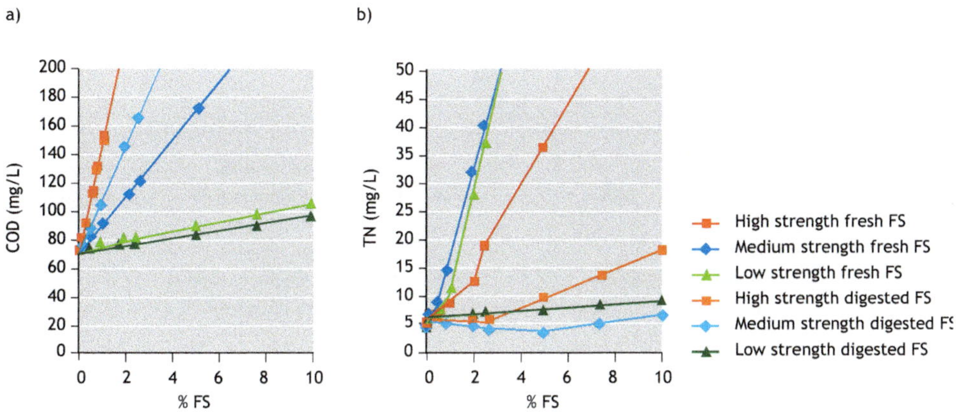

Figure 9.4 Effects of the combined discharge of municipal wastewater and faecal sludge (expressed as a percentage of the total influent discharged to the plant) on: (a) COD and (b) TN concentrations in the effluent of an activated sludge wastewater treatment plant.

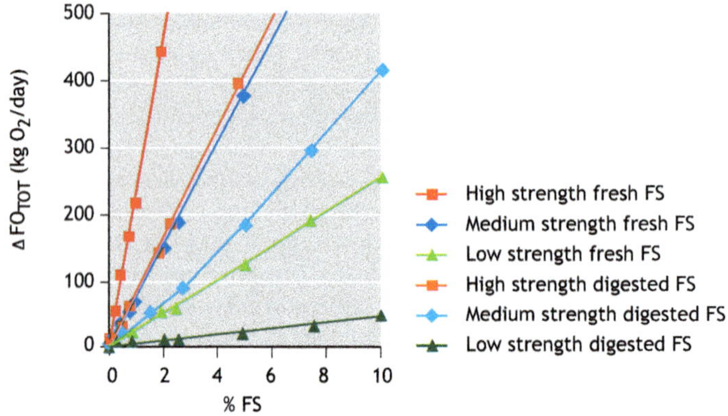

Figure 9.5 Relative increase in oxygen demand in an activated sludge wastewater treatment plant as a function of the combined discharge of municipal wastewater and different faecal sludge volumes (expressed as a percentage of the total influent discharged to the plant) (Dangol, 2013).

9.3.3 Impact on sludge generation

The accumulation of TSS is the limiting parameter for the co-treatment of FS. If the increase exceeds the maximum capacity, the plant can experience serious operational problems ranging from overloading of aeration and secondary settling tanks (with associated solid-liquid separation problems) to a considerable decrease in the oxygen transfer efficiency (which can lead to insufficient aeration and therefore to oxygen limiting conditions). As illustrated in Figure 9.6, at FS influent volumes as low as 0.5% for medium- and high-strength FS and of 2.5% for low-strength, the plant is overloaded and exceeds the maximum concentration of 5 kgTSS/m^3 recommended for the operation of aeration tanks (Metcalf and Eddy, 2003).

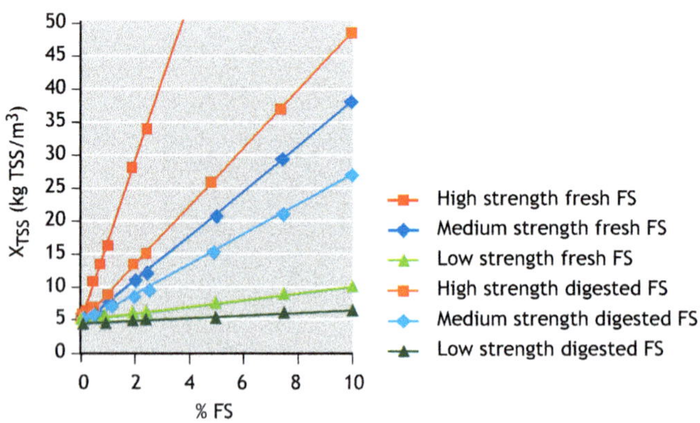

Figure 9.6 Increase in total suspended solids concentrations in the aeration tank of an activated sludge wastewater treatment plant expressed as a function of the combined discharge of municipal wastewater and of different volumes of faecal sludge (expressed as a percentage of the total influent discharged at the plant).

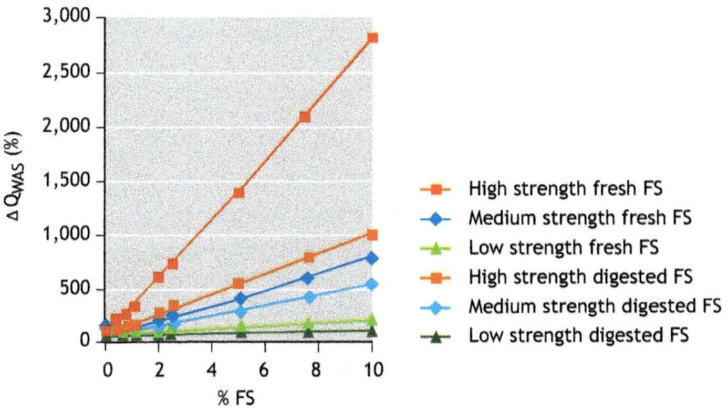

Figure 9.7 Increase in sludge production in an activated sludge wastewater treatment plant as a function of the combined discharge of municipal wastewater and of different volumes of faecal sludge (expressed as a percentage of the total influent discharged at the plant) (Dangol, 2013).

In addition, the increase in TSS and mixed liquor suspended solids (MLSS) concentrations will also result in increased volumes of waste sludge. There must be sufficient capacity in the sludge handling and disposal/enduse facilities of the plant to deal and cope with the higher sludge volumes generated, which frequently is not the case (Still and Foxon, 2012). For example, as shown in Figure 9.7, with a 100% increase in sludge production (ΔQ_{WAS}) the sludge handing facilities need to double their capacity for the co-treatment of 10% low-strength (by volume), 1% medium-strength, and 0.5% high-strength FS (Dangol, 2013).

9.3.4 Impact on aeration requirements

The increased accumulation of solids from co-treatment of FS can also lead to a reduction in the oxygen transfer efficiency. This will further increase the aeration requirements and reduce the aeration capacity of the plant. If the aeration capacity is exceeded, this will lead to oxygen limiting conditions, the creation of unaerated sections and serious operational problems. As shown in Figure 9.8, if the influent contains 2% high- or medium-strength FS by volume the demand on the aeration capacity will increase by 200%, and if it is 10% influent low-strength digested sludge this will increase by 100% (Dangol, 2013).

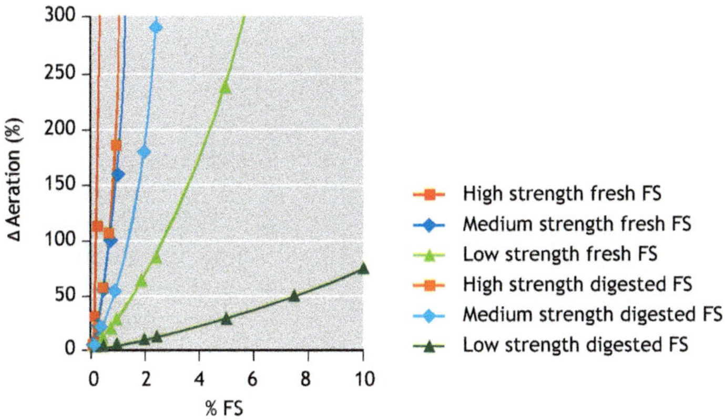

Figure 9.8 Estimation of the minimum increase in aeration requirements in an activated sludge wastewater treatment plant as a function of the combined discharge of municipal wastewater and of different volumes of faecal sludge (expressed as a percentage of the total influent discharged at the plant) (Dangol, 2013).

Potential detrimental effects caused by insufficient aeration supply include:

- Low dissolved oxygen (DO) concentrations in the aeration tank (< 1.5 mgO_2/L), or even oxygen depletion (0 mgO_2/L), resulting in incomplete oxidation of organics, a deterioration of effluent quality, high COD concentrations in the effluent, and leading to incomplete (at DO < 1 mg/L) or even cessation (at 0 mgO_2/L) of nitrification and, under extreme oxygen deficiency (for several hours), to the inactivation of bacteria.
- Creation of anaerobic pockets within the aerobic tanks resulting in the reduction of the net SRT of the aerobic system (SRT^{net}_{aer}). Such a reduction will be inversely proportional to the size of the anaerobic section(s). In particular, the SRT^{net}_{aer} drops below the minimum SRT of nitrifying organisms, this will result in the washout of nitrifying bacteria and cessation of nitrification.
- Proliferation of filamentous bacteria if the DO concentrations are below 1.5-2.0 mgO_2/L, to the detriment of desired heterotrophic and nitrifying bacteria (Martins *et al.*, 2004). Filamentous bacteria also lead to bulking sludge that does not settle well, and affects the biomass retention capacity in the secondary settling tanks. This results not only in a major increase in effluent TSS and VSS concentrations, and therefore reduced effluent quality, but also in several sludge loss from the system via the effluent. It could ultimately affect the whole operation of the treatment plant if the actual SRT drops below the minimum required values for biomass growth (Ekama, 2010).
- Partial nitrification of the high N load in FS could also result in accumulation of high nitrite concentrations (>100 $mgNO_2$-N/L) due to the oxygen limiting conditions, which would be toxic to desired heterotrophic and nitrifying bacteria. High nitrite concentrations can also have significantly negative impacts on the receiving water body where the plant effluent is discharged.

9.3.5 Impact on secondary settling tanks

Increased TSS resulting from co-treatment of FS can also overload secondary settling tanks (clarifiers). This results in problems with solids-liquid separation, solids being washed out in the effluent, and reduced biomass within the system making it difficult to maintain a stable SRT. As illustrated in Figure 9.9, the minimum surface area (A^{MIN}_{SST}) for settling tanks increases considerably with the addition of FS. 1-2% FS by volume of high- and medium-strength FS, either fresh or digested, can result in an increase in required area of more than 300% (Dangol, 2013). For low strength FS at 5 to 10%, the required area is 200% larger. Prior to co-treatment with FS, it is very important to evaluate the A^{MIN}_{SST} to determine if an adequate area is available, without assuming deterioration in sludge settleability (Ekama and Marais, 1986, 2004; Ekama *et al.*, 1997).

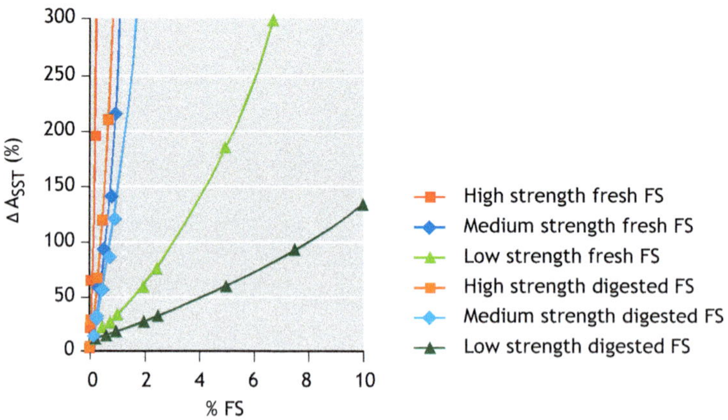

Figure 9.9 Estimation of the minimum area of the secondary settling tank required for an activated sludge wastewater treatment plant as a function of the combined discharge of municipal wastewater and of different volumes of faecal sludge (expressed as a percentage of the total influent discharged at the plant).

Table 9.5 Maximum faecal sludge volumes that can be co-treated under steady- and dynamic conditions in an activated sludge plant performing biological nitrogen removal without causing any process disruption or effluent deterioration (Dangol, 2013)

Faecal sludge type and strength	Under steady state conditions (%)	Under dynamic conditions (%)	Approximate ratio between maximum allowable faecal sludge volumes under steady-state to dynamic conditions
Digested FS			
Low-strength FS	3.75	0.64	6.0
Medium-strength FS	0.375	0.375	1.0
High-strength FS	0.25	0.25	1.0
Fresh FS			
Low-strength FS	0.375	0.125	3.0
Medium-strength FS	0.25	0.025	10.0
High-strength FS	0.125	0.025	5.0

9.3.6 Effects of the dynamic discharge of faecal sludge

Another complication when co-treating FS is the very dynamic nature of the influent FS. FS flow rates will tend to be much more dynamic than wastewater because they are not just dependent on diurnal patterns, they are also dependent on factors such as the working schedule of service providers, the customer demand for collection services, and the season. The result is heavy peak loads during the busiest times that can overload the treatment plant. Based on modelling, Dangol (2013) concluded that, under dynamic conditions, the maximum volumes that can be co-treated in an activated sludge plant without causing any process disruption or (effluent) deterioration sometimes need to be up to 10 times lower than those allowable under steady-state conditions (Table 9.5). Dangol (2013) conducted further modelling to evaluate the discharge of FS during off-peak hours (e.g. following a similar dynamic discharge pattern during the night) and the potential contribution of primary sedimentation tanks. The modelling showed no improvement in plant performance under dynamic conditions. This illustrates the importance of equalisation tanks to ensure a more even loading, and the need to distribute influent FS evenly through the entire day to approach steady-state conditions.

9.4 PRACTICAL CONSIDERATIONS FOR CO-TREATMENT OF FAECAL SLUDGE IN ACTIVATED SLUDGE SYSTEMS

Overall, the co-treatment of FS in activated sludge WWTPs can lead to severe operational problems at FS influent volumes as low as 0.025% of the total influent wastewater flow rate (which is equivalent to only one tanker of 5 m^3 per day). Thus, if the co-treatment of FS is to be employed, a very careful evaluation of the WWTP capacity needs to be made to determine which unit operation is the bottleneck of the plant (out of aeration, secondary settling tanks or sludge treatment) and how the plant is likely to fail. This will require a careful assessment and the implementation of defined measures to avoid any process disruption and deterioration of the plant. Considerations that need to be taken into account include:

- Required effluent standards. To estimate the minimum effluent COD and TN concentrations to verify the compliance with the required effluent standards.

- Maximum TSS concentrations in aeration tanks. To calculate the maximum expected TSS to evaluate if the aeration tanks will be overloaded.
- Maximum sludge production. To evaluate if the sludge handling and disposal facilities have the capacity to deal with the increase in sludge waste generation.
- Maximum installed aeration capacity. To estimate the aeration requirements based on the increase in oxygen demand and decrease in oxygen transfer efficiency. For existing plants, the DO concentration needs to be carefully monitored to maintain a concentration of at least 2 mgO_2/L.
- Secondary settling tanks. To determine the minimum surface area required for the operation of the settling tanks for the observed sludge settleability (in terms of the sludge volume index -SVI- or any other similar parameter).
- Existence and performance of equalisation tanks. To allow an even discharge of FS to the sewage plant for the longest period possible (e.g. over 24 h).

For new WWTPs that expect to receive certain volumes of FS or that are *a priori* designed to co-treat FS, the previous aspects can be used and applied to adapt the design depending upon the discharge volumes, type and strength of the FS. However, the design will probably lead to larger tank volumes, larger settling tanks, and higher installed capacity for aeration and sludge handling, treatment and disposal. For instance, compared to municipal wastewater treatment alone, for 1% FS co-treatment (regardless of the strength), the tank volumes will need to be 300% larger, the aeration capacity at least 200% higher, the secondary settling tanks 5 times larger and the sludge facilities 4 times larger. These aspects will undoubtedly considerably increase the capital and operational costs of the plant, along with the operational capacity. These considerations should be weighed carefully alongside other less expensive and more robust options presented in this book.

Case Study 9.1: Co-treatment in activated sludge wastewater treatment plants in eThekwini, South Africa

(Adapted from Still and Foxon, 2012 and Wilson and Harrison, 2012)

In spite of the apparent relatively low volumes of FS from pit latrines, two activated sludge WWTPs located in eThekwini, South Africa experienced serious operational problems caused by the high loads of organics, nitrogen compounds and suspended solids (Wilson and Harrison, 2012). A complete inactivation of the nitrification process was observed in one of the plants, which took several months to recover (Still and Foxon, 2012). A hypothesis suggests that the excessive nitrogen load discharged into the plant was the main reason (Still and Foxon, 2012). Although the causes of the problems are unclear, it cannot be discounted that the aeration capacity was exceeded as a consequence of the high loads discharged, resulting in the cessation of the nitrification process as discussed in this chapter. At the other plant under study, the high solids overloading made it practically impossible to remove the excess sludge generated as it was equal to the sludge volume produced in a month. Sludge removal was limited by the number of truckloads that could be removed, increasing associated operational costs and even the willingness of the receiving landfill to accept the material (Still and Foxon, 2012). The (digested) FS from the pits rapidly accumulated in the system and, because mixed sludge waste could not be removed at the required rate, it was retained for an extended period affecting the operation of the plant. This phenomenon resembles the excess sludge increase displayed in Figure 9.7. As Still and Foxon (2012) point out, it was clearly a case of taking one solids problem and turning it into another solids disposal problem, indicating that co-treatment in an activated sludge wastewater treatment plant can rarely be sustainable or successful.

Case Study 9.2: Co-treatment of septic tank sludge in an activated sludge wastewater treatment plant on Saint Marten, Netherlands Antilles

(Adapted from Lake, 2010 and Lopez-Vazquez, 2008)

Although the high concentrations of solids, organics and nitrogen compounds in FS attract most attention, the higher concentrations of unbiodegradable compounds and low biodegradability of organics can also hinder compliance with the effluent limits. On the island of Saint Marten, a popular tourist destination in the Caribbean, there was around 10% sewerage coverage until 2010 (Lake, 2010). Wastewater and septic tank sludge (brought by tankers to the plant) (Figure 9.11) were discharged into the existing Illidge Road WWTP, located in the Cul-de-Sac district. The plant consisted of an Imhoff tank with a volume of 154 m^3 as well as a buffer tank, secondary settling tank and sludge drying beds. The plant capacity was considerably exceeded by the wastewater flow rate (of at least 65 m^3/h) and the high FS volumes that in a typical working day accounted for an equivalent of about 175 m^3/day (Lopez-Vazquez, 2008). Since the plant was obsolete, a Modified Bardenpho (A2O) process design was proposed to achieve strict discharge standards for COD, N, P and TSS (of 125, 10, 2 and 20 mg/L, respectively). Based on local space-planning development plans, different scenarios were evaluated through mathematical modelling where the effects of the expansion of the sewer network (from 10 to 85% coverage) and population growth were taken into account (Lake, 2010). This approach helped to assess their effects on wastewater composition and WWTP performance through an estimated life span of 25 years (Lake *et al.*, 2011). Due to the loads of unbiodegradadable particulate organic matter and unbiodegradable soluble organic nitrogen from the digested FS, the study suggested that the proposed plant would only be able to comply with most of the discharge limits when the FS volumes comprise of no more than 2.8% of the influent (Lake *et al.*, 2011). However, as a consequence of the high nitrogen load and slow biodegradability of biodegradable organics (highlighted in Table 9.1), the study speculates that the nitrogen limits will probably not be met at the new plant (Lake, 2010).

Figure 9.10 Faecal sludge discharge at Illidge Road wastewater treatment plant at the buffer tank.

Case Study 9.3: Co-treatment impact on the Albireh wastewater treatment plant, Palestine

(Adapted from Al-Sa'ed and Hithnawi, 2006)

Following a similar approach as on Saint Marten, an assessment using mathematical modelling was carried out at the Albireh WWTP, located in the West Bank, Palestine. The purpose was to find an explanation for the occurrence of filamentous bulking sludge and the high effluent COD and TN concentrations that did not allow the corresponding discharge limits of 90 mgCOD/L and 18 mgTN/L to be met (Al-Sa'ed and Hithnawi, 2006). Like other plants in the region, since 2000 Albireh WWTP has been co-treating septic tank sludge from some of the 35% households not connected to the sewerage network. The modelling study indicated that, when the volumes of low-strength digested FS reached 6.6% of the total influent, the plant capacity was exceeded, requiring about 50% larger tank volumes, 50% higher oxygen requirements and generating a similar percentage of excess sludge (Al-Sa'ed and Hithnawi, 2006). The higher oxygen requirements and solids overloading might have favoured the proliferation of filamentous bacteria due to difficulties in keeping adequate aerobic conditions.

Case Study 9.4: Co-treatment of FS in Manila, Philippines

(Adapted from Robbins *et al.*, 2012)

In spite of the unsatisfactory experiences with FS co-treatment in aerobic treatment plants, activated sludge systems have recently been chosen in the Philippines as the main biological treatment process for FS treatment. Manila Water's FS operations with septic tank sludge currently utilise a FS treatment with activated sludge in the Manila South septage treatment plant (Robbins *et al.*, 2012). The plant is able to treat up to 814 m^3 per day of FS. Currently, the plant handles about 40-50% of its maximum capacity, indicating that there is still room for growth. In addition, the septage management system of the Baliwag water district has decided to build a septage treatment plant that utilises a sequencing batch reactor (a variant of the activated sludge process) as a secondary treatment process (http://watsanexp.ning.com). The full operation of the plant is scheduled by the second half of 2013. This project intends to serve as a model for water district-led septage management in the Philippines. These experiences indicate that co-treatment of FS in aerobic biological systems can be feasible and satisfactory if the design is adequate to cope with the FS influent, there is adequate operator capacity and competence, and an appropriate management scheme is implemented.

9.5 ANAEROBIC CO-TREATMENT OF FAECAL SLUDGE

The co-treatment of FS and wastewater in anaerobic processes is an alternative for sludge stabilisation, volume reduction and increased dewaterability. Possibilities include upflow anaerobic sludge blanket reactors (UASB), anaerobic digesters and anaerobic ponds. Anaerobic treatment can offset treatment costs through the production of biogas, which can be used for heating or for the generation of electricity. Pathogen reduction can also be achieved with thermophilic digestion (Metcalf and Eddy, 2003).

The characteristics of FS need to be carefully considered, as fresh or less stabilised FS will have higher concentrations of biodegradable organics but possibly also of inhibiting compounds (as discussed below). Although the biogas production and utilisation is an attractive benefit, there are currently limited applications and technologies. Therefore, further research is needed for the development of anaerobic systems for the co-treatment of high strength FS (Strauss *et al.*, 2006). FS from septic tanks (digested FS) may not be appropriate for anaerobic co-treatment, depending on the level of stabilisation it has undergone. In this case, the low concentrations of biodegradable organics in digested FS will lead to low biogas production but high solids accumulation resulting in significant operational costs with limited benefits (Still and Foxon, 2012).

9.5.1 COD overloading

As explained in Chapter 3, anaerobic digestion relies on complex interactions and dependencies among diverse bacterial groups, which makes the process susceptible to variations in influent loading rates. This is particularly important when managing FS, which by nature is highly variable. Anaerobic degradation has four stages: hydrolysis, acidogenesis, acetogenesis and methanogenesis (both acetoclastic and hydrogenotrophic). The growth rate of the fermentative bacteria that carry out acidogenesis is 10 to 20 times higher than methanogens, and their process rates five times faster (van Lier, 2008). If reactors are overloaded, the faster rate of acidogenesis will result in an accumulation of acids, as the methanogenic bacteria are not able to utilise them as fast. Depending upon the buffer capacity of the system (which depends on the nitrogen content of the organics since the hydrolysis of organic nitrogen results in an alkalinity increase), this can lead to a significant drop in pH, which inhibits the growth of methanogens, and thereby results in an even greater accumulation of acids (van Lier, 2008). This results in digester failure, and is referred to as 'souring'. In this regard, Moosbrugger *et al.* (1993) developed a simple 5-point titration method to measure both the VFA and alkalinity for anaerobic digestion control and early detection of instability to avoid 'souring'.

Anaerobic treatment processes are disrupted by overloading of COD, ammonia inhibition, pH variations, and sulfide inhibition. Therefore, these factors need to be carefully monitored, and controlled, to ensure proper operation of co-treatment of FS in anaerobic treatment systems. Each of these factors is explained below, and also how they affect appropriate FS loading rates.

UASB

To prevent overloading, the maximum COD or VSS design loading rates must not be exceeded, and reactors must have consistent and uniform feeding (Metcalf and Eddy, 2003). Figure 9.11 presents the effects of loading different FS volumes and types as a percentage of the total influent on an UASB system designed for 100,000 p.e. and operated at 25°C. The design values that were used for the UASB were medium strength municipal wastewater as described by Henze and Comeau (2008) and used by Dangol (2013). The organic loading rate (OLR) was 3 kgCOD/m^3/d and the upflow velocity 0.83 m/h. The maximum OLR for UASB systems treating wastewater with high concentrations of particulate biodegradable organics is around 6 kgCOD/m^3/d (van Lier, 2008), which suggests that, in principle, the 100,000 p.e. plant used in this study has enough spare capacity. As illustrated in Figure 9.11, the UASB reactor can handle feedings of up to 7.5% by volume of low strength fresh FS (1,500 m^3/d equivalent to the organic load of up to 180,000 p.e.), but only 0.25% high strength fresh FS due to the high COD content (10 tankers of 5m^3 per day but with an organic load equivalent to approximately 139,000 p.e.). This means that the 100,000 p.e. UASB system, as well as other UASB plants of different capacities, could handle low strength FS but are prone to overloading with high strength FS.

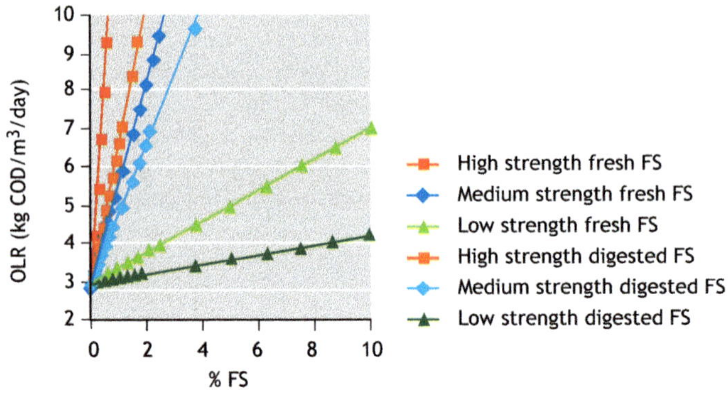

Figure 9.11 Effects of faecal sludge discharge (expressed as a percentage of the total influent discharged to the plant) on the organic loading rates of a UASB reactor designed for an average and a maximum oxygen loading rate of 3 kgCOD/m3/day and 6 kgCOD/m3/day, respectively.

Anaerobic digesters

Figure 9.12 illustrates the effect of the co-treatment of different FS types at different solids loading rates (SLR) as a percentage by volume of the total influent on an anaerobic digester. The anaerobic digester was designed to treat sludge from an activated sludge treatment plant under mesophilic conditions (35°C), with a SRT of 10 d, and a total volume of 13,750 m^3. As shown, the SRT decreases proportionally to the amount of FS being fed. Although the maximum recommended value for SLR is 4.8 $kgVSS/m^3/d$ (Metcalf and Eddy, 2003), this needs to be monitored carefully so that FS addition does not result in a drop in the SRT below the minimum recommended, causing the reactor to fail. For example, if the anaerobic digester feeding is 1% (138 m^3/d, equivalent to 28 collection and transport trucks with a 5 m^3 volume), but it contains 10% FS, an approximate 10% reduction in the operating SRT of the digester could occur (Figure 9.12).

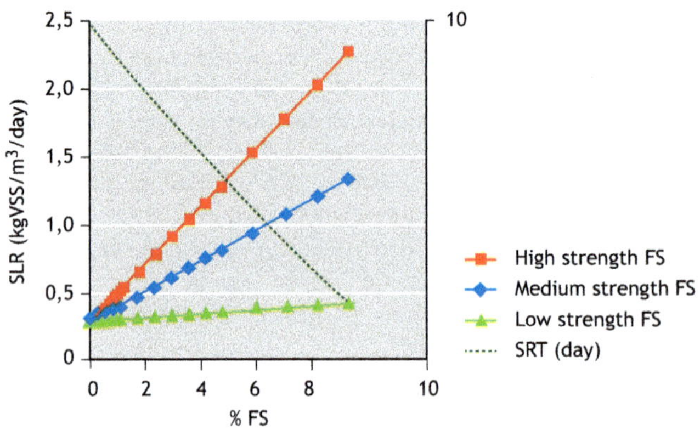

Figure 9.12 Effects of faecal sludge discharge (expressed as a percentage of the total influent discharged to the anaerobic digester) on the solid loading rates of an anaerobic digester of 13,750 m^3 designed with a sludge retention time of 10 days.

For anaerobic co-treatment in digesters, it is recommended that the feeding, including FS, is always lower than one twentieth of the digester volume (ATV, 1985). This approach would mean a maximum 5% FS loading, regardless of its strength, to prevent overloading or significant reduction in the SRT. This ratio is also based on an SRT value of 20 d commonly used for the design of anaerobic digesters (Metcalf and Eddy, 2003), which increases the reliability of the recommended approach.

Ponds

Usually, anaerobic ponds can be regarded as low loaded anaerobic systems with operational loading rates of 0.025-0.5 kgCOD/m^3/d and depths of 4 m (van Lier, 2008). For FS applications, Fernandez *et al.* (2004) suggest the pre-treatment of FS in waste stabilisation ponds (WSP) operated at maximum loading rates of 0.6 kgBOD_5/m^3/d, particularly to reduce the generation of ammonia. However, these systems may have odour problems, fast sludge accumulation (0.010 to 0.020 m^3 sludge accumulated/m^3 FS) and therefore require frequent sludge removal (Heinss *et al.*, 1998; Fernandez *et al.*, 2004). Furthermore, the loss of methane into the atmosphere, which has a 21 times higher greenhouse impact than CO_2, is an environmental impact that needs to be considered if it is not captured (van Lier, 2008). Moreover, the effluents require further polishing prior to discharge into the environment, and tend to contain high ammonia concentrations that can affect post-treatment processes in pond systems or even within the same anaerobic ponds (Strauss *et al.*, 2000). Thus, the application of anaerobic ponds for FS treatment needs to be carefully evaluated, particularly when dealing with high strength FS. This is also covered in Chapter 5.

9.5.2 Ammonia inhibition

The anaerobic co-treatment of FS can be inhibited by the high concentrations of ammonia present in FS (Still and Foxon, 2012). Among the bacteria in anaerobic reactors, methanogenic bacteria are the most sensitive to ammonia inhibition (Chen *et al.*, 2008). Inhibition of the methanogens results in lower biogas yields in spite of the availability of soluble biodegradable organics (Angelidaki *et al.*, 1993; Chaggu, 2004). Reported values for inhibition of methanogens is quite variable, and 50% reduction in activity has been observed at total ammonia concentrations between 1.7 to 14 g/L (Chen *et al.*, 2008). The broad range is due to the influence of different factors such as pH, carbon source, temperature and biomass acclimation and adaptation (Chaggu, 2004; Chaggu *et al.*, 2007; Chen *et al.*, 2008). In this regard, free ammonia (NH_3) and not ammonium (NH_4^+) has been suggested as the actual toxic agent at concentrations of 100-200 mg/L for unadapted methanogenic populations (Henze and Harremoes, 1983).

To prevent process disruptions and deterioration, Heinss and Strauss (1999) recommend limiting the volume of co-treated FS based on a total influent ammonia concentration of less than 2 g/L. However, Doku (2002) recommends limiting the maximum FS volume to reach an influent that contains less than 200 mg NH_3-N/L based on potential variations in pH (Henze and Harremoes, 1983).

Based on the total nitrogen concentrations expected in co-treatment of wastewater and fresh FS (Figure 9.3), the nitrogen concentrations would probably be higher than 200 mg/L, indicating that their volumes need to be limited to no more than 2, 5 and 8% for high-, medium- and low-strength FS, respectively.

9.5.3 pH variations

In anaerobic systems, the pH needs to be carefully monitored and kept between 7.0 and 7.5 (Chen *et al.*, 2008). The alkalinity and buffer capacity of the anaerobic systems need to be monitored to ensure that the pH remains stable (Metcalf and Eddy, 2003). pH higher than 7.5-8.0 can lead to an accumulation of free ammonia, and extreme pH levels (e.g. higher than 10.0) can fully inhibit the anaerobic biological degradation process (Chaggu, 2004; Chen *et al.*, 2008). pH values lower than 7.0 can reduce the methanogenic activity. A pH of 7.0-7.5 helps to maximise the biomass activity and reduce the potential inhibiting and toxicity effects of parameters such as ammonia and volatile fatty acids (Chen *et al.*, 2008).

Thus, to monitor, and if possible adjust, the alkalinity levels and buffer capacity of the system can help to reduce pH fluctuations and maintain an adequate pH range. However, certain practices, such as gradual feeding and the controlled addition of external compounds (including charcoal ashes to enhance pathogen removal and nutrient recovery), need to be carefully performed (Chaggu, 2004; Metcalf and Eddy, 2003). Otherwise, they may lead to pH decreases due to VFA accumulation (when overloading) or to extremely high pH levels (when overdosing alkaline or basic compounds) (Chaggu, 2004; van Lier, 2008).

9.5.4 Sulphide inhibition

Sulphide gas (H_2S) is generated during the anaerobic digestion of sludge that is rich in proteins, and due to (saline) groundwater intrusion or infiltration into the onsite sanitation system (Metcalf and Eddy, 2003; Lopez-Vazquez *et al.*, 2009). Sulphide is toxic to all living organisms and can easily affect the anaerobic digestion processes. Methanogenic bacteria are rather sensitive to sulphide leading to lower methane production, low quality biogas, bad smell, corrosion problems and high COD effluent concentrations (van Lier, 2008).

50% methanogenic activity has been observed at sulphide concentrations between 50-250 mgS/L, but H_2S is usually present in the gas phase (Metcalf and Eddy, 2003). Because the pK_{S1} value for H_2S to HS- dissociation is around 7.0, the pH should be maintained above 7.0 to keep the H_2S concentration low. Although relatively low sulphate concentrations can be expected due to the low FS volumes co-treated, the potential sulphide generation cannot be ignored since the anaerobic process may be prone to disruption at sulphide concentrations as low as 50 mgS/L depending upon other operational conditions (e.g. pH). However, data regarding the sulphate concentrations contained in FS is rare and therefore they need to be studied to assess their potential influence on the anaerobic processes when co-treating FS.

9.6 PRACTICAL CONSIDERATIONS FOR CO-TREATMENT OF FAECAL SLUDGE IN ANAEROBIC SYSTEMS

For any anaerobic treatment process, probably the most important operational aspect is the feeding. It needs to be supplied gradually and if possible continuously to avoid overloading and shocks (Heinss and Strauss, 1999; Metcalf and Eddy, 2003; van Lier, 2008).

For FS co-treatment in UASB reactors, the maximum OLR of design (including both wastewater and FS) must not be exceeded in order to avoid the overloading of the system. In particular, high strength FS needs to be carefully handled since the high organic content can easily overload the system. In this study, 0.25% high strength fresh FS (approximately 10 tankers of 5 m^3 per day) had an organic load equivalent to around 139,000 p.e. that led to the overloading of a 100,000 p.e. UASB plant.

Anaerobic digesters appear to be more robust for the anaerobic co-treatment of FS. Permissible loading rates for mesophilic digesters (operated at 35°C) depend on the operational conditions but can reach up to 1.6-2.0 kgVSS/m^3/d (Heinss and Strauss, 1999; Metcalf and Eddy, 2003). Also, the feeding, including FS, needs to be limited to the maximum daily feed rate of design which depends on the applied SRT. Thermophilic anaerobic digesters (49-52°C) are an alternative that can lead to faster hydrolysis rates (the rate limiting step in anaerobic digestion of wastewater and FS) resulting in higher biogas yields (Angelidaki *et al.*, 1993). However, they are susceptible to small temperature variations and also operating and maintenance costs are higher compared to mesophilic digesters, which make them unattractive for low-income countries (Heinss and Strauss, 1999).

Ponds appear to be cost-effective technologies for FS co-treatment when operated as low loaded systems (0.6 $kgBOD_5/m^3/d$). However, their implementation needs to be carefully evaluated because the initial investment and operational costs could be high since they have substantial land requirements and high operational costs as a consequence of the frequent desludging needed. Moreover, they can involve important environmental issues if methane is lost into the atmosphere.

Case Study 9.5: Treatment of faecal sludge in Dar es Salaam, Tanzania

(Adapted from Chaggu, 2004)

The detrimental effects of high ammonia concentrations and high pH levels need to be avoided to ensure a satisfactory performance of anaerobic digestion systems. Chaggu (2004) carried out a literature research on excreta handling in Dar es Salaam, Tanzania. He found that 50% of the filling up of pits in Dar es Salaam City was the result of a high water table and that almost 16,131 kgCOD/day from pit-latrines reached the groundwater sources. As such, he proposed to use a 3,000 L plastic tank as an experimental improved pit-latrine without urine separation for a 10-person household in Mlalakuwa settlement in Dar es Salaam City. The influent to the reactor consisted of urine and faeces in a 1.3:1 ratio. The results obtained revealed that, after 380 days of use as a daily pit-latrine, the reactor content was not yet stabilised, and 8,000 mg/L dissolved COD (but only 100 mgCOD/L as volatile fatty acids) were still present. Part of this dissolved COD was biodegradable, indicating the need for further stabilisation of the reactor content. The slow conversion of dissolved COD was assumed to be related to the inoculation of anaerobic sludge not adapted to the high ammonia concentration of 3,000 mg N/L. In the same research project, a short survey revealed that high pH values occur (up to pH 10.4) in Ecosan toilets due to addition of charcoal ashes to enhance the reduction of 'E-Coli and Ascaris eggs', but the high pH levels inhibited the anaerobic biological degradation of FS.

Case Study 9.6: Co-treatment of septage in a lab-scale UASB reactor in Ghana

(Adapted from Doku, 2002)

Although full-scale experiences are limited, Doku (2002) concluded that it is feasible to treat FS in a laboratory scale UASB reactor in Ghana provided (i) the sludge is diluted appropriately to avoid reaching high concentrations of inhibitory compounds (e.g. ammonia) and (ii) the FS is gradually and continuously fed into the reactor. Doku (2002) executed the experiments using a UASB reactor with a working volume of 50 L, operated at a mean HRT of around 12 h, and at ambient temperatures between 23.0 and 31.2 °C. The OLR was between 12.5 to 21.5 kg $COD/m^3/d$, with a relatively low upflow velocity of 0.14 m/h. The FS was diluted to a 1:6 ratio, resulting in an average total nitrogen concentration of 300 ± 50 mg/L. The average removal efficiencies were: 71% for COD, 61% for total solids, 74% for total volatile solids TVS and 73% for TSS. The calculated volume of methane in the biogas collected ranged from 4-8 L/kgCOD, not accounting for practical losses. Overall, the removal efficiencies were comparable to those obtained for a UASB reactor treating domestic sewage. However, the effluent COD concentration is too high for direct discharge and hence a form of post-treatment would be necessary. Nevertheless, full-scale studies are needed to validate the observations from this research.

9.7 CONCLUSIONS

The discharge of FS for its co-treatment in WWTPs can lead to severe operational problems when even low volumes of high-strength fresh FS are discharged (e.g. 0.25% of the total influent). This is mainly caused by the relatively higher strength of FS compared to that of municipal wastewater, which can easily lead to higher loads exceeding the plant capacity. The most common problems are the overloading of solids, COD or nitrogen compounds. They can lead to serious operational problems ranging from incomplete removal of organics to cessation of nitrification, which can take several weeks to recover. Also, the excessive solids accumulation may lead to unexpectedly highly sludge generation that can compromise the operation of the plant and increase the operational costs. Moreover, aerobic treatment systems may experience a lack of aeration capacity and severe overloading of secondary settling tanks leading to solids loss. Meanwhile, anaerobic systems are prone to inhibition by the presence of inhibitory compounds such as ammonia and pH variations. In addition, the high concentrations of soluble unbiodegradable organics and nitrogen compounds can have a serious effect on the treated effluent quality, which may hinder compliance with the required effluent standards.

If in spite of the apparent limited benefits, FS co-treatment is to be practiced in municipal WWTPs, the allowable FS volumes will probably need to be restricted to low volumes so that WWTPs do not get overloaded with total suspended solids, high COD and nitrogen loadings or high concentrations of toxic or inhibitory compounds. Moreover, FS loadings need to be added gradually and as slowly as possible to avoid overloads and shocks.

All the previous aspects need to be carefully addressed but, overall, the benefits do not seem to be attractive enough to support the co-treatment of FS with wastewater in municipal WWTPs, particularly when dealing with digested FS from septic tanks which contains low concentrations of biodegradable compounds but high concentrations of solids that will tend to overload the treatment systems. It is possible that anaerobic co-treatment of fresh FS offers certain opportunities when considering the potential recovery of resources, but further research is still needed for the development of reliable and cost-effective technologies.

9.8 BIBLIOGRAPHY

Al-Sa'ed, R.M.Y., Hithnawi, T.M. (2006). Domestic Septage Characteristics and Co-Treatment Impacts on Albireh Wastewater Treatment Plant Efficiency. Dirasat: Engineering Sciences, 33(2), p.187-198.

Andreadakis, A.D. (1992). Co-treatment of septage and domestic sewage for the greater Athens area. Water Science and Technology, 25(4-5), p.119-126.

Angelidaki, I., Ellegard, L., Ahring, B.K. (1993). A mathematical model for dynamic simulation of anaerobic digestion of complex substrates: focusing on ammonia inhibition. Biotechnology and Bioengineering 42, p.159-166.

ATV (1985). Treatment and Disposal of Sludge from Small Sewage Treatment Plants. Gesellschaft zur Förderung der Abwassertechnik e.V. (GFA), St. Augustin.

Chaggu, E.J. (2004). Sustainable environmental protection using modified pit-latrines. PhD, Wageningen University, The Netherlands.

Chaggu, E.J., Sanders, W., Lettinga, G. (2007). Demonstration of anaerobic stabilization of black water in accumulation systems under tropical conditions. Bioresource technology, 98(16), p.3090-3097.

Chen, Y., Cheng, J.J., Creamer, K.S. (2008). Inhibition of anaerobic digestion process: a review. Bioresource Technology 99, p.4044-4064.

Corominas, L., Rieger, L., Takács, I., Ekama, G., Hauduc, H., Vanrolleghem, P.A., Oehmen, A., Gernaey, K.V., van Loosdrecht, M.C.M., Comeau, Y. (2010). New framework for standardized notation in wastewater treatment modelling. Water Science Technology 61(4), p.841-857.

Dangol, B. (2013). Faecal sludge characterization and co-treatment with municipal wastewater: process and modeling considerations. UNESCO-IHE Institute for Water Education. Delft, The Netherlands.

Doku, I.A. (2002). Anaerobic treatment of nightsoil and toilet sludge from on-site sanitation systems in Ghana. PhD thesis, University of Leeds. U. K.

Dold, P.L., Ekama, G.A., Marais, G. (1980). A general model for the activated sludge process. Progress in Water Technology 12(6), p.47-77.

Ekama, G. (2010). The role and control of sludge age in biological nutrient removal activated sludge systems. Water Science and Technology 61(7), p.1645–1652.

Ekama G.A., Barnard, J.L., Günthert, F.W., Krebs, P., McCorquodale, J.A., Parker, D.S., Wahlberg E.J. (1997). Secondary Settling Tanks: Theory, Modeling, Design and Operation. IAWQ Scientific and Technical Reports #6, IAWQ London.

Ekama G.A. (2008). Organic matter removal. In: Biological Wastewater Treatment: Principles, Modelling and Design. Henze, M, van Loosdrecht, M, C, M., Ekama, G, A., Brdjanovic, D. eds. ISBN: 9781843391883. IWA Publishing. London, UK.

Ekama, G.A., Dold, P., Marais, G.v.R. (1986). Procedures for Determining Influent COD Fractions and the Maximum Specific Growth Rate of Heterotrophs in Activated Sludge Systems. Water Science and Technology 18(6), p.91-114.

Ekama, G.A., Marais, G.v.R. (2004). Assessing the applicability of the 1-D flux theory to full-scale secondary settling tank design with a 2D hydrodynamic model. Water research 38(3), p.495-506.

Ekama, G.A., Marais, G.v.R. (1986). Sludge settleability and secondary settling tank design procedures. Water Pollution Control 85(1), p.101-113.

Elmitwalli, T., Leeuwen, M.V., Kujawa-Roeleveld, K., Sanders, W., Zeeman, G. (2006). Anaerobic biodegradability and digestion in accumulation systems for concentrated black water and kitchen organic-wastes. Water Science & Technology 53(8), p.167-175.

Elmitwalli, T., Zeeman, G., Otterpohl, R. (2011). Modelling anaerobic digestion of concentrated black water and faecal matter in accumulation system. Water Science and Technology 63(9), p.2039-2045.

Fernandez, R.G., Ingallinella, A.M., Sanguinetti, G.S., Ballan, G.E., Bortolotti, V., Montangero, A., Strauss, M. (2004). Septage treatment using WSP. In: Proceedings of the 9th International IWA Specilaist Group Conference on Wetland Systems for Water Pollution Control and to the 6th International IWA Specialist Group Conference on Waste Stabilization Ponds. September 27th - October 1st, 2004. Avignon, France.

Gaillard, A. (2002). Waste(water) characterisation and estimation of digestion kinetics. MSc. Thesis, Wageningen University, The Netherlands.

Halalsheh, M.M., Noaimat, H., Yazajeen, H., Cuello, J., Freitas, B., Fayyad, M. (2011). Biodegradation and seasonal variations in septage characteristics. Environmental Monitoring Assessment 172(1-4), p.419-26.

Harrison, E.Z., Moffe, M. (2003). Septage quality and its effects on field life for land applications. JAWRA Journal of the American Water Resources Association 39(1), p.87-97.

Heinss, U., Larmie, S.A., Strauss, M. (1998). Solids separation and pond systems for the treatment of septage and public toilet sludges in tropical climate - lessons learnt and recommendations for preliminary design. EAWAG/SANDEC, Report No. 05/98.

Heinss, U., Strauss, M. (1999). Co-treatment of faecal sludge and wastewater in tropical climates. SOS - Management of sludges from on-site sanitation. EAWAG/SANDEC.

Henze, M., Comeau, Y. (2008). Wastewater characterization. In: Biological wastewater treatment: principles, modelling and design. Henze, M, van Loosdrecht, M.C.M., Ekama, G.A., Brdjanovic, D. eds. ISBN: 9781843391883. IWA Publishing. London, UK.

Henze, M., Grady, C.P.L. Jr., Gujer, W., Marais, G.v.R., Matsuo, T. (1987). Activated Sludge Model No. 1. IAWQ Scientific and Technical Report No. 1, London, UK.

Henze, M., Harremoës, P., la Cour Jansen, J., Arvin E. (2002). Wastewater treatment: Biological and chemical processes, 3rd ed., Springer-Verlag, Berlin.

Henze, M., Harremoes, P. (1983). Anaerobic treatment of wastewater in fixed film reactors—a literature review. Water Science and Technology 15(8-9), p.1-101.

Hooijmans, C.M., Dangol, B., Lopez-Vazquez, C.M., Ronteltap, M., Brdjanovic, D. (2013). Assessing the feasibility of faecal sludge co-treatment in sewage treatment plants - A practical guide. 3rd IWA Development Congress. October 14th-17th, 2013. Nairobi, Kenya.

Ingallinella, A.M., Sanguinetti, G., Koottatep, T., Montangero, A., Strauss, M. (2002). The challenge of faecal sludge management in urban areas- strategies, regulations and treatment options. Water Science and Technology, 46(10), p.285-294.

Koné, D., Strauss, M. (2004). Low-cost options for treating faecal sludges (FS) in developing countries-challenges and performance. In: Liénard, A., Burnett, H., eds. Proceedings of the 6th International Conference on Waste Stabilization Pond and 9th International Conference on Wetland Systems, Avignon, France. p.213-219.

Lake, O.A. (2010). Integrated water quality modelling: A case study on St. Maarten. MSc thesis. UNESCO-IHE, Institute for Water Education. Delft, The Netherlands.

Lake, O., Lopez-Vazquez, C.M., Hooijmans, C.M., Brdjanovic, D. (2011). Steady-state models as cost-effective tools for design and assessment of wastewater systems in developing countries. 2nd IWA Development Congress, November 21-24th, 2011, Kuala Lumpur, Malaysia.

Lopez-Vazquez, C.M. (2008). St. Maarten Island - Illidge Road Wastewater treatment plant: Wastewater characterization Final Report The Netherlands. UNESCO-IHE, Institute for Water Education. Delft, The Netherlands.

Lopez-Vazquez, C.M., Hooijmans, C.M., Chen, G.H., van Loosdrecht, M.C.M., Brdjanovic, D. (2009). Use of saline water as secondary quality water in urban environments. 1st IWA Development Congress, November 15th-19th, Mexico City.

Lopez-Zavala, M.A., Funamizu, N., Takakuwa, T. (2004). Modeling of aerobic biodegradation of faeces using sawdust as a matrix. Water Research 38(5), p.1327–1339.

Luostarinen, S., Sanders, W., Kujawa-Roeleveld, K., Zeeman, G. (2007). Effect of temperature on anaerobic treatment of black water in UASB-septic tank systems. Bioresource technology, 98(5), 980-986.

Martins, A.M., Pagilla, K., Heijnen, J.J., van Loosdrecht, M. (2004) Filamentous bulking sludge-a critical review. Water Research, 38(4), 793-817.

Melcer, H. (2003). Methods for Wastewater Characterization in Activated Sludge Modeling. Water Environment Research Foundation. ISBN-10: 1843396629 | ISBN-13: 9781843396628.

Metcalf and Eddy (2003). Wastewater Engineering: treatment, disposal, reuse. Tchobanoglous, G., Burton, F.L. eds. McGraw-Hill Book Company.

Montangero, A., Belevi, H. (2007). Assessing nutrient flows in septic tanks by eliciting expert judgement: A promising method in the context of developing countries. Water Research 41(5), p.1052-1064.

Moosbrugger, R.E., Wentzel, M.C., Ekama, G.A., Marais, G.v.R. (1993). A 5 pH Point Titration Method for Determining the Carbonate and SCFA Weak Acid/Bases in Anaerobic Systems. Water Science and Technology 28(2), p.237–245.

Robbins, D., Strande, L., Doczi, J. (2012). Sludge management in developing countries: experiences from the Philippines. Water 21. December 2012.

Roeleveld, P.J., van Loosdrecht, M.C.M. (2002). Experience with guidelines for wastewater characterisation in The Netherlands. Water Science & Technology 45(6), p.77-87.

Still, D., Foxon, K. (2012). Tackling the challenges of full pit latrines. Vol. 1: Understanding sludge accumulation in VIPs and strategies for emptying full pits. Water Research Commission Report No. 1745/1/12. ISBN 978-1-4312-0291-1.

Strauss, M., Kone, D., Saywell, D. (2006). Proceeding of the 1st Int. Symposium and Workshop on Faecal Sludge Management (FSM) Policy. EAWAG/SANDEC IWA.

Strauss, M., Larmie, S.A., Heinss, U., Montangero, A. (2000). Treating faecal sludges in ponds. Water Science and Technology, 42(10), 283-290.

Sötemann, S.W., Ristow, N.E., Wentzel, M.C., Ekama, G.A. (2005). A steady state model for anaerobic digestion of sewage sludges. Water S.A. 31(4), p.511-527.

USEPA (1984). Handbook: Septage Treatment and Disposal. EPA 625/6-84-009, U.S. Environmental Protection Agency, Center for Environmental Research Information, Cincinnati, Ohio.

USEPA (1994). Guide to septage treatment and disposal. EPA/625/R-94/002, U.S. Environmental Protection Agency, Office of Research and Development. Washington, D.C.

van Lier, J.B. (2008). Anaerobic wastewater treatment. In: Biological Wastewater Treatment: Principles, Modelling and Design. Henze, M., van Loosdrecht, M.C.M., Ekama, G.A., Brdjanovic, D. eds. ISBN: 9781843391883. IWA Publishing. London, UK.

Wilson, D., Harrison, J. (2012). eThekwini pit latrine program emptying program - The contract, the pitfalls and solutions. International Faecal Management Conference. October 29-31st, 2012. Durban, South Africa.

End of Chapter Study Questions

1. What are the common technical problems likely to be experienced in the co-treatment of FS in wastewater treatment plants?

2. Why is it important to determine the oxygen demand of the FS prior to co-treatment?

3. Explain the reason why the accumulation of TSS is a limiting parameter for the co-treatment of FS.

Technology

Chapter 10

Enduse of Treatment Products

Ives Kengne, Berta Moya Diaz-Aquado and Linda Strande

Learning Objectives

- Understand general concerns with resource recovery and how to ensure adequate protection of human and environmental health.
- Know key considerations for determining appropriate methods and rates of the land application of sludge.
- Be able to determine appropriate uses and discharge options for liquid fractions in faecal sludge management.
- Understand the wide range of potential resource recovery opportunities from faecal sludge, and key criteria in selecting the most appropriate options.

10.1 INTRODUCTION

The previous chapters covered how stabilisation, drying, and pathogen reduction of faecal sludge (FS) can be achieved with different treatment technologies, and combinations of these various technologies. Each treatment technology results in endproducts which need to be further treated, disposed of, or harnessed for some type of resource recovery. Endproducts, for example dried or partially dried sludge, compost, leachate, and biogas, each have an intrinsic value, which can turn treatment from merely a method for environmental and public health protection to resource recovery and value creation. This chapter focuses on the endproducts produced from the various FS treatment processes, addresses potential difficulties and restrictions with their enduse, and discusses additional steps that can or should be applied to turn a treatment endproduct into a valuable asset.

Historically, the most common resource recovery from sludge has been as a soil conditioner and organic fertiliser, as excreta contain essential plant nutrients and organic matter that increases the water retaining capacity of soils. There are however several other treatment options that allow for resource recovery. For example, biogas can be produced during anaerobic digestion of FS, with the remaining sludge also being used as a soil conditioner. Novel developments are underway to recover endproducts as a biofuel, for example pyrolysis, gasification, incineration and co-combustion or as resource recovery of organic matter through the growth of Black Soldier flies for protein production.

Table 10.1 Summary of potential resource recovery options from faecal sludge

Produced Product	Treatment or Processing Technology
Soil conditioner	Untreated FS Sludge from drying beds Compost Pelletising process Digestate from anaerobic digestion Residual from Black Soldier fly
Reclaimed water	Untreated liquid FS Treatment plant effluent
Protein	Black Soldier fly process
Fodder and plants	Planted drying beds
Fish and plants	Stabilisation ponds or effluent for aquaculture
Building materials	Incorporation of dried sludge
Biofuels	Biogas from anaerobic digestion Incineration/co-combustion of dried sludge Pyrolysis of FS Biodiesel from FS

This chapter addresses resource recovery options from different FS technologies, both from a biological and an energy point of view, and presents established processes as well as promising innovations.

10.2 RESOURCE RECOVERY OPTIONS

There are a wide range of FS treatment technologies that can be combined in many different ways. All treatment processes result in endproducts which are either treated further, disposed of, or harnessed in some way for resource recovery. The potential use of endproducts should be considered from the initial design phase of any complete FS management (FSM) system, as the treatment technologies used are intrinsically linked to the quality of endproducts generated. A summary of resource recovery options covered in this chapter is provided in Table 10.1.

10.3 GENERAL CONCERNS

With the implementation of resource recovery, it is important to evaluate constituents that may impact both humans and the environment. These include the presence of pathogens and heavy metals. Social factors such as acceptance in using products from FS treatment and market demand also need to be taken into account in order to ensure uptake of the intended enduse.

10.3.1 Pathogens

FS contains large amounts of microorganisms, mainly originating from the faeces. The microorganisms can be pathogenic, and exposure to untreated FS constitutes a significant health risk to humans, either through direct contact, or through indirect exposure. Pathogens are transmitted and spread through an infection cycle, which includes different stages and hosts. The faecal-oral route of transmission is shown in Figure 10.1.

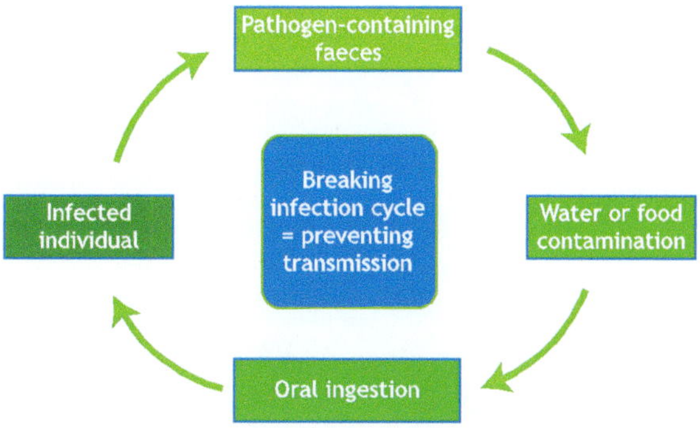

Figure 10.1 Faecal-oral transmission cycle of pathogens.

The transmission cycle of pathogens can be interrupted by putting barriers in place to block transmission paths and prevent cycle completion. FS needs to be treated to an adequate hygienic level depending on the enduse or disposal option. For example, exposure pathways are very different for treated sludge discharged to the environment, compared to sludge used in agriculture, or incinerated. The World Health Organization (WHO) guidelines for safe agricultural practice published in 1998 specified one or less helminth egg/g total solids (TS) for unrestricted irrigation (WHO, 1998). However, the more recent 2006 WHO Guidelines for the Safe Use of Wastewater, Excreta and Greywater in Agriculture and Aquaculture places less emphasis on treatment thresholds but rather highlights a multi-barrier approach where lower levels of treatment may be acceptable when combined with other post-treatment barriers along the sanitation chain. This concept of a multi-barrier approach combined with a risk assessment and risk management system to protect public health is explained in detail in the WHO's Safe Use of Wastewater and Excreta, which is available for free via their website (www.who.int/en/) (WHO, 2006).

The first barrier for beneficial use is provided by the level of pathogen reduction achieved through treatment of FS. A selection of further post-treatment barriers may include restriction of use on crops that are eaten raw, withholding periods between application and harvest to allow pathogen die-off, drip or subsurface irrigation methods, restricting worker and public access during application, use of personal protective equipment and safe food preparation methods such as thorough cooking, washing or peeling. When considering risk of infection all potential exposure groups should be accounted for which can be broadly categorised as workers and their families, surrounding communities and product consumers.

10.3.2 Heavy metals

Heavy metals are a concern due to their toxicity and long-term negative effects on soils. Heavy metals should be evaluated on a case by case basis, but are only a major concern if FS is mixed with industrial effluents that are not adequately pretreated. Heavy metals can also enter the system at the household level through the relatively common practice of improper disposal of wastes containing heavy metals (e.g. batteries, solvents, paints) into the system. The total metals concentration in the sludge differs from the bioavailable metals concentration, as the organic matter in sludge can bind metals in a form that is not biologically available. Because of this effect, sludge is also used for the remediation of metals contaminated sites.

Table 10.2 Regulations for trace element concentrations in the US and Europe for the land application of treated wastewater sludge (biosolids)

Parameters	Concentration limits (mg/kg)		
	Exceptional quality (EQ) biosolids (US EPA, 1999)	Eco label compost (Hogg *et al.*, 2002)	Use of biosolids in Spain (Hogg *et al.*, 2002)
AS	41		
Hg	17		
Fe	nm	-	-
Pb	300	100	750
Ni	420	50	300
Cr	1,200	100	1,000
Cd	39	1	20
Cu	1,500	100	1,000
Zn	2,800	50	2,500
Se	36	-	-

The US Environmental Protection Agency (USEPA) has set limits for heavy metal concentrations for the land application of wastewater sludge based on what is considered a "worst-case scenario" of metal accumulation after 100 years of application. Conservative threshold metal concentrations have been set for the protection of human and environmental health; however, these limits are less conservative than regulations that exist in Europe. An overview of regulatory limits is provided in Table 10.2.

10.3.3 Social factors

Different societies and cultures have different reactions and approaches to the management of human excreta that have to be taken into consideration when evaluating the best enduse for FS. Some cultures reject the use of excreta altogether, whereas others have a long history of excreta use in agriculture. The use of treated FS is however typically perceived differently from excreta, and has a higher acceptance based on its appearance, smell and health impacts. In a society where the use of FS is strictly taboo, other solutions such as co-treatment with other waste streams, use in building materials, or as a fuel might be more appropriate and accepted technologies. This highlights the need for evaluating the market demand of potential endproducts prior to deciding on a treatment and enduse scheme (Diener *et al.*, 2014).

10.4 USE OF FAECAL SLUDGE AS A SOIL CONDITIONER

The use of FS as a soil conditioner can range from deep row entrenchment of untreated FS, to bagged compost that is sold as a commercial product for household level use in horticulture. Using FS as a soil amendment has many benefits over using chemical fertilisers alone (Strauss, 2000). Organic matter in FS can increase soil water holding capacity, build structure, reduce erosion and provide a source of slowly released nutrients. As mentioned above, when using FS as a soil conditioner, the fate of and exposure to pathogens and heavy metals needs to be taken into consideration, and social acceptance can be closely linked to potential commercial value. Other factors that need to be considered include nutrients, which may or may not be available in the ratio required by soil and crop systems.

Table 10. 3 Nutrient content of urine and faeces and mass of nutrients required to grow 250 kg of cereals from Drangert (1998)

Nutrients	Urine[1] (kg)	Faeces[2] (kg)	Total (kg)	Nutrients needed for 250 kg cereals (kg)
Nitrogen (N)	4.0	0.5	4.5	5.6
Phosphorus (P)	0.4	0.2	0.6	0.7
Potassium (K)	0.9	0.3	1.2	1.2
Total amount of N+ P + K	5.3	1.0	6.3	7.5

[1] 500 L/capita/year; [2] 50 L/capita/year

10.4.1 Nutrient content

Theoretically, the quantity of FS produced yearly by a human contains nearly enough plant macro- and micro-nutrients to grow the quantity of food they require in a year (taken as 250 kg of cereals), as shown in Table 10.3.

It is important to determine the appropriate agronomic rate for the land application of treated sludge to maximise benefits, and to prevent environmental contamination from excessive application of nutrients. Nutrients in sludge are present in both organic and inorganic forms. Inorganic forms are more readily available than organic nutrients for plants and microbes to assimilate (e.g. NH_4^+/NH_3, NO_3^-/NO_2^-). Nutrients bound to organic matter are slowly released over time through mineralisation to become biologically available. If nitrogen is applied in excess of plant and soil microbial demand, ammonia can be lost due to volatilisation, and nitrates by leaching through the soil profile. Leaching can lead to the eutrophication of surface waters, and nitrate contamination of drinking water (e.g. resulting in methemoglobinemia).

Many countries have set limits for the land application of FS (e.g. South Africa and China). However, these are typically the maximum allowed rates (i.e. the volume of FS allowed per land area). Estimates for rates of land application can be based on experience; for example, it is estimated that 56 m^3 of FS are required to fertilise one hectare of land when cultivating cereal crops such as maize, millet and sorghum in tropical climates (Asare *et al.*, 2003). However, there are also methods for calculating application rates based on plant nutrient demand, for example, the "Nitrogen Balance" method that is employed with wastewater sludge as illustrated in Figure 10.2 (Henry *et al.*, 1999). Firstly, the amount of nitrogen taken up by plants is calculated by estimating the amount of nitrogen present in the final harvested products. The amount of nitrogen already present in the soil from natural sources is then quantified. The nitrogen required in the land application is the difference between the amount of nitrogen taken up by plants, and the amount supplied by the local natural environment (Henry *et al.*, 1999).

Other research has shown different reactions of crops to nitrogen application rates with compost and co-compost depending on the growth phase of the crops. During the vegetative phase (first 6 weeks) the transpiration efficiency of a maize crop increased up until rates of 150 kgN/ha but then decreased when the nitrogen concentration was increased to 210 kgN/ha (Adamtey *et al.*, 2010). On the other hand, during the reproductive phase (after week 8), the plant's transpiration efficiency increased with increasing nitrogen application rates (Adamtey *et al.*, 2010). These observations did not apply to soils treated with inorganic fertilisers.

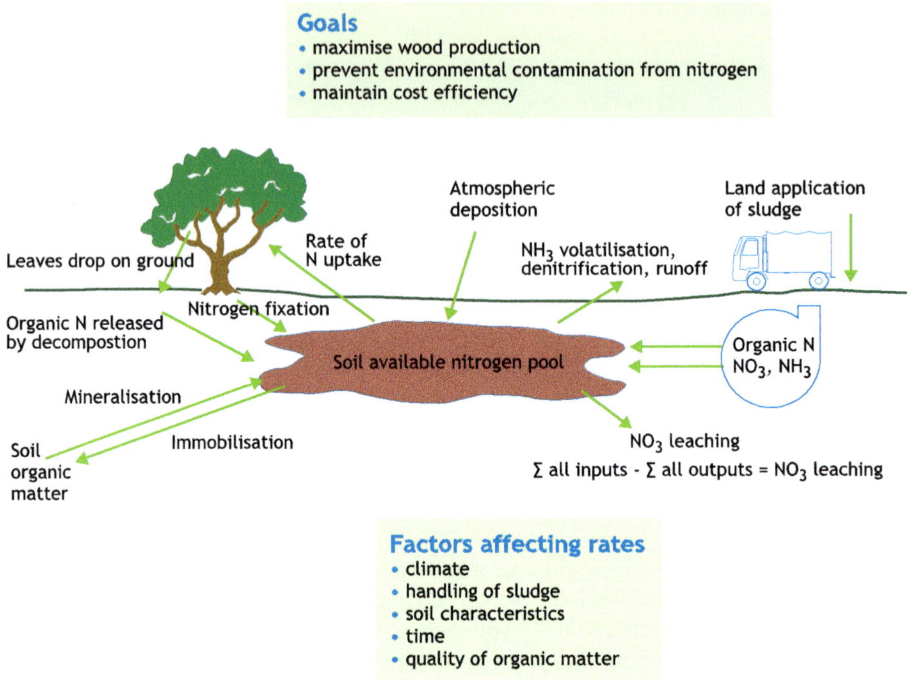

Figure 10.2 Nitrogen balance for the land application of sludge (figure: Linda Strande).

10.4.2 Untreated faecal sludge

Although it is recommended that FS is treated prior to use, some alternatives do exist for the safe disposal and use of FS directly from onsite systems. These options are dependent on the availability of adequate land area, and are therefore not generally appropriate for urban areas. Adequate barriers to pathogen exposure to protect human health are also required.

Deep row entrenchment

One possibility for the direct use of raw FS is deep row entrenchments in forestry applications. By burying sludge in deep ditches, odours are eliminated and the risk of exposure to pathogens is reduced. Trees with a high nitrogen demand are then planted on top of the buried sludge. Deep row entrenchment therefore increases the volume of sludge that can be applied at one time compared to more conventional methods such as spraying on trees or spreading on the soil surface. As with other forms of land application, the appropriate loading of nutrients needs to be considered to prevent environmental contamination. Research on deep row entrenchment of FS is being conducted in forestry trials in South Africa. The research has found that tree growth was improved, and there was no evidence of groundwater contamination (Still and Taylor, 2011). However, the effect on groundwater needs to be further studied, and considered on a case by case basis to ensure environmental protection when using this method. Important factors to consider are soil type and porosity, ground water depth, proximity to drinking water sources, and background nutrient concentrations. There is also long-term experience with deep row entrenchment of wastewater sludge in the US, and this method has been used for the remediation of gravel mining sites for use as tree farms (Kays *et al.*, 2000).

Land application

The direct use of FS has been a long-term practice in parts of China, South-East Asia and Africa. This type of application has the highest level of risk for human health impacts, and is therefore generally not recommended. This practice is best applied in arid to semi-arid regions. It must be ensured that adequate barriers are in place and that there is sufficient land area available. Raw sludge is spread out on farm fields during the dry season, and then incorporated into fields when crops are planted during the rainy season (Cofie *et al.*, 2005). A pit method is also used, where FS is buried with other crop residues and left to mature for a few months prior to use. In areas where this is practiced, there is a large demand for FS. For example, in Northern Ghana, 90% of FS is used as a fertiliser and farmers perceive the competition for FS as one of the main constrains in using it for their crops (Cofie *et al.*, 2005).

10.4.3 Treated faecal sludge in land application

Treatment and processing technologies such as drying beds, composting, and pelletising processes produce treated FS endproducts that can be used as a soil conditioner. The level of remaining pathogens will depend on the selected treatment technology, or combination of technologies.

Sludge from drying beds

The sludge resulting from treatment with planted and unplanted drying beds have very different characteristics, and therefore, different concerns with regards to land application. The majority of helminth eggs are retained in the sludge layer (Cofie *et al.*, 2006). The short retention time of sludge on unplanted drying beds (i.e. weeks) means that further sludge treatment or storage is required if pathogen reduction is to be achieved. The longer retention time of planted drying beds (i.e. years) means that significant pathogen reduction can be achieved, but this needs to be evaluated on a case by case basis. In a study evaluating helminth eggs in planted drying bed sludge, it was found that of 127 eggs/g TS, 6 eggs/g TS were still viable after 7 years (Koottatep *et al.*, 2005a). However, in another study, total eggs decreased from 78.9 to 7.5 eggs/g TS after six months, and viable Ascaris eggs decreased from 38.5 to 4.0 eggs /g TS (Kengne *et al.*, 2009). In addition, due to the long retention time, the treated sludge has similar properties and nutrient content to mature compost.

Information on treatment and removal of sludge from drying beds is covered in Chapters 7 and 8. The amount of sludge that accumulates depends on factors such as the solids content of FS, the loading frequency and the organic loading of the drying bed. With unplanted drying beds, sludge application rates of 100-200 kg TS/m^2/year resulted in a sludge accumulation of 25-30 cm with a 15 day retention time (Cofie *et al.*, 2006). With planted drying beds, sludge application rates of 100, 200 and 300 kg TS /m^2/year resulted in 30-40, 50–70 and 80–113 cm/year respectively (Kengne *et al.*, 2011). Higher loading rates are not recommended, as loadings above 500 kg TS/m^2/year reduce the treatment performance and result in plant wilting (Koottatep *et al.*, 2005b). There is however room for innovation with mixing regimes, application depth and loading rates, and solar/thermal processes.

Co-composting

Co-composting refers to composting of FS together with other waste streams such as municipal solid waste (Figure 10.3). FS with low solids content should be dewatered prior to composting, for example with settling tanks or drying beds. Pathogen reduction is achieved during the composting process through high temperatures, and/or length of time. The properly treated endproduct is a stabilised organic product that may be safely handled, stored and applied to land according to above guidelines for use, without associated concerns of pathogen transmission (Banegas *et al.*, 2007; Koné *et al.*, 2007). For more information on co-composting, refer to Chapters 3 and 5. Although composting is a proven technology to produce a safe to use soil amendment, the local market demand for compost products should be evaluated as compost frequently does not have a significantly high market value. However, other benefits are realised through resource recovery and offsetting disposal costs (Diener *et al.*, 2014).

Figure 10.3 Co-compost of faecal sludge and municipal waste by Sanergy in Nairobi, Kenya (photo: Linda Strande).

Danso *et al.* (2002) evaluated farmers' willingness to pay for compost from municipal solid waste and FS in Ghana. They interviewed 700 farmers in three different cities including both compost users and non-users. Results showed that compost is recognised as a useful resource by most farmers (all compost users and 80% of non-users), and that barriers are more likely to be economic or technical ones rather than cultural. All farmers showed a willingness to pay but at a moderate price, which was too low to achieve a profitable venture if the compost was to be sold at the prices quoted by the farmers. The prices farmers said they were willing to pay varied between 0.1 and 3 USD per 50 kg bag of compost, whereas production costs ranged between 4 and 7 USD per bag (Danso *et al.*, 2002).

Vermicomposting

With vermicomposting, worms breakdown larger organic particles, stimulate microbial activity, and increase the rate of mineralisation, thereby converting FS into humic like substances with a finer structure than normal compost (Alidadi, 2005). Vermicomposting should be operated at a maximum temperature of 35°C in order to maintain the viability of worms. This temperature is not high enough to ensure pathogen inactivation, so if this is necessary, vermicomposting should be combined with another approach such as storage, or a combination of thermophilic composting and vermicomposting should be used to achieve pathogen reduction (Ndegwa and Thompson, 2000).

Pellets

Dried pellets can be an attractive option for FS processing, producing an endproduct that is easy to transport, has reliable characteristics for enduse, and depending on the level of treatment, is safe for handling. Resource recovery options include use as a soil amendment, or for combustion as a bio-fuel. One example is the LaDePa (latrine dehydration and pasteurisation) process that has been developed in South Africa, and is currently operating in a pilot-scale implementation. The LaDePa process can be used for drier sludges (e.g. dry pit latrines, dewatered sludge), and can also be combined with wastewater sludge that has not had polymers added (Chapter 5). The process involves removal of detritus, followed by drying and infrared radiation, to produce small pellets that can be sold to consumers as a soil amendment (Harrison and Wilson, 2011). Another pellet process being developed in Ghana produces dried pellets which are enriched with urea, so the endproduct has similar fertilising properties of poultry manure. The process involves drying, composting or irradiation for hygienisation, enrichment with urea, and then pelletisation with a binder (Nikiema *et al.*, 2012). One possible concern with the use of dried pellets as a soil amendment is the availability of organic matter and nutrients when applied to soils, but Nikiema *et al.* (2012) have found that the cassava binder they are using is effective for transport stability, and also readily breaks down in the soil.

10.5 USE OF LIQUID STREAMS

Liquid streams from treatment processes can be used for agricultural and horticultural irrigation, or other forms of water reclamation (e.g. non-recreational water features, industrial processes), depending on the quantity produced and the level of treatment. Water reclamation can be beneficial in areas where water resources are limited, and also for nutrient recovery. The main consideration with reclamation of liquid streams is to ensure that the treatment quality is appropriate for its intended use. The same concerns are present as for the use of FS as previously presented. This involves undertaking a human health and environmental risk assessment, followed by a multi-barrier approach to enduse to ensure adequate risk management. With water reclamation, major distinctions are made between planned, unplanned, direct, and indirect usage (Jiménez *et al.*, 2000). Indirect usage implies a diluted waste stream, for example if wastewater or FS has been discharged to a river that is used for irrigation. Direct usage implies it is being obtained directly from the waste source, for example discharging a vacuum truck onto an agricultural field. Planned usage refers to the intended and conscious use, whereas unplanned refers to unknowing or unintentional usage.

This section focuses on the enduse of untreated liquid FS, as well as effluents from FS treatment processes. The concerns associated with liquid FS are slightly different from those with wastewater, as constituents in FS effluents are 10-100 times more concentrated than wastewater. For information specifically on the use of wastewater and wastewater treatment effluent for irrigation, it is recommended that the reader refer to texts such as Wastewater Irrigation and Health (Dreschel *et al.*, 2000) and the WHO's Safe Use of Wastewater and Excreta, both of which are available as free downloads on their websites (www.iwmi.cgiar.org, www.who.int/en/).

10.5.1 Untreated liquid faecal sludge in irrigation

Untreated liquid FS and wastewater are commonly used directly for irrigation in many regions in the world (Figure 10.4). By 'untreated' liquid FS, it is meant liquid streams that are being used directly (e.g. vacuum truck discharge, septic tank effluent), or indirectly where FS cannot be separated from wastewater (e.g. urban areas where excreta and wastewater are discharged directly to water conveyance networks). This practice can provide an essential source of water and nutrients, and is reasonably safe if carried out under controlled conditions. However the possibility for pathogen exposure is high, especially in cases of unplanned, direct use.

Research has been conducted in Ghana on farm-based measures for reducing microbiological health risks (Keraita *et al.*, 2010). Currently, irrigation is being carried out by using untreated water, that is polluted with wastewater, to the direct use of black water, resulting in pathogen contamination of uncooked food crops. The size of agricultural applications ranges from small backyard-scale to medium- to larger-scale vegetable production. Examples of types of on-farm treatment technologies include channels, ponds, wetlands, and filtration (filtration technologies range from sand filtration to passage through cloth media). Appropriate treatment solutions will obviously vary depending on the source and pollution level of water, available land area, climate, tenure of property, and intended use of water (e.g. type of crop, irrigation method). Drip irrigation provides an example of the significant impact that formal irrigation methods can have. Benefits include reduced water usage, and increased yield and human health protection. However, drip irrigation is also one of the more expensive irrigation methods. Areas for future research into the use of liquid FS include understanding the removal of pathogens and recycling of nutrients with different onsite treatment technologies so that treatment outcomes can be reliably predicted and appropriate solutions implemented.

Figure 10.4 Crops grown with faecal sludge settling tank effluent in Yaoundé, Cameroon (photo: Linda Strande).

10.5.2 Treated effluent enduse and disposal

The effluents from FS treatment processes may still contain many constituents of concern, and therefore require further treatment prior to discharge to the environment, or careful evaluation and consideration prior to direct use. Effluents are typically high in nitrogen which is beneficial for the recovery of nutrients, but which can also pose an environmental hazard. Other concerns include pathogens, heavy metals, and salinity.

An example of detrimental effects from high nutrients is demonstrated by pond systems for treatment of settling tank effluent in Ghana, where there was such a high ammonia concentration that algal growth was inhibited (ammonia toxicity to algae starts to occur at 40-50 mg NH_3-N/L). Depending on the influent, loading rates, and operations, the effluent that is discharged from waste stabilisation ponds can have similar characteristics to that achieved with more conventional wastewater treatment processes. Other examples of high constituent concentrations in effluent from FS treatment are unplanted drying bed leachate in Dakar measured at 3,600 mg COD/L, 870 mg BOD/L, 260 mg NH_3-N/L, 370 mg/L TKN, 170 mg NO_3-N/L (Koné *et al.*, 2007), and planted drying beds in Thailand 100-2200 mg COD/L, 6-250 mg TKN/L, and 5-200 mg NH_3-N/L (Koottatep *et al.*, 2005a).

Salinity can interfere with plant growth and have long-term impacts on the soil. The electrical conductivity (EC) observed in effluent from settling tanks in Ghana was 8-10 mS/cm, and leachate from planted drying beds in Thailand 2-5 mS/cm (this high conductivity is mainly due to ammonia). The maximum tolerable conductivity for tolerant plants is at most 3 mS/cm (Koné *et al.*, 2007). Using effluents for irrigation will therefore always increase the salinity of soil in the long-term, and it is recommended that salinity control practices are adopted such as soil washing, providing appropriate soil drainage and controlling salt inputs into the wastewater (WHO, 2006).

Quality standards for treated effluent enduse exist in most countries, but are not necessarily enforced in low- and middle-income countries due to economic limitations. For example, in China, a 95% reduction of helminth eggs needs to be achieved for wastewater effluent enduse, and in Ghana, The Environmental Protection Agency has stipulated more than 90% BOD and faecal removal prior to enduse (Heinss *et al.*, 1998). As discussed for the use of FS as a soil amendment, a better approach for reusing effluent in irrigation is the multi-barrier approach recommended by the WHO. A range

Figure 10.5 Crop irrigation with untreated wastewater is still practiced in many low- and middle-income countries Yaoundé, Cameroon (photo: Linda Strande).

of protective measures depending on the level of treatment should be adopted for human health protection. These include crop restrictions, irrigation technique, harvesting periods, food preparation measures and exposure control (WHO, 2006).

Koottatep *et al.* (2005b) examined the effect of using leachate from planted drying beds for irrigation in a six year study in Thailand where sunflowers were grown under different irrigation conditions. Different plots were irrigated with water containing varying amounts of leachate and the results on plant growth were observed. The experiment showed that sunflower plant growth was not hindered by the leachate, and was in fact improved. The seed and oil yields increased when leachate was used for irrigation (Figure 10.6), the best ratios being 20% and 50% leachate. Slightly lower yields were observed with 100% leachate, probably due to the high salinity leachate.

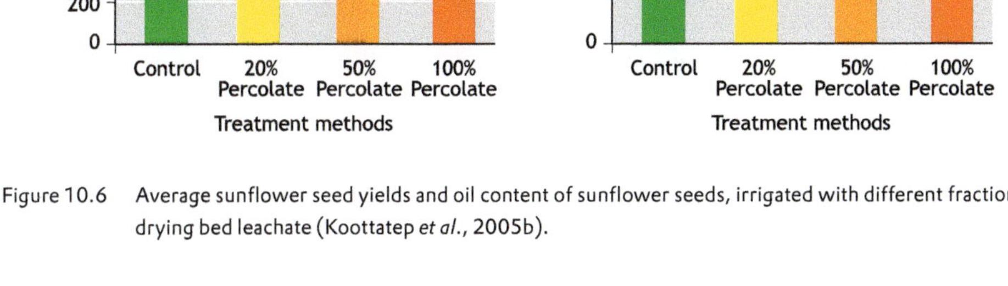

Figure 10.6 Average sunflower seed yields and oil content of sunflower seeds, irrigated with different fractions of drying bed leachate (Koottatep *et al.*, 2005b).

10.6 ADDITIONALFORMS OF RESOURCE RECOVERY

In addition to using endproducts from FS management as soil amendments and for water reclamation, there are many other opportunities for resource recovery depending on the type of treatment and processing technologies. Possibilities include food and agricultural uses (e.g. protein, fodder, fish), or energy reclamation (e.g. biofuels).

10.6.1 Protein

The Black Soldier fly (BSF) larvae (*Hermetia illucens*) can be used as a conventional protein and fat source for poultry and fish feed, and could readily replace fishmeal as a key component of animal feed (St-Hilaire *et al.*, 2007). The larvae grow while feeding on organic matter, such as FS and organic wastes. The last larval stage, the prepupa, has a high protein and fat content. The risks are very low that the BSF is a vector for disease transmission, as it does not feed during the adult stage when it can fly (Sheppard *et al.*, 1994). The use of FS as a feed source for fly larvae has been successfully demonstrated (Nguyen, 2010). However, a mixture of FS and organic municipal waste can achieve higher and faster larvae biomass production (Diener *et al.*, 2009). BSF larvae have the potential of reducing the volume of organic waste by about 55% and the residue remaining after digestion can be composted or anaerobically digested to produce a soil conditioner. It will however be lower in nitrogen and phosphorus than raw organic wastes (Diener *et al.*, 2009). BSF larvae grown only on FS with a dry matter content of 40%, can convert one ton of FS into 20 kg of dried prepupa, with a protein content of 35-44% (Nguyen, 2010). This research is still in developmental stages, and needs to be evaluated on a case by case basis to determine whether the treatment option would be appropriate, if there is a market for endproducts, and if other factors such as climate and availability of organic matter are conducive to the growth of BSF.

10.6.2 Fodder and plants

The plants used in drying beds should be harvested regularly to aid in sludge removal, but can also be harvested more frequently because they have a commercial value that can generate additional revenue. Uses of plants from drying beds include ornamental arrangements, use in compost, or as fodder for livestock (Case Study 10.1). The choice of plants to be grown on the beds should be selected taking the local conditions and market into consideration. In this way, the species that will grow well in drying beds, and which have the potential to generate the most income will be planted (for more information refer to Chapter 8). The growing of plants on drying beds have also been shown to have an increased productivity compared to traditional growing methods. For example, more than 900 shoots/m^2 of Echinochloapyramidalis have been reported with full scale planted dewatering beds in Dakar after 21 weeks of growth (Tine *et al.*, 2009), and in Cameroon, up to 150 dry tons/ha/year (approximately 750 fresh tons /ha/yr) of E. pyramidalis were reported (Kengne *et al.*, 2008).

Case Study 10.1: Market value of fodder grown in drying beds in Cameroon

A socio-economic survey conducted in three cities of Cameroon (Douala, Yaoundé and Garoua) to assess the market potential of *E. pyramidalis* has shown that the daily quantities of marketed forage vary between 5 and 8 tons of fresh weight in the dry and rainy season respectively (Figure 10.7). This fodder plant is used by breeders to feed horses, goats, sheep, dairy cows, rabbits, greater cane rats (*Thryonomys swinderianus*) and guinea pigs (Figure 10.8).

Figure 10.7 Antelope grass is a highly prized fodder in urban and peri-urban areas (photo: Ives Kengne).

E. pyramidalis is marketed in the urban and peri-urban centres and the price varies throughout the year according to its quality (dry or fresh), quantity and availability. Prices obtained varied according to the seasons from 0.1–0.2 USD to 0.2–0.3 USD/kg fresh weight, generating a daily income varying between 500–1,000 USD and 1,600–2,400 USD in dry and rainy seasons, respectively.

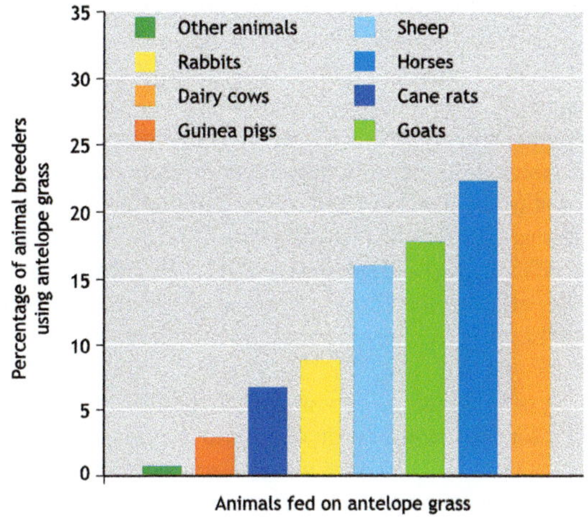

Figure 10.8 Range of animals being fed on *E. pyramidalis* (antelope grass) in Yaoundé, Cameroon.

10.6.3 Fish and plants

The nutrients in FS can be harnessed for use in aquaculture, when growing fish in stabilisation ponds with the effluent from FS treatment plants. The nutrients can increase the growth of plankton, or increase the growth of aquatic plants such as duckweed, water spinach, or water mimosa. Plankton can be harvested for use as fish feed in aquaculture, and aquatic plants can be harvested for animal feed or human consumption. Fish bred in ponds with FS can be used as animal feed, and are also sometimes used for direct human consumption. In the case of direct consumption, certain precautions have to be taken to prevent pathogen transmission and adverse health effects.

Although fish are not susceptible to human pathogens, they can be carriers of them. Faecal bacteria can accumulate in the internal organs and gills of fish. Protective barriers to prevent transfer to humans include cooking fish thoroughly before consumption, transferring the fish into clean water ponds for 2 to 3 weeks before consumption, or maintaining a faecal coliform count of less than 1,000 FC/100mL (WHO, 1998). Fish can also act as intermediate hosts to *Helminths*, which is a concern with FS. In areas where *schistosomiasis* is endemic, fishery workers may be exposed to snails, which are vectors of the disease. Preventive measures include using treated FS, wearing protective clothing such as boots, and removing vegetation on the banks of the ponds to reduce snail growth (Cairncross and Feachem, 1983).

Another concern is that there is inadequate knowledge of the technical aspects of using FS or wastewater in aquaculture, thereby making control of operating parameters more of an art than a science, and leading to potential problems such as the rapid eutrophication of ponds due to excess nutrients.

10.6.4 Building materials

Dried FS can be used in the manufacturing of cement and bricks, and in the production of clay-based products. This resource recovery option captures the material and chemical properties of FS, at a trade-off of their nutrient value not being utilised. The presence of pathogens is less of a concern as human contact is reduced, and high manufacturing temperatures result in the killing off of pathogens.

Dried wastewater sludge and FS have been shown to have similar qualities to other traditional raw building materials such as limestone and clay materials (Jordán *et al.*, 2005; Lin *et al.*, 2012). FS is commonly used in cement production in Japan as an alternative fuel in the kiln, and/or by incorporating the ash resulting from FS incineration into the cement (Taruya *et al.*, 2002).

Another possible method of integrating FS into cement manufacturing is to stabilise and dry the sludge through treatment with lime. Rodriguez *et al.* (2011) describe a process where 20 to 30% lime (CaO) is added to wastewater sludge, which triggers degradation of organic matter and hydration of the CaO. The reaction between lime and the sludge is exothermic and therefore favours sludge drying (the temperature increases from 20°C to 100°C). The product obtained after lime treatment has a powder-like texture with a particle size smaller than 40 μm and can be used as a raw material instead of limestone in cement manufacturing (Rodríguez *et al.*, 2011).The authors claim that this form of sludge dewatering is more energy efficient than other processes due to the exothermic reaction of lime and sludge generating enough heat to promote evaporation without requiring the use of fossil fuels.

FS can also be used in the manufacture of ceramics. Experiments carried out by Jordan *et al.* (2005) showed promising results for the incorporation of FS into the ceramic processing mixture. Amounts of dried sewage sludge varying from 1-10 wt.% were incorporated in the manufacture of clay. Results showed that the addition of wastewater sludge increased the permeability of clay, and reduced its bending strength (Jordán *et al.*, 2005).

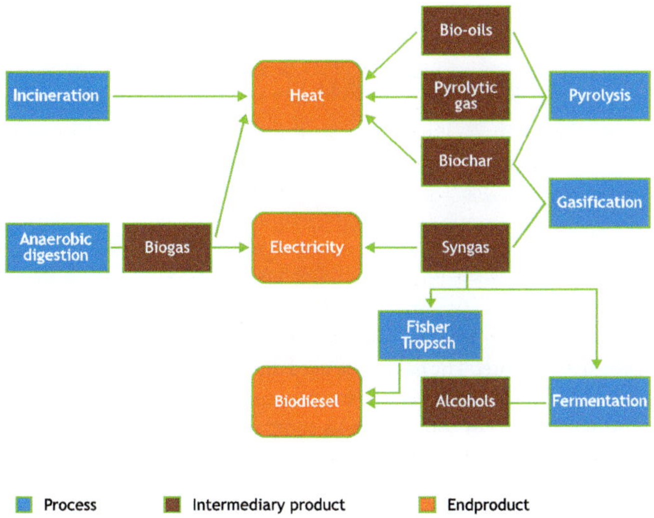

Figure 10.9 Energy recovery options from faecal sludge.

10.6.5 Biofuels

As summarised in Figure 10.9, there are several biological and thermal options for the production of energy from FS. These technologies have been receiving increased interest due to the considerable demand for sustainable biofuels. Possible technologies include anaerobic digestion, which yields biogas, heat, and digestate (sludge); pyrolysis or gasification, which yields biochar, oils and gasses; biodiesel, which can be produced through fermentation or successive chemical reactions; and incineration or co-combustion of dried FS. Energy recovery harnesses the energy potential of organic matter in FS, but frequently at a trade-off of nutrient recovery.

Biogas

The anaerobic digestion of FS produces a mixture of gaseous compounds, commonly referred to as 'biogas'. The mixture of gasses that is commonly produced is presented in Table 10.7. The mixture and amount of gas produced depends on operating parameters such as stability of the sludge, COD of the sludge, and temperature. Biogas has a high energy content due to the high calorific value of methane

Table 10.7 Gases produced during anaerobic digestion (adapted from Bates, 2007)

Substance	Symbol	Percentage (%)
Methane	CH_4	50-70
Carbon dioxide	CO_2	30-40
Hydrogen	H_2	5-10
Nitrogen	N_2	1-2
Water vapour	H_2O	0.3
Hydrogen sulphide	H_2S	Traces

Table 10.8 Biogas fuel equivalents at 15°C and atmospheric pressure (adapted from Bates, 2007)

Energy Source	Equivalent to 1m³ of biogas
Petrol	0.53 - 0.75 L
Diesel	0.48 - 0.68 L
Firewood	1.50 kg
Electricity	1.51 kW/h
LPG	0.46 kg

and can therefore be used as a fuel. The gas can be used directly for applications such as cooking fuel, but if used in engines should be 'scrubbed' prior to use to remove the hydrogen sulphide to avoid corrosion. Table 10.8 presents equivalents of common fuels compared to 1 Nm^3 of biogas. Large and small scale anaerobic digestion facilities have the same equipment requirements, the cost of which can sometimes be prohibitive for small scale applications. Smaller scale applications also tend to be more sensitive to shock loadings and process changes than larger scale plants, making the latter easier to manage. Electricity generation from biogas is not always practical on a small scale. For example, the anaerobic digestion of organic waste generates about 100-200 Nm^3 of biogas per ton of municipal organic waste (Claassen *et al.*, 1999). Taking a 25% conversion efficiency of biogas into electricity, $1m^3$ of biogas can generate 1.51 kWh (Cuéllar and Webber, 2008). Therefore the anaerobic digestion of one ton of municipal organic waste could produce a maximum of about 320 kWh, which is sufficient to operate one 100 W light bulb for around 132 days, or 3,200 100 W light bulbs for 1 hour. In this case it would be more feasible to use the biogas as a cooking gas (or as a fuel) for local district lighting.

The solids fraction remaining after anaerobic digestion can also be utilised for any of the enduses described for FS, but may require further treatment depending on the choice. The degree of pathogen destruction during anaerobic digestion depends on the operating temperature. Thermophilic digestion (>50°C) will result in significant pathogen reduction, but mesophilic conditions (30-38°C) do not guarantee pathogen inactivation. Maintaining a well-mixed reactor also increases the degree of pathogen deactivation as it prevents the formation of dead zones in the reactor (Smith *et al.*, 2005).

Incineration/co-combustion

Incineration is the complete combustion of organic matter at high temperatures, and can either be a disposal mechanism, or provide a way to generate electricity or heat. Incineration of wastewater sludge is relatively common in Europe and US. Incineration reduces sludge to ash (10% of its initial volume) which is mainly composed of remaining inorganic material, and at the same time destroys all pathogens due to the high processing temperatures (Werther and Ogada, 1999). Several methods of incineration and co-combustion are possible with FS and these are summarised in Figure 10.10. Ashes remaining after incineration can either be disposed of, or utilised as raw materials for the manufacture of construction materials.

The calorific value of wastewater sludge typically ranges from 10-29 MJ/kg, while the calorific value of FS is reported to be 17 MJ/kg solids; compared to an average coal value of 26 MJ/kg (Murray Muspratt *et al.*, 2014). Sludge can be co-combusted with coal in coal-fired power plants or other industrial applications such as cement kilns (Figure 10.11; Rulkens, 2008). The direct injection of dewatered FS can reduce NO_x emissions from a cement kiln by 40% and produces 30% less CO_2 emissions compared to when sludge is incinerated (Taruya *et al.*, 2002). The use of FS as a fuel will only be financially sustainable if the financial gains outweigh the economic and environmental costs of sufficient drying prior to combustion.

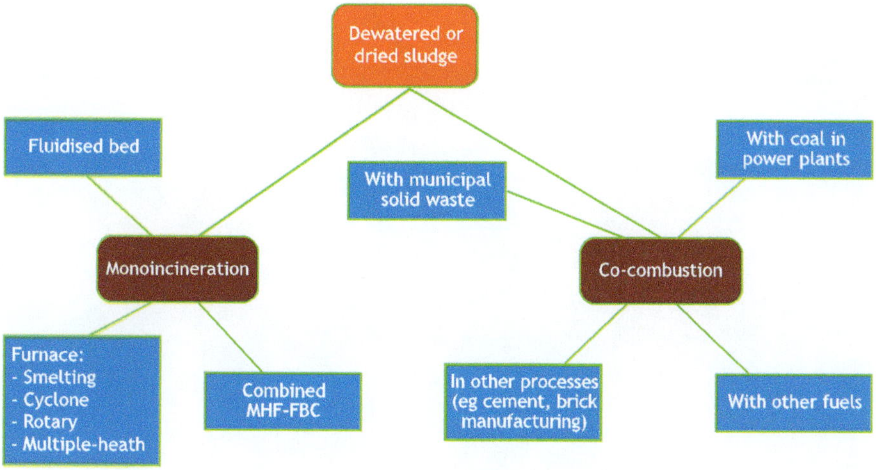

Figure 10.10 Different options for the combustion of sludge (adapted from Werther and Ogada, 1999).

Incineration can produce gases that contain pollutants which can enter the atmosphere. A gas treatment system for the removal of pollutants prior to off-gassing is typically very expensive (Rulkens, 2008). Despite the high nitrogen content of FS, it has been shown that the emissions of nitrous oxides are in fact lower from sludge incineration than from coal incineration. The emissions of dioxins and furans from sludge incinerators are also lower than from waste incinerators (Werther and Ogada, 1999).

Pyrolysis/gasification

Pyrolysis is based on the principle of heating in an oxygen-depleted environment. The absence of oxygen prevents combustion from occurring, and hence yields carbon-based endproducts that are different from those produced during incineration. These endproducts include (bio)char, oils and gases, the quantity of each depending on the processing temperature and presence of gasifying agents. At temperatures above 700°C gasification occurs, which favours the production of syngas (H_2 and CO), whereas temperatures between 350-500°C results in pyrolysis, thereby yielding a larger quantity of char, and gas with more compounds (e.g. CO_2 and CH_4). Both endproducts can be used as fuels, and the gasses produced can also be recovered (Rulkens, 2008). Reported calorific values for syngas from the gasification of wastewater sludge are similar to that produced from coal (7-9.5 MJ/m^3) (Domínguez *et al.*, 2006).

Figure 10.11 FaME (Faecal Management Enterprises) project pilot scale kiln for combustion of faecal sludge to heat oil in industrial processes, Thies Polytechnical University, Senegal (photo: Linda Strande).

Char can be used in furnaces and kilns in the same way as coal, but an energy analysis should be considered to ensure that the production of char from wet sludge has a net positive energy gain. Char can also be used as a soil conditioner; however, there is still some debate around the benefits. As char is a highly porous material, it is thought that this will increase the surface area in soils, and hence improve water retention and aeration capacity (Chan *et al.*, 2007). This technique is commonly compared to the 'terra preta' soils in the Amazon resulting from usage patterns of ancient civilisations. However, char does not provide available organic matter and nutrients present in compost as these are lost in the pyrolysis or gasification process. Growing trials with char have shown both plant yield suppression and plant yield increases. Char can also potentially deplete nutrients in the soil if they are absorbed (Brown, 2011). It therefore appears that it is more beneficial to use char as a fuel than a soil conditioner. There is however a need for additional research to further characterise the properties of char, the dependence on manufacturing conditions, and the effects on soil (Manyà, 2012). To date, information is only available based on wastewater sludge (biosolids) and not with FS, although research is currently being conducted as part of the Reinventing the Toilet Challenge (RTTC) programme of BMGF.

Conventional pyrolysis is carried out with relatively dry materials (Figure 10.12), whereas hydrothermal carbonisation (HTC) is a different type of pyrolysis that allows handling of wet materials. Hydrothermal carbonisation or hydrous pyrolysis is the thermal degradation of biomass in the presence of subcritical water and in the absence of oxygen (Libra, 2011). The solid yielded from this process is referred to as hydrochar to distinguish it from char obtained from dry pyrolysis. Hydrochar is reported to have a highly porous nanostructure, which can be utilised for ion binding, pollutant or water absorption, or as a scaffold for particle binding of catalysts (Titirici *et al.*, 2007). Berge *et al.* (2011) produced hydrochar from anaerobically digested wastewater sludge and found that its carbon content was lower than the initial feedstock, indicating an ineffective carbonisation. Possible reasons reported for this are an incomplete initial hydrolysis step, the slightly basic pH of digested sludge and its stabilised state, and being less prone to changes in carbon content (Berge *et al.*, 2011). Further research is required in the field of HTC and its applications to biomass degradation. Overall, there is less literature available on HTC than dry pyrolysis and char, probably due to the intense research interest that the discovery of the 'terra preta' soils instigated in char research (Berge *et al.*, 2011), and the high energy and pressure requirements of HTC.

Gasification is made up of a series of chemical and thermal steps: drying, pyrolysis, oxidation and reduction (Dogru *et al.*, 2002). This process mainly produces a synthetic gas, or syngas, which is made up of carbon monoxide (CO), carbon dioxide (CO_2), hydrogen gas (H_2) and other trace elements. Syngas has a high energy content and can be either directly used for electricity generation in gas engines and turbines, or it can be further processed to obtain liquid fuel. It has been reported that gasification yields 37% more energy than pyrolysis (Nipattummakul *et al.*, 2010). Dogru *et al.*, (2002) obtained gas with a calorific value of 4 MJ/m^3 in a bench-scale experiment with a fixed bed downdraft gasifier. The use of this type of gasifier is limited to small scale applications as it cannot be easily scaled up, whereas circulating fluidised bed configurations, most commonly used for coal applications, are planned to be used on an industrial scale for wastewater sludge gasification (Ferrasse *et al.*, 2003).

Hydrogen gas is potentially a valuable renewable fuel, which has the potential to power hydrogen fuel cells or hydrogen engines without greenhouse gas emissions. Under the right operating conditions, hydrogen can make up a substantial portion of the syngas that is produced, and research efforts are focused on optimising processing conditions to maximise hydrogen gas yield. Greater volumes of hydrogen gas can be obtained with higher reactor temperatures, and three times as much hydrogen can be obtained with steam gasification of sewage sludge than with air gasification (0.076 g gas/sample at 1,000°C) (Nipattummakul *et al.*, 2010).

Figure 10.12 Iiribogo gasification project utilising corn husks and sawdust, located in Muduuma Sub-county, Mpigi District, Uganda (photo: Linda Strande).

Other alternatives for biofuel production include the processing of syngas into transportation fuel. Syngas can be fermented to produce alcohols such as ethanol. This fermentation is mediated by microorganisms, which convert syngas into hydrocarbons. These microorganisms are mesophilic and the gases therefore need to be cooled down b efore the fermentation step. Heat recovery during the cooling process is possible (Henstra *et al.*, 2007). Another option is to apply the Fischer Tropsch process to syngas to obtain biodiesel, which involves a chain of chemical reactions aided by a metal catalyst (e.g., cobalt, iron, ruthenium). This process is complex, and applications of producing liquid hydrocarbons from biomass are only in the first stages of commercialisation (Srinivas *et al.*, 2007).

Biodiesel

Biodiesels are produced from oils and fats, and therefore the lipids contained in FS have to be harvested through extraction processes. Once lipids are isolated, they undergo a base- or acid-catalysed trans esterification process using alcohol. The resulting compounds are fatty acid alkyl esters (i.e. methyl, propyl or ethyl), which make up the biodiesel. The difficulty in maximising the extraction of lipids from sludge and the associated costs are the main barriers to producing biodiesel from FS (Kargbo, 2010).

Biodiesel can be used in similar applications to conventional fossil fuel-based diesel. Biodiesel has a slightly lower heat of combustion than petroleum-based diesel, resulting in about a 10% reduction in power when using biodiesel. Also It does however have benefits compared to conventional diesel such as increasing engine life and producing less exhaust gas emissions (Demirbas, 2009).

Figure 10.13 Screenings from the Niayes faecal sludge treatment plant in Dakar, Senegal (photo: Linda Strande).

10.7 GRIT SCREENINGS

Screening at the influent of treatment plants is essential to prevent clogging of pumps and machinery, and to prevent detritus in endproducts (Figure 10.13). Unfortunately, there are not many options for resource recovery from these screening solids. The screenings contain a large number of pathogens, are odorous, haves a high water content, and a high density and weight. Organic decomposable wastes represent the largest constituent of screenings from FS, as also observed for municipal solid wastes in low-income countries (Troschinetz and Mihelcic, 2009). Screenings also contain rocks, sand, iron, wood, textiles and plastics in various proportions.

The most common form of disposal is landfilling. Incineration is usually not an option due to the presence of non-decomposable materials in the screenings (e.g. rocks, sand). Composting is an option to treat the organic decomposable fraction, potentially co-composted with domestic household solid wastes to provide sufficient readily degradable matter (Koné *et al.*, 2007, Niwagaba, 2009).

10.13 BIBLIOGRAPHY

Adamtey, N., Cofie, O., Ofosu-Budu, K.G., Ofosu-Anim, J., Laryea, K.B., Forester, D. (2010). Effect of N-enriched co-compost on transpiration efficiency and water-use efficiency of maize (zea mays L.) under controlled irrigation. Agricultural Water Management 97, p. 995-1005.

Alidadi, H., Parvaresh, A.R., Shamansouri, M.R., Pourmoghodas, H., Najafpoor, A.A. (2005). Combined compost and vermicomposting process in the bioconversion of sludge. Iranian Journal of Environmental Health Science & Engineering 2, p.251-254.

Asare, I., Kranjac - Berisavjevic, G., Cofie, O. (2003). Faecal Sludge Application for Agriculture in Tamale. Urban Agricultural Magazine 10, p.32-33.

Banegas, V., Moreno, J, L., Garcia, C., Leon, G., Hernandez, T. (2007). Composting anaerobic and aerobic sewage sludges using two proportions of sawdust. Waste Management 27, p.1317-1327.

Bates, L. (2007). Technologies that really work. Boiling Point 53. Available from http://stoves.bioenergylists.org/taxonomy/term/790.

Berge, N, D., Ro, K. S., Mao, J., Flora, J. R.V., Chappell, M. A., Bae, S. (2011). Hydrothermal carbonization of municipal waste streams. Environmental Science & Technology 45(13), p.5696-5703.

Brown, S.L. (2011). Climate change connections: Real Solutions Fill The Vacuum. Biocycle 52(1), p.51-56.

Cairncross, S., Feachem, R. (1983). Health aspects of waste re-use. Environmental Health Engineering in the Tropics. Second edition edn. p.205-213.

Chan, K.Y., van Zwieten, L., Meszaros, I., Downie, A., Joseph, S. (2007). Agronomic values of greenwaste biochar as a soil amendment. Soil Research 45(8), p.629-634.

Claassen, P.A. M., van Lier, J.B., Lopez Contreras, A.M., van Niel, E. W. J., Sijtsma, L., Stams, A.J.M., de Vries, S.S. Weusthuis, R.A. (1999). Utilisation of biomass for the supply of energy carriers. Springer Berlin/Heidelberg.

Cuéllar, A, D., Webber, M,E. (2008). Cow power: the energy and emissions benefits of converting manure to biogas. Environmental Research Letters 3.3.

Cofie, O.O., Agbottah, S., Strauss, M., Esseku, H., Montangero, A., Awuah, E., Koné, D. (2006). Solid–liquid separation of faecal sludge using drying beds in Ghana: Implications for nutrient recycling in urban agriculture. Water Research 40(1), p.75-82.

Cofie, O.O., Kranjac-Berisavljevic, G.,Drechsel, P.(2005).The use of human waste for peri-urban agriculture in Northern Ghana. Renewable Agriculture and Food Systems 20(2), p.73.

Danso, G., Fialor, S.C., Drechsel, P.(2002). Farmers' perception and willingness-to-pay for urban waste compost in Ghana. In: Almorza, D., Brebbia, C., Sales, D., Popov, V., eds, Waste Management and the Environment. Southampton: WIT Press, p.231-241.

Demirbas, A. (2009). Progress and recent trends in biodiesel fuels. Energy Conversion and Management 50(1), p.14-34.

Diener, S., Semiyaga, S., Niwagaba, C., Muspratt, A., Gning, J.B., Mbéguéré, M., Ennin, J.E., Zurbrugg, C., Strande, L. (2014). A value proposition: resource recovery from faecal sludge – can it be the driver for improved sanitation? Resources Conservation & Recycling (in press).

Diener, S., Zurbrügg, C., Tockner, K. (2009). Conversion of organic material by black soldier fly larvae: establishing optimal feeding rates. Waste Management & Research 27(6), p.603-610.

Domínguez, A., Menendéz, J.A., Inguanzo, M., Pís, J.J. (2006). Production of bio-fuels by high temperature pyrolysis of sewage sludge using conventional and microwave heating. Bioresource technology 97(10), p.1185-1193.

Dogru, M., Midilli, A., Howarth, C.R. (2002). Gasification of sewage sludge using a throated downdraft gasifier and uncertainty analysis, Fuel Processing Technology 75, p.55–82.

Drangert, J-O. (1998). Fighting the urine blindness to provide more sanitation options. Water SA 24(2) p.157-164.

Dreschel, P., Scott, C., Raschid-Sally, L., Redwood, M., Bahri, A. (2000). Wastewater Irrigation and Health: Assessing and Mitigating Risk in Low-income Countries. Earthscan, IWMI, and IDRC.

Ferrasse, J.-H., Seyssiecq, I., Roche, N. (2003) Thermal gasification : a feasible solution for sewage sludge valorisation? Chemical Engineering and Technology 23 (9), p.941-945.

Harrison, J., Wilson, D. (2011). Towards Sustainable Pit Latrine Management Through LaDePa, WISA 2012, 2011.

Heins, U., Larmie, S.A., Strauss, M. (1998). Solids Separation and Pond Systems for the Treatment of Faecal Sludges in the Tropics.05/98. Dubendorf, Switzerland: EAWAG/SANDEC.

Henry, C., Sullivan, D., Rynk, R., Dorsey, K., Cogger, C. (1999). Managing nitrogen from biosolids. Seattle: Washington State Department of Ecology.

Henstra, A.M., Sipma, J., Rinzema, A., Stams, A.J. (2007). Microbiology of synthesis gas fermentation for biofuel production. Current Opinion in Biotechnology 18(3), p.200-206.

Hogg, D.,Barth, J., Favoino, E., Centemero, M., Caimi, V., Amlinger, F., Devliegher, W., Brinton, W., S. Antler, S. (2002). Comparison of compost standards within the EU, North America and Australasia. The Waste and Resources Action Programme, Oxon, UK, p.1-97.

Jimenez, B., Chavez, A., Barrios, J. E., Perez, R. (2000). Impact and potential of reused water in the Mezquital Valley, Water 21, p.34-36

Jordan, M. M., Almendro-Candel, M. B., Romero, M., Rincón, J. M. (2005). Application of sewage sludge in the manufacturing of ceramic tile bodies. Applied Clay Science 30(3–4), p.219-224.

Kargbo, D.M. (2010). Biodiesel Production from Municipal Sewage Sludges. Energy and Fuels 5, p.2791-2794.

Kays, J.S., Flamino, E.J., Felton, G., Flamino, P.D. (2000). Use of deep-row biosolids applications to grow forest trees: a case study. In C.L. Henry, R.B. Harrison, R.K. Bastian (Eds.). The Forest Alternative: Principles and Practice of Residuals Use, p.105-110. Seattle, WA: University of Washington College of Forest Resources.

Kengne, I.M., Kengne, E.S., Akoa, A., Bemmo, N., Dodane, P.-H., Koné, D., (2011). Vertical-flow constructed wetlands as an emerging solution for faecal sludge dewatering in developing countries. Journal of Water, Sanitation and Hygiene for Development 1(1), p.13-19.

Kengne, I.M., Akoa, A.A., Soh, E.K., Tsama, V., Ngoutane, M.M., Dodane, P.H., Koné, D. (2008). Effects of faecal sludge application on growth characteristics and chemical composition of Echinochloa pyramidalis (Lam.) Hitch. and Chase and Cyperus papyrus L. Ecological Engineering 34(3), p.233-242.

Kengne, I.M., Akoa, A., Koné, D. (2009). Recovery of biosolids from constructed wetlands used for faecal sludge dewatering in tropical regions. Environmental Science and Technology 43(17), p.6816-6821.

Keraita, B., Konradsen, F., Drechsel, P. (2010). Farm-based measures for reducing microbiological health risks for consumers from informal wastewater-irrigated agriculture. In: Drechsel, P., Scott, C., Raschid-Sally, L., Redwood, M., Bahri, A. Wastewater Irrigation and Health: Assessing and Mitigating Risk in Low-income Countries. London: Earthscan. p.189-208.

Koné, D., Cofie, O., Zurbrugg, C., Gallizzi, K., Moser, D., Drescher, S., Strauss, M. (2007). Helminth eggs inactivation efficiency by faecal sludge dewatering and co-composting in tropical climates. Water Research 41(19), p.4397-4402.

Koottatep, T., Polprasert, C., Hadsoi, S. (2005a). Nutrient Recycling and Treatment of Faecal Sludge through Constructed Wetlands and Sunflower Plant Irrigation, Proceedings of the International Forum on Sustainable Techniques for Wastewater Management between Thailand and Taiwan ROC, p.26-29.

Koottatep, T., Surinkul, N., Polprasert, C., Kamal, A.S.M., Koné, D., Montangero, A., Heinss, U., Strauss, M. (2005b). Treatment of septage in constructed wetlands in tropical climate: lessons learnt from seven years of operation. Water Science and Technology 51(9), p.119-126.

Libra, J.A., Kyoung, S.R., Kammann, C., Funke, A., Berge, N.D., Neubauer, Y., Titirici, M.M., Fühner, C., Bens, O., Kern, J., Emmerich, K.-H. (2011). Hydrothermal carbonization of biomass residuals: a comparative review of the chemistry, processes and applications of wet and dry pyrolysis, Biofuels 2(1), p.89–124.

Lin, Y., Zhou, S., Li, F., Lin, Y. (2012).Utilization of municipal sewage sludge as additives for the production of eco-cement. Journal of Hazardous Materials, 213–214, p.457-465.

Manyà, J. J. (2012). Pyrolysis for Biochar Purposes: A Review to Establish Current Knowledge Gaps and Research Needs. Environmental Science and Technology 46 (15), p.7939–7954.

Murray Muspratt, A., Nakato, T., Niwagaba, C., Dione, H., Baawuah, N., Kang, J., Stupin, L., Regulinski, J., Mbéguéré, M., Strande, L. (2014). Fuel Potential of Faecal Sludge: Calorific Value Results from Uganda, Ghana, and Senegal. Journal of Water Sanitation and Hygiene for Development, in press.

Ndegwa, P.M., Thompson, S.A. (2000). Effects of C-to-N ratio on vermicomposting of biosolids, Bioresource Technology 75(1), p.7-12.

Nguyen, H.D. (2010). Decomposition of organic wastes and fecal sludge by black soldier fly larvae. School of Environment, Resources and Development. Asian Institute of Technology, Thailand, p.75.

Nikiema, J., Cofie, O., Impraim, R., Dreschel, P. (2012). Fortified Excreta Pellets for Agriculture. Conference Proceedings – 2nd International Faecal Sludge Management Conference, Durban, South Africa, October 29-31, 2012.

Nipattummakula, N., Ahmeda, I., Kerdsuwan, S., Guptaa, A. K. (2010). High temperature steam gasification of wastewater sludge. Applied Energy 87(12), p.3729-3734.

Niwagaba, C. (2009). Treatment technologies for human faeces and urine. PhD Thesis, Doctoral Thesis No. 2009: 70. Swedish University of Agricultural Sciences, Uppsala, Sweden. ISBN 978-91-576-7417-3.

Rulkens, W. (2008). Sewage Sludge as a Biomass Resource for the Production of Energy: Overview and Assessment of the Various Options. Energy & Fuels 22(1), p.9-15.

Rodríguez, N.H., Granados, R.J., Blanco-Varela, M.T., Cortina, J.L., Martínez-Ramírez, S., Marsal, M., Guillem, M., Puig, J., Fos, C., Larrotcha, E., Flores, J. (2011). Evaluation of a lime-mediated sewage sludge stabilisation process. Product characterisation and technological validation for its use in the cement industry. Waste Management 32(3), p.550-60.

Sheppard, C.D., Newton, L.G., Thompson, S.A., Savage, S. (1994). A value added manure management system using the black soldier fly. Bioresource Technology 50(3), p.275-279.

Smith, S.R., Lang, N.L., Cheung, K.H.M., Spanoudaki, K. (2005). Factors controlling pathogen destruction during anaerobic digestion of biowastes. Waste Management 25(4), p.417-425.

Srinivas, S., Malik, R.K., Mahajani, S.M. (2007). Fischer-Tropsch synthesis using bio-syngas and CO_2. Energy for Sustainable Development 11(4), p.66-71.

St-Hilaire, S., Sheppard, C., Tomberlin, J.K., Irving, S., Newton, L., McGuire, M.A., Mosley, E.E., Hardy, R.W., Sealey, W. (2007). Fly Prepupae as a Feedstuff for Rainbow Trout, Oncorhynchus Mykiss. Journal of the World Aquaculture Society 38(1), p.59-67.

Still, D., Taylor, C. (2011). Simple sludge disposal with benefits? Deep-row entrenchment with agroforestry. What happens when the pit is full? FSM Seminar report, 14-15 March 2011, WINSA, p.29-30.

Strauss, M. (2000). Human Waste (Excreta and Wastewater) Reuse. ETC/SIDA Bibliography on Urban Agriculture. Duebendorf, Switzerland: EAWAG/SANDEC.

Taruya, T., Okuno, N., Kanaya, K. (2002). Reuse of sewage sludge as raw material of Portland cement in Japan. Water Science and Technology 46(10), p.255-258.

Tine D., 2009. Traitement de boues de vidange de systèmes d'assainissement autonome à Dakar (Sénégal) : Etude d'une phase d'acclimatation de deux espèces utilisées pour le traitement des boues de vidange domestiques. Mémoire de DEA en Sciences de l'environnement. Institut des Sciences de l'Environnement, Université Cheikh Anta Diop de Dakar

Titirici, M.M., Thomas, A., Antonietti, M. (2007). Back in the black: hydrothermal carbonization of plant material as an efficient chemical process to treat the CO_2 problem? New Journal of Chemistry 31(6), p.787-789.

Troschinetz, A.M., Mihelcic, J.R. (2009). Sustainable recycling of municipal solid waste in developing countries. Waste Management 29(2), p.915-931.

US EPA (1999). Biosolids Generation, Use, and Disposal in The United States. U.S. Environmental Protection Agency Municipal and Industrial Solid Waste Division Office of Solid Waste EPA 530-R-99-009. Available from http://www.epa.gov/compost/pubs/biosolid.pdf.

Werther, J.,Ogada, T. (1999).Sewage sludge combustion. Progress in Energy and Combustion Science 25(1), p.55-116.

WHO (2006). WHO Guidelines for the Safe Use of Wastewater, Excreta and Greywater - Volume IV: Excreta and greywater use in agriculture. World Health Organization (WHO), Geneva, Switzerland

Yadave, L.S., Hesse, P.R. (1981): The Development and Use of Biogas Technology in Rural Areas of Asia (A Status Report 1981). Improving Soil Fertility through Organic Recycling. Food and Agriculture Organization (FAO) and United Nations Development Program (UNEP).

End of Chapter Study Questions

1. Identify at least six resource recovery options for FS, the associated treatment technologies, and their advantages and disadvantages.

2. Describe different options for composting and their advantages and disadvantages.

3. Char from FS could be used as a soil conditioner or as a fuel. List advantages and disadvantages for both resource recovery options.

Management

Chapter 11

Operation, Maintenance and Monitoring of Faecal Sludge Treatment Plant

Magalie Bassan and David M. Robbins

Learning Objectives

- Understand the importance and role of operations and maintenance for faecal sludge treatment plants.
- Understand critical operations and maintenance factors to include starting with the design and planning phases.
- Be able to design an effective monitoring and operations and maintenance plan to ensure treatment performance.
- Understand the role of administrative management in the long-term operation of faecal sludge treatment plants.

Management

11.1 INTRODUCTION

Faecal sludge treatment plants (FSTPs) require ongoing and appropriate operation and maintenance (O&M) activities in order to ensure long-term functionality. O&M activities are at the interface of the technical, administrative, and institutional frameworks that enable sustained FSTP function. "Operation" refers to all the activities that are required to ensure that a FSTP delivers treatment services as designed, and " maintenance" refers to all the activities that ensure long-term operation of equipment and infrastructure (Bräustetter, 2007). Proper O&M of FSTPs requires a number of crucial tasks to be carried out regardless of the size of the plant, and complexity of the technological setup (Figure 11.1). Having skilled workers perform these tasks in a timely manner and in accordance with best practices will maximise the value of the FSTP and ensure its long-term performance.

Figure 11.1 Maintenance worker cleans mechanical screens, a critical activity to be performed each shift to keep the system operational, Dakar, Senegal (photo: Linda Strande).

Many FSTPs fail following construction, regardless of the choice of technology or the quality and robustness of the infrastructure. Reasons for failure are not always investigated, but the most frequent explanations given are low operational capacity (Fernandes *et al.*, 2005; Lennartsson *et al.*, 2009; Koné, 2010; HPCIDBC, 2011), and the lack of financial means to accomplish O&M tasks (Koné, 2002). Lessons learned from these failures are that O&M must be considered as an integral component of the full life cycle costs of a facility, and that ongoing training and capacity building is essential for the operators. In addition, the O&M plan must be incorporated into the design process and receive appropriate review and approvals along with the engineering plans. This helps to ensure that O&M is fully integrated into the facility once construction is complete and operation has begun.

Financial, technical and managerial inputs are needed to ensure the continuous operation of even the simplest of FSTP systems. The procedures that establish how the treatment facility and equipment are utilised, are documented in several O&M plans, monitoring programmes, reports and log books, and health and safety plans, which outline the step-by-step tasks that employees are required to carry out in order to ensure the long-term functioning of the FSTP. While many O&M activities are process-specific, others are common to all facilities and all O&M Plans should therefore include information on:

- the procedures for receiving and off-loading of faecal sludge (FS) at the FSTP;
- the operation of specific technologies such that they function as designed;
- maintenance programmes for plant assets to ensure long-term operation and to minimise breakdowns;
- the monitoring and reporting procedures for the FSTP O&M activities as well as the management of treatment endproducts;
- management of health and safety aspects for protection of the workers and the environment;
- the organisational structure, distribution of and the management of administrative aspects; and
- procedures for the onsite storage of FS and the off-site transportation.

The level of organisation required at any given FSTP is a function of its size and treatment capacity. Small systems that receive a few loads of FS a week may only need one operator, and therefore have relatively simple O&M plans, while large municipal systems that receive FS loads around the clock are more complex and require more staff with different levels of operators and maintenance personnel. This chapter discusses the O&M planning process as well as the specific components of the O&M Plan. It references the procedures and tasks that are common to all FSTP facilities, as well as considerations for technology specific tasks.

11.2 INTEGRATING O&M INTO THE FAECAL SLUDGE TREATMENT PLANT PLANNING PROCESS

There are several important factors that need to be considered when planning FSTPs which will have a direct impact on O&M and monitoring. They encompass both classical engineering aspects of technology integration, as well as other issues concerning the institutional management that defines the FSM programme. Since O&M aspects are important for the overall long-term success of the programme, O&M planning, including the financial provision of funds, should be included in the terms of references for the design of each FSTP (Fernandes *et al.*, 2005; Lüthi, 2011). Furthermore, the O&M plan should be reviewed and approved along with engineering designs and specifications, including the following considerations:

- location of the FSTP and its proximity to residential areas;
- volumes and schedules of FS collection;
- availability of local resources;
- degree of mechanisation of technologies; and
- final enduse or disposal of endproducts.

11.2.1 Location of the faecal sludge treatment plant

The location of a FSTP is a crucial aspect when designing an O&M plan. FSTPs are often associated with nuisances such as odors, flies and mosquitoes, and noise. Facilities located close to residential areas must therefore install preventative controls, all of which have O&M implications. Examples include FSTPs that utilise waste stabilisation ponds located near to residential areas, where mosquito control is an important requirement. For FSTPs located such that access roads cross residential areas, reduction of noise and dust produced by trucks needs to be regulated.

Other site specific factors that might influence O&M activities and costs include:

- soil conditions, such as soil depth and bearing capacity, that might have impact on equipment selection and installation;
- groundwater level and proximity of the FSTP that could result in pollution of water resources or infiltration of groundwater into treatment tanks, directly impacting on the pumping and solids handling equipment; and
- surface waters and flooding risks, which might inhibit site access during rainy seasons, adversely affect or undermine equipment due to scouring or erosion.

11.2.2 Volumes and schedules of faecal sludge delivery

The volume of FS that is collected and delivered to the treatment plant, as well as the operational times of the FSTP will have a significant influence on the O&M costs and requirements. Cultural habits or events can influence the volumes that are discharged at the FSTP at different times of the year. Similarly, seasonal variability of waste volumes will impact O&M staffing requirements. Larger systems that operate on a daily basis have very different staffing requirements to those that operate intermittently.

Figure 11.2 Maintaining the fleet of faecal sludge vacuum trucks in Dumaguete City, Philippines, (photo: David M. Robbins).

The distribution of the FS volume received at the plant throughout the day is critically important in the planning process, as low or high flows that exceed the design of the treatment system can have a significant impact on the operational efficiency. The initial planning phase must therefore ensure that the chosen technology is appropriate for local conditions, and that it is correctly sized to accommodate the expected volumes and related fluctuations. Institutional arrangements that closely coordinate activities between facility owners and those responsible for the FS collection and transportation can help to address these issues.

11.2.3 Availability of local resources

The availability of local resources impacts not only those aspects that determine the cost of construction such as technology selection and building materials but also on the costs of O&M requirements. Local resource issues that must be considered from the O&M perspective include:

- the availability of spare parts and tools;
- the availability of consumables (e.g. chemicals for flocculation);
- the availability and reliability of local utilities including water and power;
- the availability of trained human resources to properly operate the facility;
- the availability of local laboratory resources that may be required for monitoring programs; and
- the availability of local contracting firms to assist with periodic tasks that may be labor intensive, or require very specific skills.

Ideally, equipment that can be maintained and repaired within the country should be used. If no local supplier is available, fast delivery and repair services need to be ensured, or adequate replacement components must be stocked at the plant. For example, the powerful vacuum trucks that are needed to empty settling-thickening tanks require specific maintenance skills, which are often not locally available in mechanical workshops (Figure 11.2). It is therefore recommended that contracts be prepared during the equipment acquisition process whereby conditions for the repair services, for example, the annual maintenance of vacuum trucks, is defined. When designing FSTPs that require the addition of consumables for the treatment process (e.g. lime or chlorine), the costs and availability of these needs to be assessed, as well as the requirements for safe storage. Other aspects that impact on O&M costs include emergency operation procedures during power or water outages, and any shipping or transportation charges for delivery of samples requiring laboratory analysis. The choice of technology should therefore not only be made based on installation costs, but also O&M costs.

11.2.4 Degree of mechanisation of technologies

The degree of mechanisation of the FSTP depends on the availability of spare parts, electrical power and trained operators. Where this is limited, passive technologies such as drying beds and stabilisation ponds might be better technology choices. If power availability is intermittent, technologies that utilise manual systems should be chosen over mechanical ones whenever possible. For example, screenings can be removed manually or by a mechanical rake, dried sludge can be transported with a mechanical shovel or with a wheelbarrow, and small composting piles can be mechanically aerated, while compost heaps need to be turned manually.

11.2.5 Final enduse or disposal of treatment products

The enduse or disposal of the treatment endproducts has an influence on the technologies and processes needed to achieve the required level of treatment (Chapter 10). This in turn, has a significant impact on the costs and skill levels required to operate and maintain equipment. In a simple FSTP where sludge is dried for disposal in a landfill or for enduses such as combustion, both of which do not require high pathogen reduction, less rigorous treatment and lower O&M costs are involved compared to a system that produces endproducts for use on food crops that are directly ingested without cooking (e.g. salad greens). Determining if the value associated with the enduse activities is outweighed by the technology and O&M costs needed to achieve the required levels of treatment is a key driver for FSTP technology design. Understanding the costs associated with the specific O&M and monitoring tasks for identified enduse activities assists in the planning of a FSM programme.

11.3 RECEIVING FAECAL SLUDGE AT THE TREATMENT PLANT

It is important to take the traffic patterns and the management of truck traffic in and out of FSTPs into consideration in order to maximise the efficiency of the receiving and off-loading processes. Receiving FS loads at the FTSP involves:

- traffic control; and
- approving the FS for discharge into the facility.

These aspects are discussed in the following sections.

11.3.1 Traffic control

At facilities which are used infrequently, traffic control is rarely an issue. In most cases, the employees at these facilities is mainly required for discharge approval and direction of trucks in the FSTP. On the other hand, at busy facilities, where vacuum or sludge delivery trucks and other vehicles may be competing to discharge their loads, operational employees can help facilitate rapid unloading by providing direction and assistance to drivers, and thereby avoiding accidents.

Traffic control is simplified through a well-designed facility layout. Access roads that allow vehicles to drive through after discharge rather than turn around are not only more efficient, but also safer. Mechanised unloading stations that record the drivers identification and discharge volume can also reduce O&M costs at busy facilities. The turning radius and weight of the largest trucks that will utilise the facility should be considered when planning roads and driveways. In addition, off-loading and truck parking areas should be level, and access roads should not have more than a 3% gradient.

11.3.2 Approving faecal sludge for discharge

Wastes from different sources can have widely differing characteristics, which may impact upon the operation of the FSTP. Residential FS (e.g. from pit latrines or septic tanks) is often relatively free of toxic chemicals. Restaurant FS, however, may have significant quantities of fats, oil and grease, especially if grease traps or interceptors are absent or not functioning properly. Similarly, FS from auto repair shops, dry cleaning establishments, hospitals, or other commercial or institutional settings may contain toxic materials that are detrimental to the treatment process. In areas with a large number of commercial facilities, it is recommended that FSTP have parallel treatment trains, one that can accommodate residential sludge, and another for commercial wastes.

Depending on the institutional framework, and the arrangement between the stakeholders in charge of the collection, transport and treatment, a manifest system can be utilised to record the origin, volume and special characteristics of FS. A form can be completed at the origin of the FS and signed by the owner (Figure 11.3). Where the trucks frequently contain FS from several onsite technologies, the form should include this information. The manifest is then carried by the driver and presented at the FSTP for review by operations employees prior to off-loading. Once the load is approved, the manifest is then signed by the operator and returned to the driver as proof that the waste load was discharged into the facility.

Manifest Form

Sludge / septage origin

Name (Household unit owner) ____________________

Address ____________________

____

Date and time of collection ____________________

Source and volume of sludge/septage

Source	Check one	Volume (cubic metre)
Residential		
Commercial / industrial		
Institutional		
Wastewater treatment plant		

Commercial / industrial waste must be sampled and tested before it is offloaded at the treatment facility to ensure that the material will not contaminate the treatment process. Contamination can be caused by grease, oil, metals and chemicals.

Description of commercial / industrial waste:

____

____

Excavator / transporter

Operator / company	
Address	
Type of vehicle	
Plate number	
Name of driver	
Signature	
Driver's license number	
Name of other personnel	

Approved by authorised representative

____

(Name and signature)

Figure 11.3 Manifest form identifying the origin of the load, waste volume and driver's name adapted from the Philippines Department of Health (2007).

Operators of FSTPs should be trained in the physical inspection of sludge samples. If there is any doubt as to the origin of the load, samples should be drawn and inspected for color, odor, and presence of grease or oil. FS from residential sources has a distinct visual appearance, as do loads contaminated with excessive oil and grease. Loads that do not conform to standards that have been established for the treatment process should be rejected if segregation is not possible.

11.4 OPERATION & MAINTENANCE PLANS

The O&M plan provides details on the tasks, materials, equipment, tools, sampling, monitoring and safety procedures which are necessary to keep the plant running properly, all of which have cost implications that must be carefully considered.

11.4.1 Operational procedures

FSTPs require clear operational procedures. Therefore, the O&M plans should include an operation manual, containing the following information:

- the engineering drawings and FSTP specifications;
- the manufacturer's literature and equipment operation guidelines;
- the responsible person for each task;
- the frequency of each activity;
- the operation procedures and tools required to perform the task;
- the safety measures required; and
- the information that is to be monitored and recorded.

If chemicals or other consumables are required for the operation of a specific component, they should also be listed together with the name of the supplier and information on how they are to be used and stored. If some operational activities require the use of external companies, or if a transport company is needed to discharge the endproducts, their contact and description should also be given in the operation manual. The operation manual must also have a special section for emergency or non-routine operations requirements. Procedures should be planned for specific cases such as extreme climatic events, power shortages, overload, degradation of a pump, basin or canal, and other accidents. All procedures provided in the operation manual must be prepared in order to ensure conformance with the local laws and standards.

The treatment technologies described in Chapters 5 to 9 all require the control of the following aspects:

- screenings removal;
- load (quantity, quality and frequency);
- processing (e.g. mixing compost pile, chemical addition for mechanical drying);
- residence time;
- extraction, further treatment or disposal of endproducts;
- collection and further treatment or disposal of liquid endproducts; and
- storage and sale of the endproducts.

The operational procedures should take the climate and the other context-dependent variables into account. The drying time or retention time may vary greatly during intensive rain periods or droughts. Rain events may also increase FS volumes delivered to the FSTP if the onsite sanitation systems were not built adequately, due to runoff or a rise in the groundwater table. The operational activities at the FSTP can then be planned to take these aspects into account. For example, macrophytes of planted drying beds can be weeded during a dry season, when there is potentially less FS to treat, and there is a shorter drying time.

The operational procedure also needs to take the FS characteristics (e.g. viscosity, amount of waste, fresh or partly stabilised sludge), and the required level of treatment into account. The information collected though the monitoring system also needs to be considered in order to improve the operational procedure and planning. For example, the frequency of sludge extraction from a settling-thickening tank or from a waste stabilisation pond can be adjusted based on the observed quantity of sludge accumulated over time.

11.4.2 Maintenance procedures

There are two main types of maintenance activities: preventative maintenance and reactive maintenance. Well-planned preventative maintenance programs can often minimise reactive interventions to emergency situations, which are frequently more costly and complex. Component breakdowns at FSTPs can result in wider system failure, or non-compliance. Therefore, each component at the FSTP has specific preventative maintenance requirements that need to be described in detail in a maintenance plan including the tasks, frequency of actions, and step-by-step procedures for accomplishing the tasks, including inspections. Physical inspections conducted at scheduled intervals are important, where operators look for specific indicators such as cracked wires, broken concrete and discolored and brittle pipes in order to identify preventative maintenance needs.

The maintenance plan should be guided by the local context, the climate, and the asset-specific monitoring information. Coastal FSTPs, for example may require more frequent painting and corrosion control due to the salt air compared to the same plant located inland. The task details include the equipment, tools and supplies needed to accomplish the task and the amount of time it should take to complete. Once completed, the task details should be entered into the equipment maintenance log book or database, along with any difficulties encountered.

Frequent maintenance tasks include:

- corrosion control – scraping rust, painting metal surfaces, and repairing corroded concrete;
- sludge and coarse solids extraction from the basins and canals;
- repacking and exercising valves (i.e. locating and maintaining fully operational valves);
- oiling and greasing mechanical equipment such as pumps, centrifuges or emptying trucks; and
- housekeeping activities including picking up of refuse and vegetation control.

11.5 ASSET MANAGEMENT

Asset management is a holistic approach to FSTP maintenance in order to maximise long-term effectiveness of the facility at the lowest possible cost. Cost items that are included in the full lifecycle costs of an asset include:

- capital cost of purchasing and installation;
- labour required for operation and maintenance;
- spare parts for repairs;
- essential consumables, such as grease or chemicals; and
- replacement costs once the component has reached the end of its useful life.

Integral to the full lifecycle costs are the stocks of tools and supplies that are required for long-term operational needs. These should ideally be available at each FSTP site (Lüthi *et al.*, 2011). If several FSTP rely on the same technology or equipment, centralised stocks can be organised.

Asset management is crucial for large FSTPs and the following aspects should be included in the maintenance plan (USEPA 2012):

- the current state of the assets;
- the required 'sustainable' level of service;
- the assets which are critical to sustained performance;
- the minimum life-cycle costs; and
- the long-term funding strategy.

Without an asset inventory, no comparison can be made on the cost of equipment or the importance of the asset. Components that are crucial for the operation of the FSTP should be highlighted, and once used, replenished immediately. In these cases it is therefore important to have a reputable provider and agreements drawn up to ensure swift service. Case Study 11.1 provides an example of a FSTP failure due to the lack of permanent employees and the pump not being listed as a key component.

Case Study 11.1: Example of treatment plant failure
(Adapted from Bassan, 2009)

A FSTP was constructed with one screening channel, two parallel settling-thickening tanks, nine unplanted drying beds and a pipe conveying the liquid fraction to the waste stabilisation ponds of the wastewater treatment plant located nearby.

In 2009, after less than 5 years of operation the FSTP was out of order for some months, despite the selection of robust technologies. This was partly due to the design process that had resulted in the selection of pumps that were not powerful enough to extract the thickened sludge from the tanks, but also due to insufficient sludge extraction by the vacuum trucks. As a result, the settling-thickening tanks were not emptied for several months, the sludge was not dried on the beds, and the waste stabilisation ponds were saturated with high loads of suspended solids. Additionally, no maintenance was carried out on the beds and the filter media, resulting in degradation of the walls and the valves. Consequently, significant resources were needed to remove the weeds and to once again ensure good treatment performance.

This situation was the result of a weak human resource (HR) strategy, a lack of precise procedures for O&M, and a rigid administrative system. There were no permanent employees at the treatment site, and daily workers were often hired without any training. This mode of recruitment does not encourage accountability which is necessary for careful maintenance, and nor does not allow for continuous operational activities. Additionally, no skilled mechanical technician was hired to repair the pump. Once this information was communicated to the head office, the required repair and maintenance work was carried out, and the FSTP was again able to operate efficiently.

This example demonstrates the extent to which the priority level given to HR operating the FSTP can influence the performance and the long-term viability. It is therefore essential to have sufficient budget in order to hire skilled and permanent employees at a FSTP. It also highlights that the operation of a FSTP requires a flexible internal management process. If the hierarchical procedure is overly time consuming and complex, repairs or improvements are not possible at short notice and may result in the deterioration of the FSTP.

11.6 MONITORING

The maintenance of a FSTP involves a detailed understanding of the treatment processes and performance requirements. This understanding should not only be based on the theoretical information concerning the treatment mechanisms and the design of the technology, but also on a monitoring procedure that requires specific planning, infrastructure (e.g. laboratory), employees, and finance. The monitoring programme should be structured in order to provide the operations employees with adequate information to continuously optimise the plant performance, and to provide control over the effluent quality. Monitoring may include a range of different methods such as:

- visual or sensory inputs: this includes visual observations of plant conditions, such as scum on a treatment lagoon, the color of the sludge, or odours emanating from a pump tank;
- analysis or measurement at source: this includes test strips or kits that can be utilised in the field for measuring pH, dissolved oxygen, or temperature; and
- laboratory testing of samples (either onsite or offsite).

Monitoring is expensive and time consuming. A written monitoring plan is essential and will assist operators in collecting and organising the data that is required, relevant, and accurate. This plan is based on the following aspects:

- why the information are required;
- what information will be obtained;
- how and when the information or samples will be collected in the field; and
- who will collect them.

11.6.1 Monitoring of physical-chemical and microbiological parameters

Planning an efficient laboratory analysis programme provides the data necessary for making operational decisions and reporting findings. The more accurate and timely the information is, the better the operational decisions that can be made. For example, the load and residence time in a waste stabilisation pond or in an anaerobic digester can be adjusted based on the results of the laboratory analysis. If the laboratory analyses reveal biochemical oxygen demand and suspended solids values above the discharge standards, the residence time in the basins can be increased, and the treatment performance improved.

The 'Chain of Custody' form is the mechanism by which the sampler at the FSTP communicates with the laboratory with regards to the samples taken and analytical tests requested. It provides a written record of field sampling conditions, special instructions, and a list of who was responsible for the samples at all times. Specific information includes:

- sample identification;
- data related to the site conditions at the time of sampling;
- instructions to the laboratory as to which analytical tests to perform on each sample; and
- the date, time and signature of each person that maintains custody of the sample.

The parameters that are most often analysed include (HPCIDBC, 2011):

- the solid and suspended matter content: these analyses assist in the evaluation of the settling and solid/liquid separation performances (Figure 11.4);
- the moisture content of the endproducts: this parameter provides an estimation of the drying performances;
- the biological and chemical oxygen demand in the liquid fraction: these parameters monitor the available oxygen which has a direct impact on aquatic life;
- the nutrient content (i.e. nitrogen and phosphorus) which influences the potential for resource recovery in agriculture, as well as the risk of eutrophication of water bodies; and
- the pathogen content: this involves an evaluation of the presence and number of *E Coli*, faecal coliforms or helminthes eggs which allows control of the risks related to waterborne diseases.

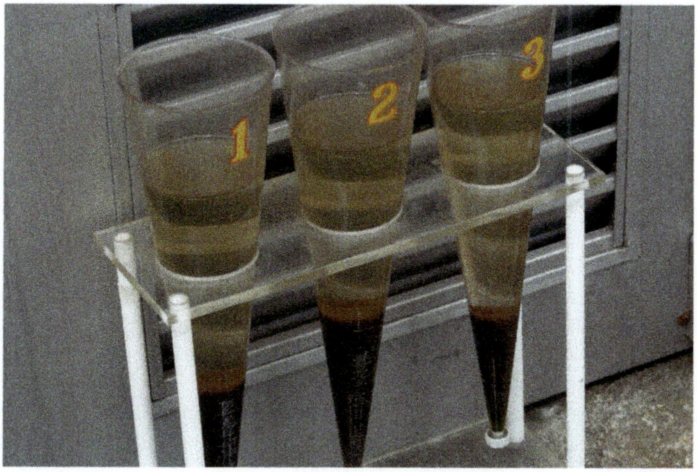

Figure 11.4 Settleability tests performed on site at the Manila Water South Septage Treatment facility in the Philippines (photo: David M. Robbins).

These monitoring parameters can be adjusted depending on the technologies used, on the local effluent discharge standards, and on the enduse objectives (see Chapters 2 and 10). For example, assessment of the pathogen content may not be necessary if the endproducts are to be used as a fuel in a cement kiln, but pH may be a very important factor for loading an anaerobic digester.

Laboratory monitoring requires strict procedures and skilled employees, as well as significant funds to operate and maintain the analytical equipment and infrastructure, and to purchase the required consumables. A specific laboratory budget is therefore required. Some technologies involve more complex laboratory monitoring to ensure an efficient process (e.g. composting, activated sludge, lime treatment), while others only require laboratory analyses to evaluate the treatment performances. Laboratories also require quality assurance and quality control (QA and QC) procedures.

Where specific analyses are required, external laboratories can be contracted to undertake these procedures. Contract laboratories are an important source of information and support to FSTPs operation. If external laboratories are to be used for the monitoring programme a clear definition of sampling techniques, preservation methods for maintaining sample integrity, and procedures for sample analysis are required. FSTPs which make use of contract laboratories may request copies of the QA and/or QC plan in order to review procedures and verify that they will meet the required needs.

11.6.2 Analysis manual

If laboratory analysis is required for a specific FSTP, an analysis manual shall be provided, encompassing the following information:

- the sampling frequency, site and procedure (e.g. grab or composite), and the conditions under which these samples should be transported;
- the storage of the samples and the chemicals (e.g. the type of container, the chemicals required and the temperature);
- the analyses protocol for each parameter; this should be based on standardised methods if possible;
- QA/QC plan for sampling and any onsite analytical activities to ensure the accuracy of the analytical data;

- requirements for split or duplicate samples, or travel blanks; and
- information on the calibration and maintenance of the laboratory and onsite equipment (e.g. probe for oxygen content and pH evaluation).

11.7 RECORDKEEPING

Effective O&M programmes for FSTPs require that accurate records be kept of all O&M activities, monitoring as well of any malfunctions. Operators frequently refer to records in order to identify previous fluctuations in the operation of the facility and operational problems that may recur periodically, review the effectiveness of mitigation measures that may have been used to correct past operating problems, and to optimise the O&M procedures. These records should therefore be easily accessible to FSTP operators.

Some examples of recordkeeping that are useful for FSTPs include:

- information on the operation of the FSTP including daily operating records, the operators log book, manifest reports (an example is provided in Figure 11.5), the treatment unit operating data sheet, and other records related to FS deliveries to the plant;
- disaster response and emergency recovery records;
- preventative and corrective maintenance records including the equipment maintenance log books and store room supply reports;
- compliance reports including field and analytical data, and correspondence from regulatory officials; and
- employee records, such as employee schedules, time sheets and injury reports.

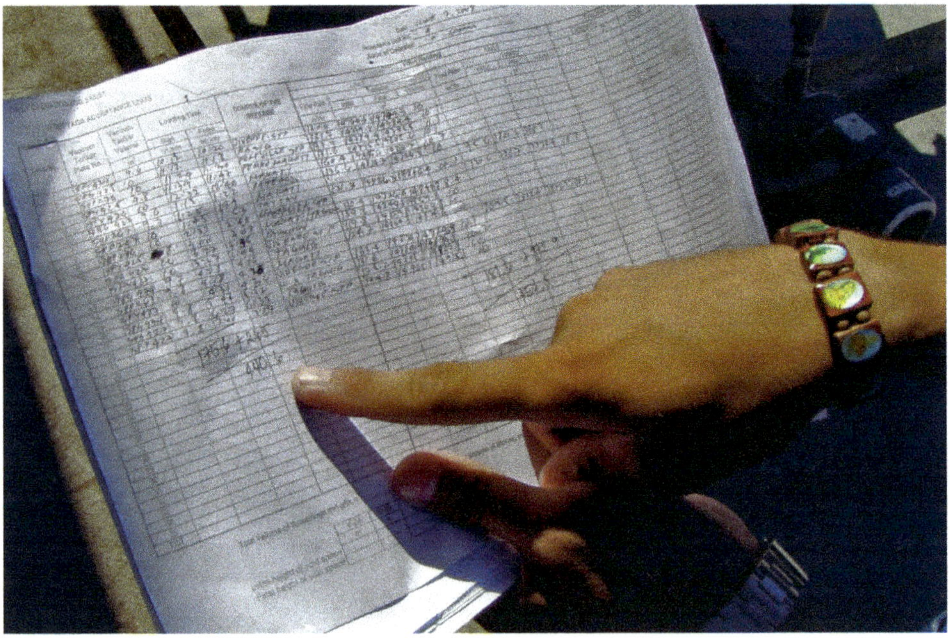

Figure 11.5 Reception reports track the total number of loads delivered, the time, date and driver's name. These records are important to maintain for all faecal sludge treatment plants (photo: David Robbins).

Management

The type of records and the length of time for which they will be retained for a particular facility will be determined by the size of the FSTP, regulatory requirements, and the technologies that are used. Since these records are tools that can be used by employees to assist in the day to day operation of the facility, a summary of the information should be used to optimise the O&M plan, as well as in the planning of any expansion to a FSTP or in the design of new FSTPs. An explanation of some of the key recordkeeping aspects is provided in the following sections.

11.7.1 Operator's log book

The operators log book is perhaps the most important record for a FSTP. This log book provides a means of communication between operators of the plant and a written record of important events. Typical entries include the names of people on duty, weather conditions, any equipment malfunctions, operating problems, important phone messages, security information and actions taken in response to unusual circumstances. An excerpt from a typical operator's log book is provided in Case Study 11.2 from the New Jersey, US Administrative Code on Wastewater Management.

11.7.2 Reception monitoring reports

Reception monitoring reports record the amount of FS received at the plant each day, the discharge fees collected, and any issues reported by drivers or employees. Maintenance of accurate reception monitoring reports is critical as it minimises fraud and assists in guaranteeing that the collected FS was delivered to the FSTP and not discharged elsewhere.

Case Study 11.2: Excerpt from the New Jersey, US administrative code on wastewater management (operator's log book)

The results of all mechanical equipment and related accessoires inspections essential to the proper O&M of the system shall either be recorded in ink and maintained in bound inspection log books or be maintained in secured-access computer databases or files or other equivalent method of recordkeeping. The log books or computer databases, or file or equivalent shall also include:

- time, date and subject of all system inspections;
- a report of all breaks, breakdowns, problems, bypasses, pump failures, occurrences, emergencies, complaints and/or intervening factors within the system that result in or necessitate deviation from the routine O&M procedures; and any situations that have the potential to affect public health, safety, welfare, the environment or have the potential to violate any permits, regulations or laws;
- a record of the remedial or follow up action and protocol taken to correct all of the above issues; and
- the date and time of each entry, and by whom it was entered.

11.7.3 Treatment unit operation sheets

Treatment unit operation sheets are used to record the quantity of FS loaded into each treatment unit, the operational activities performed (e.g. load of FS or extraction of endproducts), the operational variable applied (e.g. mixing ratio of fresh to stabilised sludge, addition of lime), the quantity of endproducts and wastes extracted, and the consumables required. The number of employees required and the relevant skills needed to perform all the activities should also be recorded, together with any difficulties encountered and potential solutions. These sheets therefore provide historical records of the maintenance carried out on each piece of equipment, the failures experienced and the solutions implemented, together with the budget and HR involved. Distinction should be made between preventative and reactive maintenance, and recommendations for optimising the planning process made.

11.7.4 Interpretation and communication of technical data

The data collected in the laboratory and from onsite monitoring (i.e. log books, reports and operation sheets) are used in conjunction with one another in order to optimise treatment performances through the adjustment of O&M procedures. For example, the volumetric load of FS on planted drying beds can be adjusted through a comparison of the laboratory results and with observations on the pollution load and residence time (Koottatep *et al.*, 2005). The optimal operating conditions can then be identified, and the treatment performances improved.

All information collected through the monitoring program and recordkeeping should be analysed, and reports prepared for internal communication. An effective communication system is crucial for the optimisation of the administrative and operational management procedures, and also ensures that all the employees have comprehensive information on the operation of the FSTP. This communication system should therefore also define the frequency of delivering reports and the decision making process that is to be followed.

To ensure that the monitoring data and reports are used, that the correct conclusions are made, and that follow up action is taken, the laboratory analysis reports should be made available to the operating employees, and the operational reports made available to the management. In order that the significance of the laboratory results are understood, both the laboratory technician and the FSTP operating employees need to be suitably trained. If the laboratory analysis data provides results which lie outside of the expected range, the laboratory technician and operating employees need to meet to discuss the necessary adjustments to the operational activities. All data recorded in the O&M monitoring sheets and in the laboratory analysis reports is then captured in a summary report or in a database which provides an overview of the FSTP performance and difficulties over the previous months and years. For example, it is important to know how often a pump fails over a period of one year in order to adjust the maintenance planning programme, and whether to install a better prescreening process or an improved pumping unit. O&M activities are also affected by the seasons and need to be considered in the O&M plan in order to optimise the operational activities under these different conditions.

11.8 PLANT SECURITY AND SAFETY

FSTPs are critical infrastructures and must therefore be secured from unauthorised entry and vandalism by fencing off of facilities and engaging security employees. Managers of FSTPs can also create a culture of security by enacting the following guidelines:

- including security as a topic in employees meetings and discussions;
- appointing a Plant Security Officer or assigning the duties to a responsible employees member;
- enforcing security policies and procedures consistently and equitably; and
- providing security training for all employees.

11.8.1 Health and safety

There are many health and safety hazards associated with the typical tasks required to operate and maintain FSTPs. Health and Safety aspects should therefore form an integral part of the O&M plan but are quite often not given adequate attention.

Figure 11.6 Safety posters and signs are good reminders to follow proper procedures (photo: David M. Robbins).

The "Health and Safety Plan" specifies the procedures, practices and equipment that should be used by employees in order to conduct activities in a safe manner. Health and safety plans are prepared specific to each FSTP but also contain aspects that are common to all FSTPs. Health and safety procedures are strictly enforced by management through the preparation of the safety plan, and also through posters and signs located in areas of risks (e.g. ponds and tanks, electrical device, confined spaces). An example of a safety notice is provided in Figure 11.6. Based on the authors' experience, the following topics should be included in the health and safety plans:

- personal protective equipment (PPE) and safety measures for O&M activities;
- infection control and hygiene measures;
- emergency contact procedures;
- protection against falling and drowning hazards;
- confined space entry protection; and
- electrical safety and the use of the '*Lock-Out Tag-Out*' procedure.

Further details and recommendations can be found on the Occupational Safety and Health Administration (OSHA) website (http://www.osha.gov/), and the following sections explain each of these aspects in more detail.

11.8.2 Personal protective equipment

Personal protective equipment (PPE) is equipment worn in order to minimise exposure to hazardous conditions, and includes:

- hard hats to provide head protection from falling items;
- eye protection such as safety glasses, goggles or face shields to protect against chemical or dust exposure;
- gloves for hand protection from chemicals or abrasion, made from rubber latex or other materials dependent upon the specific hazard;

- breathing safety devices such as respirators, dust masks or self-contained breathing apparatus (SCBA), should certain tasks require them;
- other protective clothing including foot protection, and coveralls; and
- other equipment required for task specific safety.

While the health and safety plan specifies the PPE required for each task, it is the management's responsibility to ensure that appropriate PPE is provided, that employees receives training in the proper use of PPE, and that employees are complying with the requirements regarding PPE usage.

Clear safety procedures are also required for all O&M and monitoring activities at the FSTP, including the receiving and movement of trucks; the discharge of FS, the O&M of equipment, the use, storage and disposal of chemicals, the sampling of various processes and the processing and removal of endproducts. For example, safety requirements for the receiving of trucks and FS discharge include the use of chocking wheels during off-loading or when trucks are parked, wearing personal protective clothing, and the prohibition of smoking.

11.8.3 Infection control

FS, by its nature is infectious material. It often carries disease-causing bacteria, viruses or other pathogens. Workers should have proper immunisations (e.g. hepatitis A, tetanus), and follow hygienic procedures at all times when handling equipment that might have come into contact with fecal materials. Showers and areas to wash hands should be available for workers, as well as a locker room where workers can store clothes. Infection control procedures include:

- use of appropriate PPE to protect skin from contact with faecal material;
- washing hands prior to eating or after coming in contact with faecal material;
- no eating or drinking in areas where FS or chemicals are stored or processed;
- reporting illness to plant supervisors immediately; and
- prohibition against smoking, an activity that can transmit pathogens via the fecal oral route of entry.

11.8.4 Emergency contact procedures

Emergency contact procedures provide current telephone numbers and contact information that can be used by employees in the case of an emergency. The contact list should be posted in a common area that is accessible to all employees and which has access to an operational telephone. For all FSTPs, but especially those in remote areas, first aid materials, supplies and equipment must be provided. A typical emergency procedure consists of the following actions:

- contacting the appropriate emergency personnel;
- depending on the situation (e.g. explosion, fire or chemical spill), evacuating the employees;
- contacting the plant manager if not already on site; and
- providing support to affected personnel until emergency personnel arrive and take control of the emergency situation.

Emergencies must be documented on an emergency report form that is then sent to management for investigation. All emergencies must also be fully detailed in the operators log book.

11.8.5 Protection against falling and drowning hazards

FSTPs that utilise lagoons or waste stabilisation ponds, or even large reactor tanks need to have a drowning prevention programme in place that provides safety equipment, signage and training. Plants with large lagoon cells often have boats from which O&M tasks are accomplished. In these situations, workers must make use of floatation devices, work in pairs, and be trained in proper procedures to minimise the hazard of drowning. At all FSTPs, measures should be taken to avoid slip hazards such as preventing the spilling of FS, as well as ensuring that manholes are closed in order to avoid falls.

11.8.6 Confined spaces

A confined space is defined as any place in a FSTP that is enclosed and has limited access, such as tanks and dry wells. They are potentially hazardous as the breathable atmosphere may become compromised, either by a depletion of oxygen or the presence of chemical gasses, such as chlorine or hydrogen sulphide. In order to prevent confined space accidents, a "Confined Space Entry Permit" programme is utilised at FSTPs.

The first step in this programme is for senior management to identify all confined spaces in the plant. When maintenance is required inside these areas, certain procedures can be defined in order to protect the worker. These typically include the following:

- a confined space entry permit is prepared by the worker and signed by the supervisor;
- prior to entry, the atmosphere is tested with an oxygen meter or, in the case of manholes, with a hydrogen sulphide meter; and
- the work is conducted using the buddy system, with one person entering the confined space secured with a harness attached to a safety rope, and one person located outside of the confined space ready to provide assistance if needed. When the work is completed, the permit is returned to the supervisor for signature indicating the completion of the task.

11.8.7 Electrical safety

FSTPs with electrical equipment must enact specific procedures to keep workers safe when performing O&M activities on powered devices. An example of such a safety procedure is the *Lock-out Tag-out* procedure which ensures that the breaker to the power source for the equipment that is to be repaired is turned off and locked in the off position. A tag which specifies the work to be carried out, the person doing the work, and the date and time the work will be conducted, is attached to the locked out breaker. This tag must be signed by the plant or shift supervisor and the electrician doing the work. When the task is completed, the tag is removed by the supervisor and electrician, and the lock removed. Only then can the equipment be powered up.

11.9 ADMINISTRATIVE MANAGEMENT

Effective management of a FSTP requires a well-defined management strategy specific to each FSTP. If not incorporated in the management strategy, aspects such as employees coordination, planning, supervision, and capacity strengthening, it can result in reduced treatment performances. This can be due to poor operational skills of the employees, misunderstanding of the technical priorities by the administrative employees, poor communication, or poor financial management (see Case Study 11.1). The procedures for the O&M, and monitoring of the plant, as well as the communication requirements should be strategically defined by the decision makers, and tie up with the financial and HR of the company. These aspects are described in more detail in the following sections.

11.9.1 Financial procedures

It is recommended that financial procedures are defined based on operational needs. Therefore, the operating costs should be monitored, and the budget adjusted based on the actual expenses. The various types of costs that can be incurred are discussed in Chapter 13. Special provision and administrative mechanisms should be in place in case of breakdown of equipment that is crucial for the operation of the FSTP, as well as for the replacement of old equipment. The procedures for the acquisition of tools, other stock items, and safety equipment must be rapid, and special funds should be available for small repair work in order to ensure continuous operation (e.g. repairs to a screening grid or a valve). For example, if a valve or a pump is broken, the funds need to be available immediately for the repair, not after three or six months of budget approval process.

11.9.2 Human resource management

HR management refers to the way in which employees are managed and trained, including the definition of job descriptions, chain of authority, and policies and procedures for work place activities.

While HR management can be considered as a key aspect for the successful operation of any treatment plant, very often, no financial mechanisms are defined in order to ensure that sufficient and appropriate HR is available to operate the FSTP. HR requirements can be defined based on the specifications of the design consultants, and the operational requirements observed during the startup period. In some cases, where O&M activities may involve very specific skills or resources (e.g. mechanical skills to repair centrifuge or vacuum trucks) which are not available in-house, external services can be hired. Specific provisions are then needed to ensure that the required level of service is provided (see Case Study 11.3).In this case, the service and frequency must be well defined to allow continuous operation of the FSTP.

Case Study 11.3: Outsourcing of maintenance services for treatment plants

The National Operator for Water and Sanitation in Morocco (ONEP) has the responsibility for managing the operation of several wastewater treatment plants countrywide. Due to the wide territory covered, ONEP cannot afford the equipment and employees for specific maintenance activities for all the treatment plants (e.g. mechanical repair of pumps). Private companies are therefore hired on a 5 year contract basis to provide maintenance of the treatment equipment. Each company covers one region, answers to quality standards defined by ONEP, and the employees is trained at the ONEP training center. This type of organisational structure results in optimisation of the equipment and operational costs as well as ensuring a maintenance plan for the treatment plants

Such dependency on external services must be well managed. Long-term collaboration should be encouraged, and quality standards well defined. If this external service includes the maintenance of key equipment, and cannot be planned precisely, the service must be available at short notice at any treatment site.

Irrespective of the size of the FSTP, employees should have defined roles and responsibilities in order to ensure complete understanding of specific job requirements. HR aspects of FSTPs therefore include:

- description of the lines of communication indicating who the employee reports to;
- outline of the level of authority required for making operational decisions; and
- appropriate and ongoing training to ensure that employees can carry out their responsibilities.

11.9.3 Staffing, roles and responsibilities

FSTPs can have a broad range of staffing requirements depending on the size of the plant, the treatment volume and the required level of skill.

An organisational chart that clearly specifies the roles and responsibilities of each employees member, as well as the lines of communications is a useful management and training tool which should be defined during the design and planning phase. Employees are recruited through HR management systems as described above, complete with job descriptions for each employee classification.

Smaller FSTPs may combine various job titles such as plant superintendent, safety officer, and maintenance technician into one job description. The following sections outline the key employees requirements and the respective responsibilities which are crucial for the long-term operation of FSTPs.

Plant superintendent

The FSTP superintendent forms part of the management team and is responsible for the day to day management of the FSTP. The superintendent defines the goals, objectives, policies and priorities, concerning the O&M, and is responsible for:

- all paperwork and correspondence, grounds and equipment maintenance, and supervision of personnel;
- participating in the development and implementation of goals, objectives, policies, and priorities;
- coordinating the organisation, staffing, and operational activities including assuming responsibility for critical decisions regarding operational changes, process control, maintenance priorities, scheduling, and compliance;
- identifying opportunities for improving O&M, monitoring and safety methods and procedures;
- directing, coordinating, and reviewing the work plan for O&M functions;
- directing the testing of various treatment phases, interpreting tests to determine necessary changes in treatment parameters;
- directing the adjustment and repair of equipment such as pumps, chlorinators, metering devices, electrical control panels, and treated or digested sludge dewatering;
- serving as a team member on construction project teams with construction management companies and contractors;
- selecting, training, motivating, and evaluating assigned personnel;
- overseeing safety programs for assigned sections and work groups and assisting with action planning for safety programs; and
- participating in the development and administration of assigned programme budget.

Figure 11.7 Sludge removal from drying beds at the Bugolobi Treatment Plant in Kampala, Uganda (photo: Linda Strande).

Plant engineer

The FSTP engineer serves as the chief technical employees member. Typical roles and responsibilities include:

- ensuring the overall efficiency of the plant and optimisation of the treatment process;
- controlling operating expenses;
- organising and coordinating the work carried out by subordinate teams (e.g. sludge removal from drying beds as shown in Figure 11.7);
- recommending technical solutions to problems that may be encountered;
- contributing to the monitoring and reporting on the performance of equipment and processes; and
- managing technical subcontractors and suppliers.

Plant operator

The FSTP operator is responsible for carrying out the day-to-day technical aspects of plant operations in order to ensure that equipment is operating properly and in compliance with all requirements. Typical duties include:

- performing equipment inspections, monitoring operations, and collecting samples in order to verify system performance in collaboration with laboratory employees;
- operating trucks, pumps, blowers, generators, compressors, and other machinery/equipment;
- testing, calibrating, repairing, and operating control and instrumentation systems under general supervision;
- keeping records of operational activities, degradations and failures;
- preparing field and office reports summarising the records and providing recommendations for optimising the system; and
- assisting in site environmental investigations, field surveys, and cleanups as required.

Plant maintenance technician

The FSTP maintenance technician performs routine and emergency maintenance and repairs on plant facilities, pumps, engines, motors, filters, bar screens, valves, pipes, and other equipment at the FSTP. Typical responsibilities include:

- checking, adjusting and maintaining mechanical equipment including greasing of moving parts, changing oil, and performing other routine maintenance activities;
- maintaining buildings, roads and grounds;
- performing janitorial work;
- replacing worn parts and performing routine and emergency service and repairs including replacing motors, bearings, flanges, seals and other equipment components;
- inspecting mechanical and hydraulic equipment being installed under contracts to ensure compliance with contract requirements;
- monitoring facilities and equipment in order to identify and repair leaks or other malfunctions; and
- keeping records through the logging of maintenance activities and repairs, and preparing reports summarising the main activities, malfunctions and recommendations.

11.10 COORDINATION

Communication should be encouraged between the O&M and monitoring employees of different FSTPs in the same jurisdiction, as well as with the decision makers. An effective vertical communication ensures that the administrative employees understands the constraints and needs of the O&M employees, and results in rapid acquisition of parts or repairs in order to ensure continuous operation of the FSTP. Horizontal communication between the different FSTPs allows the exchange of experiences and therefore assists in the optimisation of the procedures. Frequent (weekly or monthly) meetings should be held in order to facilitate discussions between the operating, monitoring and administrative employees on the difficulties experienced and possible solutions. If the operating company is in charge of several FSTPs, one person can be designated to ensure quality control and harmonisation of the O&M procedures over all the facilities. This would result in the adjustment of procedures and guidelines based on experiences, the standardisation of these for all similar FSTPs, and would ensure the uniform implementation of safety rules and O&M procedures.

11.11 STARTUP PERIOD

For newly built FSTPs, a transition period is necessary at the beginning of operation in order to evaluate the preliminary procedures. This allows definition of the frequency, safety measures and communication lines for the operation, maintenance and monitoring activities. During this startup period, there should be frequent communication between the operating and administrative employees in order to discuss any problems. The final procedures and documents (i.e. operation manual, information sheets, monitoring sheets, logbooks etc.) will be prepared based on the information collected during this startup period.

For some treatment technologies, the startup period may involve specific procedures. For example, biogas digesters need to be started up slowly to allow for the development of the appropriate anaerobic microorganism community, and planted drying beds need to be progressively loaded to allow the acclimatisation of plants. Even though the infrastructure and equipment may be operational within a relatively short time period (e.g. unplanted drying beds, settling-thickening tanks), the following operational aspects should be assessed and optimised during the startup period:

- quantities of FS discharged in the FSTP;
- truck circulation in and around the FSTP;
- removal frequency, and quantities of screened wastes;
- loading of the treatment unit(s);
- organisation of the activities required for the treatment process (e.g. turning the heaps in co-composting plants or in solar sludge driers);
- removal frequency, and quantity of the endproducts from the treatment unit(s);
- time and conditions required for efficient stabilisation and pathogen removal depending on the enduse goals;
- frequency and type of routine maintenance activities; and
- frequency and interpretation of the monitoring analysis and observations.

The time required for the startup period may differ depending on the technology used. For example, the acclimatisation of macrophytes on planted drying beds or lagoons (Figure 11.8) may require between 3 and 6 months until the nominal treatment efficiency is reached. For some technologies, it is also important to plan the startup period given the seasonal climatic variations, as these influence the operational activities and performance. For example, the time needed for FS to dry at the surface of unplanted drying beds may differ greatly during dry and rainy seasons in arid climates. The quantities of FS produced may also vary based on the rainfall patterns. Therefore, an assessment of the ideal loads and retention times during dry and rainy seasons, or warm and cold seasons is useful, and it is recommended that the startup period covers at least two seasons.

To ensure a successful startup period, the entire employees should be trained in order that they understand all the necessary procedures before the commissioning of the FSTP. Therefore, site visits to similar treatment plants should be organised, and basic information on the treatment mechanisms provided. During the startup period, the operator may need technical and managerial assistance from experts in the field.

Figure 11.8 Starting up period of faecal sludge lagoon system, San Fernando City, Philippines. In this case, lagoon basins were seeded with activated sludge from a nearby wastewater treatment plant (photo: David M. Robbins).

The operating hours of the FSTP and the procedures for FS discharge (e.g.. FS characteristics and discharge fees) should be monitored over several months and discussed with the collection and transport stakeholders. Similarly, the treatment efficiency of the plant, and the quantity and quality of endproducts needs to be assessed, and the enduse or disposal procedures defined and agreed upon with the relevant stakeholders.

At the end of the startup period, all the administrative, operational, maintenance, monitoring and communication procedures should be defined and well understood by the entire employees. Final versions of tools such the O&M plans and manuals, laboratory reports, monitoring sheets, and health and safety plans should be developed validated and enforced.

11.12 BIBLIOGRAPHY

Bassan, M. (2009). Méthodologie d'évaluation des facteurs de succès et d'échec des stations de traitment des eaux usées et des boues de vidange à grande échelle. A methodology to evaluate success or failure of full-scale wastewater and faecal sludge treatment systems. Master Thesis, Ecole Polytechnique de Lausanne, Switzerland; available from http://infoscience.epfl.ch/record/140429.

Bassan, M., Mbéguéré, M., Tchonda, T., Zabsonre, F., Strande, L. (2013). Integrated faecal sludge management scheme for the cities of Burkina Faso. Journal of Water, Sanitation and Hygiene for Development 3(2), p.216–221.

Bräustetter, A. (2007). Operation and maintenance of resource-oriented sanitation systems in peri-urban areas. Fakultät Umweltsicherung, Fachhochschule Weihenstephan Master.

Fernandes, A., Kirshen, P., Vogel, R. (2005). Faecal Sludge Management, St. Elizabeth, Jamaica. Impacts of Global Climate. Anchorage, AK.

HPCIDBC (2011). Status and Strategy for Faecal Sludge Management in the Kathmandu Valley. High Powered Committee for Integrated Development of the Bagmati Civilization, Nepal.

Koné, D. (2002). Epuration des eaux usées par lagunage à microphytes et à macrophytes en Afrique de l'Ouest et du Centre: Etat des lieux, performances épuratoires et critères de dimensionnement. Faculté Environnement Naturel, Architectural et Construit. Lausanne, Switzerland, Ecole Polytechnique Fédérale de Lausanne. PhD thesis.

Koné, D. (2010). Making Urban Excreta and Wastewater Management contribute to Cities' Economic Development - A paradigm shift. Water Policy 12(4), p.602-610.

Koottatep, T., Surinkul, N., Polprasert, C., Kamal, A.S.M., Koné, D., Montangero, A., Heinss, U., Strauss, M. (2005). Treatment of septage in constructed wetlands in tropical climate: lessons learnt from seven years of operation. Water Science and Technology 51(9), p.119-126.

Lennartsson, M., Kvarnström, E., Lundberg , T., Buenfil , J., Sawyer, R. (2009). Comparing Sanitation Systems Using Sustainability Criteria. EcoSanRes, Stockholm, Sweden.

Lüthi, C., Panesar, A., Schütze, T., Norström, A., McConville, J., Parkinson, J., Saywell, D., Ingle, R. (2011). Sustainable Sanitation in Cities: A Framework for Action. Rijswijk, The Netherlands, Papiroz Publishing House.

End of Chapter Study Questions

1. What are important operations and maintenance factors that should be taken into consideration when planning FS treatment plants, and why are they important?

2. List three site-specific factors that could have an impact on the operation and maintenance of FSTPs.

3. Give four examples of types of records that need to be collected in the operation and of FSTPs.

4. Explain why monitoring is critical in the ongoing operation of FSTPs.

Management

Chapter 12

Institutional Frameworks for Faecal Sludge Management

Magalie Bassan

Learning Objectives

- Be able to identify important management aspects that need to be incorporated in an institutional framework.
- Understand regulations and contracts that can be used to to ensure effective faecal sludge management.
- Understand the strengths and weaknesses of stakeholders roles in institutional frameworks.
- Obtain an overview of potential institutional arrangements for the distribution of responsibilities in the service chain.
- Understand the main advantages and drawbacks of different institutional arrangements.

12.1 INTRODUCTION

For the successful implementation of faecal sludge management (FSM) systems, an institutional framework needs to be developed based on the specifics of the local situation (Ingallina *et al.*, 2002; Koné, 2010; Lüthi *et al.*, 2011). The focus of the FSM service chain in this book is collection and transport, treatment and enduse or disposal. This service chain depends on an effective management system. Laws and strategies need to be clearly defined, including regulating and enforcing the roles and responsibilities of each stakeholder throughout the entire service chain. This comprehensive approach incorporating multiple levels of institutional aspects requires a strong commitment by the government (Strauss and Montangero, 2003) that is linked to their sanitation policy, including onsite sanitation in the short-, medium- or long-term. Therefore, the FSM institutional framework requires dedicated funding and training strategies (Strauss and Montangero, 2003; AECOM and SANDEC/EAWAG, 2010).

Adequate attention to organisational aspects is rare and unfortunately many projects only consider one aspect of the service chain (e.g. subsidising septic tanks or only building a treatment plant). There are several examples where governments have focused only on the physical infrastructure and not the organisational or financial aspects, and as a result experienced failures of their FSM systems (Koné, 2010).

The institutional framework is defined by the laws, contracts and regulatory documents that determine the relationships between the stakeholders involved in FSM, and it defines the organisation of the entire service chain. This chapter focuses on institutional aspects that ensure the sustainable management of the service chain in the following three sections:

- Success Factors (Section 12.2);
- Enabling Regulatory Environment (Section 12.3); and
- Institutional Arrangements (Section 12.4).

This chapter presents a broad overview and introduction to the topic, and related information can also be found in Chapters 13 and 17. The selection of an adequate institutional framework is part of the planning process, and it requires a detailed assessment of the situation (Chapters 14 and 15), and participatory involvement of the stakeholders (Chapter 16).

12.2 SUCCESS FACTORS

The selection of a FSM institutional framework must be driven by local socio-economic, climatic and environmental contexts, taking into account existing sanitation infrastructures, institutions and planning procedures (Ingallinea *et al.*, 2002). Important factors for success that need to be considered when defining an institutional framework are discussed below (Klingel, 2001; Pybus and Schoeman, 2001; Bolomey, 2003; Jeuland *et al.*, 2004; Moe and Rheingans, 2006; Bassan *et al.*, 2014). These factors can be considered as objectives for the different stakeholders concerned (e.g. managers, politicians, practitioners). The implementation of these objectives depends on the local context. For example, the coordination of the local stakeholders will require more effort if several private companies are in charge of different activities than if they are represented and organised in association. All these objectives can be reached in a stepwise process, with more aspects being integrated as the local experience increases:

Priority level given to FSM: The political prioritisation of FSM and its implementation through regulations, financial resources, incentives and organisational efforts is the main enabling condition for the system's sustainability and efficiency. If it is not a priority of the national and/or local government as part of its overall sanitation program, then comprehensive, effective and safe FSM is unlikely to develop.

Coordination of the stakeholders: The identification and coordination of stakeholders is crucial to get their input and commitment. To ensure this happens, frequent meetings or workshops should be organised (e.g. with municipalities, the police, utilities, private sector companies, and customers). The incentive and enforcement strategies must also be clearly defined (e.g. requirement for monitoring by laboratories for resource recovery and penalties). Committees and associations can be created to simplify the communication between the stakeholders. For example, organising workshops for all the separate private collection and transport companies requires more time and investment than if they are represented by an association (Chapter 15). Incremental solutions can be adopted to facilitate the involvement of stakeholders. For example, based on the initial involvement and skills of the stakeholders, coordination committees can first be organised with the different departments of the government involved (e.g. public works, health, environment), and then expanded to include the private sector. The coordination work can be conducted by NGOs, governments, or through associations at each step in the service chain.

Response to the needs of the whole area and population: The system must address the sanitation needs of the entire population at affordable prices. Collection and transport services should be available for all types of onsite technologies and infrastructure in the entire city area, including in densely populated areas such as informal settlements. Survey and field investigations are therefore needed to assess the existing and potential demand for collection and transport. The faecal sludge treatment plants (FSTPs) must be located and designed in order to serve the entire city (Chapter 17). The treatment and processing of endproducts also need to be designed so that they can be effectively transported. The provision of these services to the entire population can be included as a principle requirement in the regulation by the governments, who can then further distribute the responsibilities among stakeholders.

Financial, environmental and social sustainability: The institutional framework should ensure long-term financial viability. These aspects are discussed in Chapter 13. Two other crucial requirements for the institutional framework are to meet environmental protection principles and to be accepted by all local stakeholders. Therefore, provisions can be made to avoid uncontrolled discharge into the environment and incentives given to favour resource recovery. For example, transfer stations could be built if the FSTPs are far away. Financial mechanisms such as subsidies can be implemented to provide access to repair shops for collection and transport operators. This can be useful to ensure no spillage happens during the transportation. Also, agricultural areas can be established near the FSTP if compost is produced, or subsidies can be given to the industries that are able to use the treatment endproducts as fuel. Coordination committees or associations can be involved in the monitoring of these aspects.

Awareness raising and information dissemination: Efficient communication on the advantages of the FSM system on public health and the environment has a positive impact on public acceptance. Provision of information to all the stakeholders involved in FSM is crucial for demand generation, demand management and the long-term viability and acceptance of the system. Good practices should be encouraged. Raising the awareness of the population can help to increase willingness to pay realistic tariffs and commitment at all levels, including that of private companies and politicians (e.g. through visits, workshops and information campaigns) – this is discussed further in Chapter 16. NGOs, public or private utilities and governments can be involved at different levels for the awareness raising.

Figure 12.1 Project coordination meeting with universities and research institutes from five countries together with the national sanitation utility in Dakar, Senegal (photo: Linda Strande).

Development of local expertise: Collaboration with local universities, NGOs, research centres, and institutions from other countries will contribute to the emergence of local expertise. Specific curricula on FSM should be developed in training centres as part of sanitation courses. Training and exchange of information between the public and private stakeholders contributes to enhancing the global level of understanding on the requirement of the FSM service chain. Universities and governments should be involved in the implementation of new curricula. Trade associations can also be created to facilitate the exchange of practical skills and solutions.

Capacity for monitoring and optimisation of the system effectiveness and efficiency: Monitoring and evaluation of the technical operation, the financial balance and customer satisfaction must be implemented in each institution or company involved in FSM. Lessons learned from experience should be capitalised on and incorporated to improve system performance. Means for monitoring and optimisation are discussed in Chapter 11. Financial viability and efficiency is discussed in Chapter 13.

Operation and maintenance management ability: The operation and maintenance (O&M) is a priority for the entire service chain. The choice of technology should ensure that the compexity and cost related to O&M are appropriate for the local context. Spare parts need to be readily available for all the equipment. External contracts for O&M services should only be arranged if the services are available immediately when needed (e.g. a pump repair should not be delayed due to lack of an available mechanical service). Chapter 11 is dedicated to the system requirement for the O&M of FSTP. Most of the recommendations can also be applied to other equipment and infrastructures such as collection and transport trucks, transfer stations, and resource recovery plants.

Management system efficiency and flexibility: The operator(s) should try to maintain flexibility in their management of the service chain to allow for growth and innovations (e.g. in pricing procedure or technology developments). The internal decision-making process must be short and efficient. Incremental solutions can be considered by all stakeholders at all levels of the service chain. For example, if a FSTP is first built in a peri-urban area for small amounts of FS from septic tanks but change of land use results in an increased production of public toilet FS, then the operation of the treatment technologies should be changed. FS can be mixed, the residence time changed, and maybe new investment made to provide new endproducts for resource recovery (e.g. compost). In a case like this, the collection and transport operators should also adapt to answer the new demand for services. Public private partnerships often provide greater flexibility in the FSM system.

Figure 12.2 Harvesting of sludge from drying bed for use in agriculture, Dakar, Senegal (photo: Linda Strande).

Financial management ability: Sound financial management must be ensured by each organisation by means of well-defined business plans (Chapter 13). Meetings with stakeholders and authorities must include discussions on pricing, fees, tariffs and funding opportunities.

Transparency of the system: The management system must ensure transparency to strengthen the trust between the stakeholders and with the service users. Coordination between the stakeholders through meetings and committees is a good approach to facilitating transparency, as well as communicating to the customers.

Endproduct marketing and customer relations: Customer relations should include the marketing of products and services for the collection and transport of FS as well as the way in which endproducts can be used. Customers must be able to contact the organisation easily and positive information dissemination on the benefits of resource recovery, product quality and good practices must be carried out. The importance of the link between the endproduct processing and the market demand for these products is addressed in Chapter 10.

Ability to acquire land: Long-term planning should secure access to land for existing and future project developments. The authorities in charge of land planning should be involved early on in the process, together with nearby inhabitants of future FSTPs (Chapter 17).

12.3 ENABLING REGULATORY ENVIRONMENT

The national authorities need to be involved in the development, validation and dissemination of an array of policies, strategies, laws and standards that define the stakeholders' roles, the quality standards, the procedures, and penalties (Hecht, 2004). Private sector stakeholders must also be taken into account when defining the regulations, as they may offer more cost-effective services and often fill the gaps between demand and governments' ability to supply the services. Aspects that should be considered in developing regulatory texts are discussed in the following sections, which can be included incrementally in the regulation, and according to the local expertise development (Case Study 12.1) to reach the objectives described in Section 12.2.

Human and environmental health: The measures needed to protect human and environmental health from risks linked to FSM need to be clearly laid out by regulations. This includes storage, transfer and treatment infrastructures, protective equipment for employees working in contact with FS, and measures to avoid discharge into the environment (Figure 12.3).

Figure 12.3 Illegal faecal sludge discharge directly into the environment, Yaoundé, Cameroon (photo: Linda Strande).

Overall sanitation strategy: To ensure an integrated approach, a strategy for the management of sanitation services needs to be defined, and should include FSM and wastewater management. This includes the existing onsite sanitation technologies and the FS quantities. Also future strategies for the provision of sanitation at the household level need to be coordinated with FSM and wastewater management.

City-wide approach: Strategic plans for FSM need to be developed on a city-wide scale in order to define the protocols for implementation at a local level, taking into account the future urban development plans (Strauss and Montangero, 2003). The land use, population characteristics, and type of buildings need to be considered.

Complete service chain: Regulation is needed to support the management of each step of the service chain, including the storage, collection, transport, treatment, and enduse or disposal of FS.

Enforcement: Regulations need to be enforced at both national and sub-national or city level by separate decrees, decisions, standards and guidelines defining the rules and potential penalties for the following aspects:

- the authorised stakeholders for each step of the service chain, their roles and obligations, and the mechanisms responsible for the monitoring and enforcement of each activity;
- the required design and construction standards for the onsite sanitation technologies and treatment infrastructures;
- the authorised roads and traffic rules for collection and transport;
- the authorised sites for treatment and disposal;
- the access and discharge conditions for the treatment, resource recovery and disposal sites (e.g. opening hours, tariffs);
- the required standards for services and products; and
- the required enforcement and monitoring outputs.

Incentives and control means for the enforcement of regulations are needed for each step (AECOM and SANDEC/EAWAG, 2010; Figure 12.4).

Figure 12.4 Official responsible for enforcement of illegal dumping of faecal sludge, Dakar, Senegal (photo: Linda Strande).

Permits and licenses: Documents are necessary to define the role of the stakeholders involved in the service chain. Sufficient financial and human resources need to be allocated to the institutions in charge of the activity, enforcement, and periodic renewal of these documents. The administrative procedure to obtain these documents should be clearly communicated.

Coordination: There needs to be structure(s) and financial mechanisms in place for the coordination and evaluation of the entire FSM system (AECOM and SANDEC/EAWAG, 2010). The flow and frequency of communication between the stakeholders and the data required for evaluation of the system should be clearly defined in strategies and regulatory documents.

Case Study 12.1 : Institutional and regulatory framework in Malaysia
(Adapted from AECOM and SANDEC/EAWAG, 2010)

The example of Malaysia shows the extent to which government commitment can improve sanitation and FSM. This country has developed a very efficient system for the management of FS that was supported by real institutional changes and a global vision to solve sanitation issues.

In 1993 Indah Water Consortium (IWK) was created, a company that is responsible for the provision of wastewater and FS services across the country. The objectives of IWK are to build infrastructures, develop collection and transport services, and increase acceptance for scheduled FS collection and wastewater fees. In 2000, IWK was incorporated into the Ministry of Finances in order to increase the subsidies and the financial control. The Sewerage Service Act fixes the conditions for the construction and O&M of treatment systems and septic tanks, and for the collection and transport services that are undertaken both by IWK and private operators.

In 2008 a new regulatory institution was created by the Ministry of Energy. Suruhanjaya Perkhidmatan Air Negara (SPAN) is responsible for the definition of sanitation strategies, and the regulation of the water and wastewater infrastructures management. IWK thus relies on the strategies defined by SPAN, and the discharge and quality standards defined by the Ministry of Nature and Environment. Specific committees are responsible for the control of financial viability and transparency. These committees have the power to define wastewater tariffs, subsidies and taxes. Since that same year the Water Service Industry Act also allowed the federal government to collaborate with water and wastewater companies, thus supporting the management of water resources from source to disposal for the country. This Act aims to raise the efficiency of the water sector industries, and to assist in the dissemination of achievements and the sharing of best practices across the country.

Such a strong institutional setup supports the success factors discussed in Section 12.2, as FSM in the country is supported by specific regulations and is considered an integral part of the water resource management process. Additionally, collaboration with national universities ensures the development of a strong national expertise through research and training programmes. The publication of several booklets and press releases has also increased public awareness.

These changes to the institutional and regulatory framework over the last 10 years have resulted in an increase in the percentage of households connected to the sewer network from 5% in 1993 to 73% in 2005, with the remaining 27% of the population benefiting from scheduled collection of FS.

12.4 INSTITUTIONAL ARRANGEMENTS

12.4.1 Overview of the service chain organisation

One of the main reasons for the failure of FSM systems is the overlapping and unclear allocation of responsibilities and a lack of incentives for efficient operation. This situation frequently occurs where an incomplete institutional framework exists, resulting in both a lack of accountability and disagreements between stakeholders. Since the entire service chain is interlinked, each aspect influences another and it is essential that the roles and responsibilities are clearly defined. For example, stakeholders in charge of the collection and transport of FS must also participate in the organisation of the discharge of FS at the treatment plant. In turn, the FSTP managers need to coordinate their activities with the stakeholder(s) who are in charge of resource recovery and disposal of the endproducts. Thus, coordinating the link between each step in the chain is imperative to ensure a successful FSM system. This differs from wastewater management systems where the waste is transported via the sewer and typically only one stakeholder is in charge of the entire system.

As illustrated in Figure 12.5, where each block represents one stakeholder, there are many possible ways to organise the FSM service chain. Systems that have more stakeholders involved will be more complex, regardless of who the stakeholders are. In contrast, if only one stakeholder is in charge of the whole service chain, flexibility may be hard to ensure and intensive management procedures are then necessary. Thus the selection of an institutional arrangement that is appropriate for the local context is crucial. The arrangement can also be changed incrementally, based on the demand for services. All the stakeholder roles can be carried out by either the public or the private sectors.

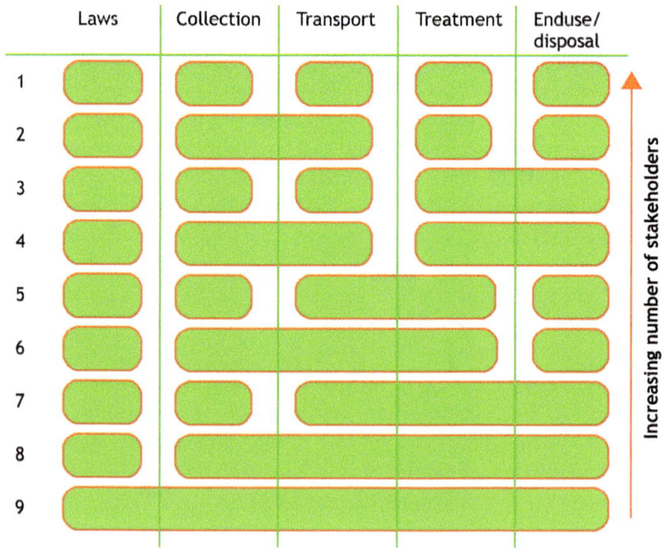

Figure 12.5 Schematic representation of different organisational arrangements for distribution of operational responsibilities among stakeholders (one block represents one stakeholder).

There are advantages and disadvantages to each of the options presented in Figure 12.5:

Option 1: Each step of the service chain is provided by a different stakeholder. This allows for organisational flexibility, but enforcement, monitoring and coordination are difficult and may result in tension at the many interfaces. The fact that the collection and transport activities are undertaken by different stakeholders favours job creation. However, a potential drawback is the fact that transfer of FS is needed after collection to transport it to the FSTP, thus involving more infrastructure and organisation (e.g. to operate transfer stations).

Option 2: Collection and transport services are operated by one stakeholder, and the treatment is carried out by a separate stakeholder. This option is preferable when mechanical collection and transport services are already available. It simplifies the financial flow and organisation of the transport of FS to the FSTPs. However, the procedure to discharge FS at a treatment plant may be complex, and it is difficult to control the qualitative and quantitative variation of the load. Solutions must also be found for densely populated areas where truck access is difficult.

Option 3: The value created through the marketing of endproducts can be used to finance the treatment infrastructure if the same stakeholder is in charge of these activities. This allows optimisation of the O&M and financial management of the treatment and resource recovery plant and endproducts are more easily decontaminated. However, in this option, FS transportation and discharging procedures at the treatment site are not optimised.

Option 4: One type of stakeholder manages all the equipment for the collection and transport of FS, while another is in charge of the infrastructures for the treatment of FS and resource recovery. In this case, the two types of stakeholder can develop specific skills for their activity. As with the previous three options, the main disadvantage is that the discharging of FS is not facilitated at the treatment plants. However, similarly to Option 3, the treatment technology can be chosen based on the resource recovery required.

Option 5: This option allows local job opportunities to be created in the communities, as well as the development of industrial processes and use of the endproducts. This system is advantageous in densely populated areas, where access by trucks is difficult. The discharging procedure of FS at the treatment plant can be optimised, and the possibility exists to improve control over the quality of sludge that is treated. However the organisation of the transfer of fresh FS between the collection and transport steps may be complex. It is also important to have clear conditions for the delivery of treated FS to the stakeholder in charge of the resource recovery.

Option 6: The management of the collection and transport equipment together with the treatment infrastructures involve highly developed managerial skills. This option has the advantage of facilitating the management of FS from the onsite technologies user to the treatment plant, and reducing the risk of unauthorised discharging. However, the financial flow between the enduse step and the rest of the service chain is not optimised.

Option 7: Similarly to options 1, 3, and 5, this option is best implemented where transfer stations exist, and an additional responsibility is assigned for the management of the transfer station. This creates local job opportunities and allows for management of FS in densely populated areas. In this option, the service chain is more complex, but resource recovery is easily organised, as there is no transfer between several stakeholders.

Option 8: Having one stakeholder in charge of the entire service chain allows easy coordination and optimisation of each component of the service chain based on the needs of the other components, but requires highly developed managerial skills and financial resources.

Option 9: This option should be avoided, as it does not allow for transparency. Regulations and enforcement should be performed by government entities, independent from the interest of companies.

12.4.2 Role distribution among the stakeholders

The selection of one of the above-mentioned arrangements depends on the characteristics of the local stakeholders. For example, a small private company might not be structured enough to manage the entire service chain, as described in Option 8. Thus, the features of each stakeholder must first be understood (Chapter 15) and then the institutional system defined.

In most of the currently existing systems, a combination of stakeholders tends to provide services in the FS service chain (e.g. Sanitation Utility, Municipal Services, Military Department, Private Entrepreneurs, Group of Economic Interest (GEI)) (Koné, 2010). Table 12.1 summarises the possible responsibilities of stakeholders. They may take charge of one or more activities within the service chain (Koanda, 2006).

Table 12.1 Different stakeholders in the faecal sludge sector and their possible involvement at different levels of the faecal sludge organisation

Stakeholder	Laws	Coordination	Collection & Transport	Treatment	Resource recovery	Enforcement	Training & Information	Monitoring
Ministries	▪	▪				▪	▪	▪
National/ municipal utilities		▪	▪	▪	▪	▪	▪	▪
Police						▪		
Private companies			▪	▪	▪			
Associations[1] /CBO[2]		▪			▪	▪	▪	
NGOs							▪	▪

1 Associations = groups of stakeholders organised around defined objectives

2 CBO = Community Based Organisations that can provide services for the community

The distribution of the responsibilities among the stakeholders should be decided taking into account the intrinsic strengths and weaknesses of each stakeholder involved in the service chain (Table 12.2). Incremental improvements can be facilitated either through capacity building or reorganisation of different stakeholders.

The police, environmental agencies and NGOs are excluded from Table 12.2 as they are only responsible for the enforcement and training aspects. The stakeholders in charge of enforcement and quality monitoring should be clearly recognised and impartial. Ideally, the national or municipal authorities should be involved in the supervision of laws, standards and guidelines (AECOM and SANDEC/ EAWAG, 2010). Consumer organisations can also be involved in discussions about prices, service requirements and quality monitoring (Klingel, 2001).

Table 12.2 Possible stakeholders, their advantages, drawbacks and needs

Stakeholder	Advantage	Drawback	Needs
Ministeries, National/municipal utilities	• Subsidies available • Easy enforcement • Possibility to manage complex technologies	• Dependency on political situation (e.g. changes of direction with political rearrangements) • Potential low priority level among government activities • Time consuming internal procedures • Low flexibility	• Capacity strengthening • Autonomous organisation from the national authorities • O&M-driven organisation
Private companies	• Service flexibility • Demand-led market • Answer to O&M needs • Easy contact with customers • Local job production	• Lack of legal enforcement • Lack of recognition • Poor management capacity • Complex coordination • Difficulty to accessing subsidies • Potential low technical skills	• Capacity strengthening • Tax reduction for the delivery of public services • Licenses and contracts needed
CBO, associations	• Service flexibility • Local job production • Involvement of local population • Possibility to inform and raise awareness of the community	• Complex coordination • Varying service fees between areas managed by different CBOs • Low accountability level • Poor management capacity • Weak human resource continuity • Difficulty in organising service delivery to customer living outside an area that is managed by the CBO	• Coordination committee • Capacity strengthening • Need simple technologies • Increasing the feeling of accountability

The advantages and drawbacks linked to the involvement of each type of stakeholder, together with documentation and contractual requirements, are discussed further in the following sections. Signing of documentation and definition of the institutional setup should take place early on in the process (Chapter 16).

12.4.3 Institutional arrangements for colection and transport

Collection and transport form the first step in the FSM service chain. Any FSM work must include consultation with the collection and transport stakeholders in order to ensure their commitment to the system thus strengthening capacity and coordination. The omission of these stakeholders may result in failure of the process (Case Study 12.2).

Case Study 12.2: Faecal sludge treatment plant built without involvement of the collection and transport operators

A FSTP built in Bamako, Mali, was implemented without the involvement of the collection and transport operators, who were not given adequate consideration in the location of the plant. It was thus built too far out of the town, and the collection and transport operators could not afford to drive to the facility between the collection at each onsite technology. As a result, the facility was never utilised, and has since been abandoned.

Different types of stakeholders can be in charge of the collection and transport, with or without transfer stations. National or municipal utilities, or private companies can undertake either collection at the onsite technology, or transport to the treatment plant (Options 1 and 3 in Figure 12.1), and combine transport and treatment activities (Options 5 to 8, Figure 12.1). CBOs commonly have a weaker management structure, and are best involved in collection at a local level. The advantages and constraints related to the involvement of these three different stakeholders include:

National or municipal government utilities: National or local departments of governments and municipal utilities (e.g. public works, environment, cleaning) can be responsible for collection and transport of FS. This can also be effective with small, local government-owned companies. In the case of Addis Ababa, Ethiopia, the Sewerage Authority provides low-cost collection and transport services, and benefits from state subsidies that would not be available to private companies (Kebbede, 2004). This option also avoids difficulties with the police who often respect the right of the Authority's trucks more than those belonging to private operators. However national and municipal utilities often lack human resources and equipment resulting in poor quality of the collection and transport service (Strauss and Montangero, 2003; Koanda 2006; AECOM and SANDEC/EAWAG, 2010).

Private companies : Private companies offer more flexibility as they often provide other services to improve their competitiveness (e.g. collection of solid wastes, construction etc.), they create employment on a local level and can rapidly adapt to the service demand (PS-Eau&Hydroconseil-Mauritanie *et al.*, 2002; Blunier, 2004; Hecht, 2004; Jeuland *et al.*, 2004; Koanda, 2006). However, if the competition is weak, profit seeking can lead to bad practices and high prices (Jeuland *et al.*, 2004). Private operators frequently lack financial viability, and have a bad reputation with the authorities and the public (Klingel, 2001; Bassan *et al.*, 2013). In Africa, some collection and transport companies are organised in associations that are legally recognised and provide an interface with the authorities, which can then adopt several measures such as tax exemption. Important collection and transport contracts that could not be undertaken by a single company can be secured through this type of association that exists in Senegal, Burkina Faso, Mali and Uganda (Bolomey, 2003; Blunier, 2004; Mbéguéré *et al.*, 2010; Bassan *et al.*, 2013). These associations improve the recognition of small operators, thereby facilitating sector formalisation, regulation and transparency. They should thus be encouraged. Licenses to provide the collection and transport services, and for the truck circulation can be provided by the local authorities.

Associations/CBOs: CBOs can take charge of the collection of FS before transfer stations, as well as the management of these stations. This structure favours job creation and also facilitates the information and awareness raising of the customers' awareness concerning the maintenance of the onsite system, as the local community is involved through the CBO. CBOs require a contractual arrangement with local authorities to define their roles, the quality of the service, and the standards for the monitoring.

As will be discussed in Case Study 12.3, the responsibility for emptying septic tanks and latrines can be assigned to the user of the onsite technology or to the service provider (Klingel, 2001; AECOM and SANDEC/EAWAG, 2010). Collecting on demand requires minimal customer management procedures and the responsibility to empty at an adequate frequency is given to the user. However, the frequency of collection cannot be controlled, and customers might only call once the system is full, or more realistically, overflowing, as people tend not to maintain systems until there is a problem. Thus, information campaigns are needed to inform users about the maintenance requirements of their onsite technology and the importance of frequently extracting FS. Another possible disadvantage is the difficulty of controlling illegal discharging of FS. This type of management system is commonly used where the operator does not have sufficient resources to manage a customer database. It is also more flexible and allows for collection and transport services to be provided by different companies.

Figure 12.6 Privately owned collection and transport trucks discharging faecal sludge at a receiving facility owned by the municipality, Kampala, Uganda (photo: Linda Strande).

Where a contractual agreement is signed between the operator and the user, the responsibility to empty the onsite technology at regularly scheduled intervals lies with the operator (i.e. on-demand service must be possible for full onsite technology). In this case, the collection and transport operator needs to have a very organised and efficient management structure in order to manage the service for all types of customers. Typically, the collection frequency is scheduled for regular intervals (AECOM and SANDEC/EAWAG, 2010). The use of a billing system that integrates the collection and transport operators' O&M costs allows for continuous income, rather than just having income on demand when services are requested. Illegal discharging is also more easily controlled. However, possible disadvantages of scheduled collection and transport services could be the limited flexibility, and the dependency on an enforcement system to compel customers to pay (e.g. no water delivery if the bills are unpaid).

Case Study 12.3: Service chain organisation in Malaysia

Under the Sewerage Service Act, the collection and transport of FS in Malaysia was fully managed by Indah Water Konsortium (IWK) who developed a database to organise scheduled collection per area. Customers were contacted by IWK prior to the FS collection and paid semi-annual wastewater bills. This system was promoted through media spots.

With the adoption of the Water Service Industry Act in 2008, the responsibility for the collection of FS was transferred to the onsite technologies users who have to organise collection and can be fined for non-compliance. Private companies also provide collection and transport services. This system is more flexible, but these changes involve a complex enforcement system for the different companies. Campaigns to strengthen the users' commitment and to raise their awareness of the importance of frequent collection are also needed.

A progressive strategy was adopted for the management of FSTP infrastructures in Malaysia. Old wastewater treatment plants were first rehabilitated and converted to enable FS treatment; then simple technologies were encouraged; and finally modern technologies were implemented in the biggest cities. Today, FS is treated depending on the type of land use in each area.

This example shows that a progressive approach can be adopted which allows the development of a well-coordinated FSM scheme. Each arrangement has advantages and drawbacks. In all cases provision of information to the population and communication among the stakeholders is crucial in order to ensure proper coordination and sustainability of the program. All the steps in the service chain need to be taken into account. Even though Malaysia has achieved great advances in FSM, the system is largely subsidised, and an important challenge is the acceptance of representative, non-subsidised collection and transport fees by the population.

12.4.4 Institutional arrangements for treatment of faecal sludge

FSTPs are important technical infrastructures that require adequate training of the personnel responsible for their management, O&M and monitoring (Chapter 11). All treatment technologies need to be managed by a well-organised and effective institution (Strauss and Montangero, 2003). Therefore, CBOs are not recommended, as the high level of technical and managerial skills required are often not available in these organisations.

Referring again to Figure 12.5, both national or municipal utilities and private companies can be in charge of only the treatment plant (Options 1 and 2), or they can combine this activity with transport and/ or enduse management (Options 3 to 8). In each case, the contractual links, the financial management, and the communication and monitoring procedures need precise definition. The monitoring of the quality of the endproducts can be done by an independent laboratory, especially in the case of private management. Agreements are useful to define the frequency of sampling and the access rights to the sampling points. The institution in charge of treatment can either own the property and infrastructure or have some type of public-private partnership. Different arrangements can exist as follows:

Direct management by national or municipal utilities: The national or municipal utility owns the FSTP. This arrangement has the advantage of facilitating the enforcement of pollutant discharge standards, and also offers the possibility of financing O&M activities through subsidies, without which the finances allocated to the FSTP O&M are often insufficient. The national or municipal utilities should be sufficiently autonomous and not suffer from long or complex internal procedures that can hinder operation activities (Bassan *et al.*, 2013). Contracts or agreements with the authorities can be signed in order to define the responsibilities.

Direct management by private companies: The FSTP is owned by a private company. Experiences in Benin, Mali and Gabon with direct private management show that operational requirements of FSTPs can be met, and that competitiveness is raised by a benefit-driven approach. Low technical and managerial skills, and limited access to subsidies are potential drawbacks to private management (Jeuland *et al.*, 2004). However, licenses or contracts can be provided by the local authorities in order to set the quality standards and the monitoring program. The potential for private sector involvement is higher when there is a financial gain from the resource recovery or from FS treatment endproducts.

Delegated management by national or municipal utilities, or private company: One potential advantage of delegated management is that the operator can be chosen by the FSTP owner based on their technical and managerial capacity. In this case, contracts need to be signed with the owner, specifying the requirements in terms of O&M. Licenses for the FSTP's O&M can also be provided by the authorities in cases where the FSTP is publicly owned.

Figure 12.7 Meeting with municipal government responsible for sludge management, research institutes, and donor agency in Bac Ninh, Vietnam (photo: Linda Strande).

12.4.5 Institutional arrangements for enduse and disposal

The institutional framework needs to promote sustainable business models for the entire service chain. Therefore, good quality endproducts must be ensured, which must also be safe to use (Chapter 10). Similarly for the treatment of FS, resource recovery from the endproducts can require a high level of skills for O&M and monitoring, depending on the choice of technology (Chapter 5). The products not only need to be sanitised and processed, but they also need to be of value to the local market. This requires a preliminary assessment of the market demand, proper marketing and the provision of a high quality of service (Klingel, 2001). A multi-barrier approach should also be adopted to protect the workers, customers and final users from health risks linked to pathogens.

Two types of management structures can be followed – direct or delegated. In the case of delegated management of publicly owned infrastructures and equipment, licenses are useful to define the O&M requirements, the quality standards, and the monitoring program. The comparative advantages and drawbacks are the same as discussed in Section 12.4.4. Three types of stakeholders may be in charge of these activities:

National or municipal utilities: A complex process can be managed by a national or municipal utility, which could also deliver the endproducts to customers. Where national or municipal utilities are in charge of the resource recovery plant they are also likely to be involved in the FSTP management, either through direct or delegated management.

Private companies: Small private companies providing services for resource recovery from waste and treatment endproducts are found worldwide (Jeuland *et al.*, 2004). Their main strengths are related to the inherent dynamism of private entrepreneurs. Capacity strengthening and close coordination of private companies are often needed to ensure efficient management and O&M of the facility (Bolomey, 2003). Contracts or agreements can be signed with the stakeholder in charge of the FSTP O&M in order to define the agreement, as well as the price and quality of the endproducts to be processed and marketed.

Associations/CBOs: CBOs or associations can be involved if the technology used to process, treat and package the endproduct is low, and if customers come to the plant to buy the products. This solution may be applicable where people are living near to the FSTP, especially if endproducts are used directly in the community (e.g. as building material or soil amendment; Klingel, 2001). The management rules of a CBO stipulate the need for sustainable O&M and transparent financial transactions, and therefore licenses can be provided by local authorities.

As for the collection and transport processes, the activities linked to resource recovery can be carried out on demand, or based on a contractual agreement outlining a scheduled sale or delivery. Where valuable endproducts can be produced over the entire year, the main advantage of the scheduled sale is the provision of a regular income that can be used for the O&M of the infrastructures.

12.4 BIBLIOGRAPHY

AECOM and SANDEC/EAWAG (2010). A Rapid Assessment of Septage Management in Asia: Policies and Practices in India, Indonesia,Malaysia, the Philippines, Sri Lanka, Thailand, and Vietnam. USAID. Bangkok, Thailand.

Bassan, M., Mbéguéré, M., Koné, D., Holliger, C., Strande, L. (2014). Success and failure assessment methodology for wastewater and faecal sludge treatment projects in low-income countries. Journal of Environmental Planning and Management (in press).

Bassan, M., Mbéguéré, M., Tchonda, T., Zabsonre, F., Strande, L. (2013). Integrated faecal sludge management scheme for the cities of Burkina Faso. Journal of Water, Sanitation and Hygiene for Development 3 (2), p. 216-221.

Blunier, P. (2004). La collecte et le transport mécanisés des boues de vidange dans la ville de Ouahigouya (Burkina Faso): Analyse du marché et propositions de réorganisation des flux financiers. M.Sc., Ecole Polytechnique Fédérale de Lausanne.

Bolomey, S. (2003). Amélioration de la gestion des boues de vidange par le renforcement du secteur privé local -Cas de la Commune VI du District de Bamako. EAWAG, Dübendorf, Switzerland.

Bolomey, S. (2003). Amélioration de la gestion des boues de vidange par le renforcement du secteur privé local: Etudes et Outils - Cas de la Commune VI du District de Bamako. EAWAG, Dübendorf, Switzerland.

Hecht, A. D. (2004). International efforts to improve access to water and sanitation in the developing world: a good start, but more is needed. Water Policy 6 (1), p.67-85.

PS-Eau and Hydroconseil-Mauritanie (2002). Les entreprises de vidange mécanique des systèmes d'assainissement autonome dans les grandes villes africaines. Etude de cas : Nouakchott (Mauritanie) I : Enquête auprès des entreprises de vidange mécanique. Rapport. P. G. D. D. D. E. DE and L. A. U.

Ingallinella, A.M., Sanguinetti, G., Koottatep, T., Montangero, A., Strauss, M. (2002). The challenge of faecal sludge management in urban areas – strategies, regulations and treatment options. Water Science and Technology 46 (10), p.285-294.

Jeuland, M., Koné, D., Strauss, M. (2004). Private Sector Management of Fecal Sludge: A Model for the Future? Focus on an innovative planning experience in Bamako, Mali. EAWAG, Dübendorf, Switzerland.

Kebbede, G. (2004). Living with urban environmental health risks: the case of Ethiopia. Hants, England, Ashgate Publishing.

Klingel, F. (2001). Nam Dinh Urban Development Project: Septage Management Study. EAWAG, Dübendorf, Switzerland.

Koanda, H. (2006). Vers un assainissement urbain durable en Afrique subsaharienne : Approche innovante de planification de la gestion des boues de vidange. PhD, Ecole Polytechnique Fédérale de Lausanne.

Koné, D. (2010). Making Urban Excreta and Wastewater Management contribute to Cities' Economic Development - A paradigm shift. Water Policy 12 (4), p.602-610.

Lüthi, C., Panesar, A., Schütze, T., Norström, A., McConville, J., Parkinson, J., Saywell, D., Ingle, R. (2011). Sustainable Sanitation in Cities: A Framework for Action. Sustainable Sanitation Alliance (SuSanA), International Forum on Urbanism (IFoU), Papiroz Publishing House, Rijswijk, The Netherlands.

Mbéguéré, M., Gning, J.B., Dodane, P.H., Koné, D. (2010). Socio-economic profile and profitability of faecal sludge emptying companies. Resources, Conservation and Recycling 54 (12), p.1288-1295.

Moe, C. L., Rheingans, R.D. (2006). Global challenges in water, sanitation and health. Journal of Water and Health 4 Suppl. 1, p.41-57.

Pybus, P., Schoeman, G. (2001). Performance indicators in water and sanitation for developing areas. Water Science and Technology 44 (6), p.127-134.

Strauss, M., Montangero, A. (2003). FS Management – Review of Practices, Problems and Initiatives. Engineering Knowledge and Research Project - R8056 Capacity Building for Effective Decentralised Wastewater Management. EAWAG, Dübendorf, Switzerland.

UNEP (2010). Africa Water Atlas. Department of Early Warning and Assessment (DEWA). Nairobi, Kenya, United Nation Environment Programme (UNEP).

End of Chapter Study Questions

1. Name five important institutional aspects that play a role in FSM, and explain why they are important.

2. Explain the role of enforcement of regulations in FSM.

3. An example of responsibilities in a service chain is collection and transport services operated by one stakeholder, and the treatment by another stakeholder. When is this way of organisation preferable? Which aspects can be challenging for this arrangement?

Chapter 13

Financial Transfers and Responsibility in Faecal Sludge Management Chains

Elizabeth Tilley and Pierre-Henri Dodane

Learning Objectives

- Understand how the different stakeholders in a service chain relate to each other from a financing point of view.
- Know which types of financial transfers play a role in faecal sludge management.
- Be able to describe different financial flow models for faecal sludge management.
- Understand the complexity involved in designing, implementing, monitoring and optimising an entire faecal sludge management system that includes all stakeholder and financial interactions.

13.1 INTRODUCTION

One of the reasons that faecal sludge management (FSM) systems have not been widely implemented is because of the financial and political complexity involved. This is not only due to the number of stakeholders who have a financial interest in the system, but also to the diversity of the interests each stakeholder has.

Unlike other types of infrastructure (e.g. electricity) where a single utility is usually responsible for the generation, delivery, operation, maintenance and billing, a faecal sludge (FS) system is more commonly a collection of stakeholders, each of whom is responsible for a different part of the treatment chain. Consequently, payments must be made each time responsibility is transferred from one stakeholder to another. Only a special set of political and financial conditions can foster an environment that allows each essential stakeholder to perform their task and permit a complete treatment chain to take form.

Figure 13.1 Servicing and billing in informal settlements is always difficult; it is exacerbated by a lack of access and tenure (photo: Linda Strande).

This chapter will examine the financial flows within various FSM systems and will illustrate and discuss the critical financial and responsibility transfer points. To understand the complete FS system, this chapter will begin by defining the various stakeholders and their roles within the FSM system. The types of financial transfers will be discussed with particular attention paid to the stakeholders between whom they are transferred. Five different FSM models, i.e. different combinations of stakeholders with various responsibilities and financial transfers are presented and examined. Finally, a short problem is presented using the business model of a small-scale collection and transport entrepreneur in order to illustrate the number and magnitude of financial transfers that affect even a minor element of a FSM system. The chapter concludes with future perspectives.

13.2 FINANCIAL MODELS

13.2.1 Stakeholders involved in financial transfers

Almost every stakeholder in a FS system is involved in some kind of financial interaction. Stakeholders are those people, institutions or enterprises that send or receive payment in exchange for taking responsibility for one or more processes in the FS treatment chain. The stakeholders and their financial responsibilities are summarised (in alphabetical order) in the following paragraphs.

Enduse industries are those stakeholders that make use of the inherent nutrients, energy potential, and bulking properties of treated FS. Enduse industries are a relatively new, but growing sector in the FS process chain. The enduse(s) of FS should be considered when designing the entire FSM service chain to ensure the appropriate design of treatment technologies; i.e. so that the best quality FS can be generated for its specific final use (Diener *et al.*, 2014).

With a growing need for low-cost, locally sourced, sustainable nutrients, the agricultural industry will likely emerge as an important enduse stakeholder. FS is also a promising sustainable energy source. In the future, the financial benefits and environmental necessity of enduse may become drivers for improved FSM and influence the design of FS systems. The demand for sludge, as well as the legal framework for its application, will have an increasingly powerful impact on how FS is managed through the entire process chain. Refer to Chapter 10 for a full range of industries and products associated with enduse.

Government authorities are responsible for the rules and regulations to which private enterprises and public utilities must adhere. Government authorities may allocate budgets to utilities and outsource work to private enterprises, but may also plan and manage their own FS programs internally. Government authorities are responsible for collecting taxes in order to cover, or partly cover their budgets. Authorities may also be recipients of foreign aid, which may be allocated to the construction, operation or maintenance of public infrastructure.

Household-level toilet users are those people who are responsible for removing FS from property that they own or rent. These people have some type of onsite sanitation technology that requires periodic FS removal. Technologies that require periodic emptying include septic tanks, pit latrines, anaerobic baffled reactors (ABRs) (for clusters of houses) or other similar, water-based storage technologies.

Non-Governmental organisations (NGOs) are enterprises that operate on a not for profit basis and which are not funded or supported directly by government, although they are often sub-contracted by government for specific tasks. NGOs operate in the social-service niches left where governments and private enterprise are unwilling or unable to operate effectively.

Private enterprises are organisations that operate on a for-profit basis by providing goods or services in exchange for payment. Private enterprises are bound by the laws of the state, and may accept contracts to work for the state. However, private enterprises are not wholly or in part, associated with government at any level and do not receive guaranteed government funding (though they may apply for subsidies, loans, etc.).

Public utilities are responsible for operating and maintaining public infrastructure (e.g. water or electricity). They are extensions of government authorities, and as such, are funded by government budgets. Depending on how well the public utility (PU) is run, and how users are billed, the PU may operate at a loss. Public utilities provide a useful service, which may not otherwise exist in a free market (e.g. sludge treatment) but have typically operated as monopolies. Increasingly however, private enterprises have recognised the financial potential of operating within the PU marketplace and as a result, PUs are no longer free from competition.

13.2.2 Financial transfers

Within a FSM system, money is exchanged for different activities (e.g. emptying, transport, processing), at different orders of magnitude (e.g. small service payments, massive construction costs), and with different frequency (e.g. daily transfer frees, annual taxes). To achieve a financially sustainable business model, a prudent selection of the transfer types must be implemented. A brief summary of the most common financial transfers, applicable to FSM, is presented below.

Budget support is the name given to cash transfers between stakeholders to partly or fully cover one stakeholder's operating budget. Typically, a government authority would provide budget support for a public utility, but foreign governments or agencies (e.g. USAID, Asian Development Bank) also provide budget support to different ministries and/or sectors. The duration of the budget support is usually long-term and non-conditional. In other words, it is not related to a specific task or output, but rather, is made to support daily budgetary requirements (conditional cash transfers have become increasingly promising since they reward outcomes and encourage transparency).

Table 13.1 Discharge fees and rates at official discharge sites in 2004 (adapted from Collignon, 2002; Jeuland, 2004)

City	Cost per discharge (€)	Percent of total discharges	Discharges per year	Destination type
Cotonou, Benin	8.6	75%	26,667	Treatment*
Kampala, Uganda	5.6	42%	7,000	Treatment
Dar Es Salaam, Tanzania	3.1	7%	100,000	Treatment
Kumasi, Ghana	2.0	95%	-	Treatment
Dakar, Senegal	1.2	74%	67,525	Discharge only

* Proper treatment cannot be guaranteed since the facility is improperly designed and overloaded

Capital investment costs are those that are paid once, at the beginning of the project to cover all materials, labour and associated expenses needed to build the facilities and associated infrastructure. Examples of capital investments could include the purchase of land for the construction of FS drying beds, the design and build of a treatment plant, the purchase of a vacuum truck for collection and transport, or the installation of a septic tank at the household level. Capital investments can be paid by any of the stakeholders listed in the previous section.

Discharge fee is a fee charged in exchange for permission to discharge FS at some type of facility. The fee is paid with the intention of transferring responsibility to a stakeholder who has the legal and technical ability to safely process and/or transfer FS to another responsible stakeholder. In theory, anyone who owns property could charge a discharge fee and allow FS to be dumped, despite the lack of appropriate safety precautions. Official discharge fees (in conjunction with enforced laws) must therefore be structured so as not to create an incentive for individuals to charge their own, unregulated, discharge fees and compete with the formal discharge fee structure. It has been argued however, that discharge fees do not correlate with illegal discharge, i.e. higher discharge fees do not result in reduced use of authorised facilities as shown in Table 13.1.

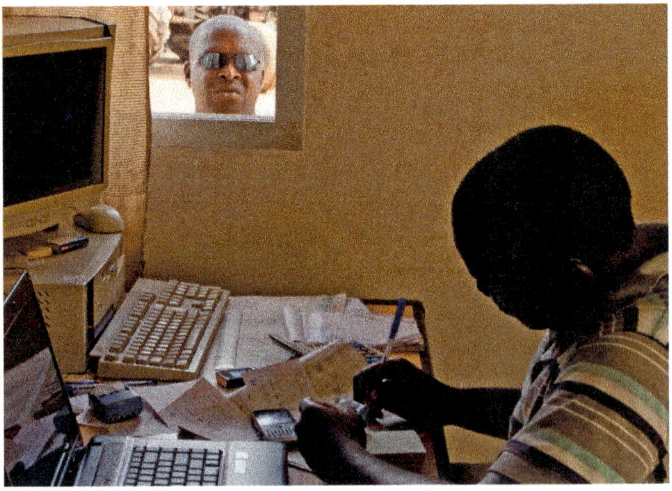

Figure 13.2 Collection of discharge fees. Good accounting is essential to understand how any business operates and how it can be improved (photo: Linda Strande).

The most equitable and financially beneficial way to charge a discharge fee is not clear. It may be charged according to the volume of sludge discharged (which may be difficult to measure, and does not take into account the density of the sludge), or per discharge event regardless of the volume (though the entire volume of the truck may be difficult to empty). Both models have consequences for the collection and treatment (C&T) business and the FS treatment plant (FSTP) in terms of how they optimise their finances. Payments based on discharge events, for example, may encourage C&T enterprises to maximise the volume of FS in each truck more efficiently, resulting in the FSTP being faced with more infrequent, highly loaded discharge events.

Discharge incentive is the opposite of a discharge fee. It is a payment used to reward the C&T business for discharge the sludge in a designated location and to disincentivise unregulated, or illegal discharge. Making payments, rather than collecting fees, means that the FSTP would require other means of meeting their costs, likely in the form of a sanitation tax. A discharge incentive of 5 USD per load of sludge was proposed for Ouagadougou, Burkina Faso to prevent illegal discharge, although the long term results of this program have not been published (SANDEC, 2006). Incentives are essentially payments made to people as rewards for performing tasks that they may not otherwise do, but that are socially desirable. Incentives are controversial because, as some argue, people should not be paid for doing what is 'right', but programs to date have been highly effective at achieving their objectives using more of the 'carrot' than the 'stick' approach, and achieving higher returns on public investment than comparable public announcement, social-pressure, or education campaigns (Gertler and Boyce, 2001; Kakwani *et al.*, 2005; Eldridge and Palmer, 2009; Banerjee *et al.*, 2010).

Discharge license is a financial instrument used to control the number and quality of C&T enterprises that are allowed to discharge FS at the FSTP. The license, in theory, is given out depending on proven quality of the service that the stakeholder is able to provide. In practice however it is often a way for the license issuer to generate revenue, and few license applicants are therefore denied. Since 1998, operators in Nairobi have been paying between 260 and 780 USD (for trucks less than $3m^3$ and greater than $7m^3$, respectively) for annual licenses. The license allows C&T enterprises to discharge into the city's sewerage network, thereby reducing their travel time and indiscriminate discharge (Water and Sanitation Program Africa, 2005). However, the licensing system may exclude smaller, less capital-rich stakeholders from operating. This could have the unwanted effect of creating a parallel, black-market system devoid of permits or licenses.

Emptying fee is the fee that is charged at the household level for removing FS from the onsite sanitation technology where it is collected and stored. Typically, the same stakeholder that is responsible for emptying is also responsible for transporting the sludge away (from where it has been emptied), although some independent operators who manually empty tanks/pits are not able to transport the FS and so leave that task to the household. Household members may also assist the C&T company with the emptying to reduce the fee. The emptying fee can be paid once the service is provided, but this type of payment model does not encourage the household to arrange for the emptying until it is absolutely necessary or long overdue. This type of emptying schedule, which may be completely unpredictable, or correlated with the seasons, causes a great deal of uncertainty for both the C&T companies and the FSTP operators. Some poorer households that cannot afford to pay the fee for emptying the entire quantity of FS may opt instead to have a small portion removed (e.g. the top metre of sludge in a pit).

Emptying fees vary depending on country, region, currency, market, volume, road condition and a host of other criteria. For example, within one informal area of Nairobi, known as Kibera, it costs 8 USD to have 0.2 m^3 of sludge emptied manually, or 196 USD for a vacuum truck that removes $3m^3$ of sludge (Water and Sanitation Program Africa, 2005).

Fines are tools used by the government, or other legal authorities to control and discourage undesirable behavior. Fines can be used to prevent the illegal discharge of sludge and provide an incentive for the less-costly behaviour of paying for a discharge license or the discharge fee. This only occurs when the fines are high enough, and enforced often enough, to present a genuine threat to illegal/informal practices. It should be noted however, that fines are only equitable when there is an alternative option available at a reasonable cost; e.g. access to a FSTP with regular hours and affordable discharge fees.

Operation and maintenance (O&M) costs are expenses that must be paid regularly and continually until the service life of the infrastructure/equipment has been reached. Equipment like pumps, trucks, hoses, etc., will wear down with use and the frequency of replacement will depend on the operating conditions and how often the parts are maintained. Although the service life of the equipment will be significantly shortened in the absence of O&M payments, more immediate needs (e.g. fuel) often take precedent. Owners of vacuum or pump trucks used for FS management face high O&M costs because of the wear that foreign material (e.g. sand, garbage) puts on the equipment. Further information on O&M is presented in Chapter 11.

Purchase price is the price paid by one stakeholder to another in exchange for becoming the sole owner of a good. A purchase fee can be paid at any point or with any frequency, as opposed to capital costs, which are only paid at the beginning of a project. The purchase price is dependent on supply, demand, and any subsidies that may be available. The agricultural industry for example, may pay a public utility a purchase price for treated FS to set up a greenhouse, in which case it would be categorised as a capital cost; a brick-making industry may buy FS weekly to use as a fuel source, in which case it would be deemed an O&M cost.

Sanitation tax is a fee collected either once, or at regular intervals, and which is paid in exchange for environmental services such as a water connection, a sewer connection / removal of FS, or any combination of these services. The benefit of a sanitation tax for the government agency is that it provides a steady source of income allowing treatment and upgrade activities to be more easily planned. However, the sanitation tax may be applied to households with no sewer connection, so although it may cover the water connection (or not) the household could still be responsible for paying an additional emptying fee (if they have an onsite technology). In this case, the household may be billed twice for sanitation services; i.e. paying the sanitation tax for a non-existent sewer connection as well as an emptying fee to desludge on onsite sanitation technology. This type of model may have the effect of charging the poor more for lower-quality service, but it may also help to cross-subsidise sanitation services. A summary of the implementation of sanitation taxes in four cities in the Philippines is provided by Robbins *et al.* (2012) and shows how a sanitation tax paid on top of water bills or property taxes was used to improve FSM, by subsidising the collection and transport of sludge from households.

The sanitation tax can however be designed in such a way that it benefits the poor and directly pays for service improvement. For example, flat-rate taxes based on a uniform per-capita FS generation rate (applied to the whole city) or as a function of water consumption, would force those using more water to subsidise those using less water (and probably requiring pit emptying) (Steiner *et al.*, 2003). Fees as low as 1 USD per person per year have been calculated to completely support a sustainable FSM system. Although monthly payments may be preferable to some low-income customers who cannot afford the high, one-time emptying fee, this type of monthly payment model requires a high degree of transparency and organisation to issue, track and receive payments.

Both O&M and capital costs are paid to a large and diverse group of stakeholders (e.g. mechanics, suppliers, banks) all of whom are not, nor could be, listed here. A more detailed list of costs is presented in Section 13.4 where the financial transfers of a small scale C&T enterprise are examined in detail.

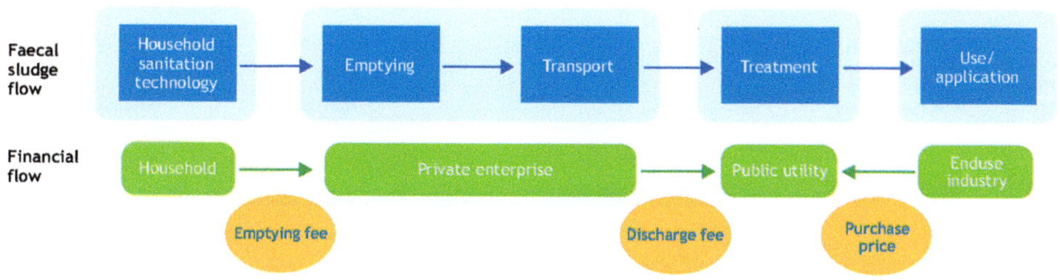

Figure 13.3 Model 1: Discrete collection and treatment model showing the responsibility of each stakeholder and the related financial transfers.

13.3 FINANCIAL FLOW MODELS

There is no single FSM model that has proven to be effective in all situations; indeed, service delivery models are constantly modified and restructured depending on the economic, legal, and environmental conditions. Furthermore, the responsibilities within the system are constantly changing and as such, the financial transfers between stakeholders can take several forms.

Various financial models for the management of FS have been proposed and an extensive list of possible configurations is summarised by Steiner *et al.* (2003). This section presents a representative selection of five different models based on existing case studies and theoretical examples. The models differ in terms of the stakeholders, the stakeholders' responsibilities, and the types of financial transfers that take place.

For the following diagrams (Figures 13.3-7) the different parts of the FSM system are shown on the upper part of the diagram in blue. The associated responsibility is indicated below in green. The type of transfer is indicated by a yellow oval. The direction of the arrow between the stakeholders indicates the direction of the payment. A dashed line indicates that the transfer is optional and may or may not occur.

Figure 13.3 illustrates a simple model of financial transfers. In this example, each of the stakeholders is responsible for a single technology in the FSM chain, and consequently, money is exchanged each time responsibility is handed over (emptying and transport are identified here as a single technology). The household-level toilet user pays a private enterprise (PE) an emptying fee to remove the sludge and the PE is responsible for the emptying and transportation of the sludge. The PE is then charged a discharge fee by the public utility for accepting, and treating the sludge. The utility is also paid a purchase price by an end-use industry in exchange for treated FS or sludge-grown products (e.g. fodder). In this model, the utility operates independently from the government authority and must cover all costs by collecting sufficient discharge and purchase fees.

This type of model has two potentially negative consequences; either, private enterprises are forced to pass the high discharge fee costs on to their customers, and thus exclude the poorest; or, the PE avoids paying the high discharge fee by illegally discharge, free of charge, on land that is not designated for FS discharge or treatment. In an effort to cut costs, and maintain a competitive advantage in the local market, the PE may also attempt to save money on O&M costs (e.g. regular maintenance of truck and pump), and as a result, limit the useful service life of the equipment, effectively putting the company out of business. In addition, because the utility is operating without direct financial support from the government authority, it is less likely to be subjected to administrative supervision and the quality of treatment, and the adherence to regulations may suffer as a result.

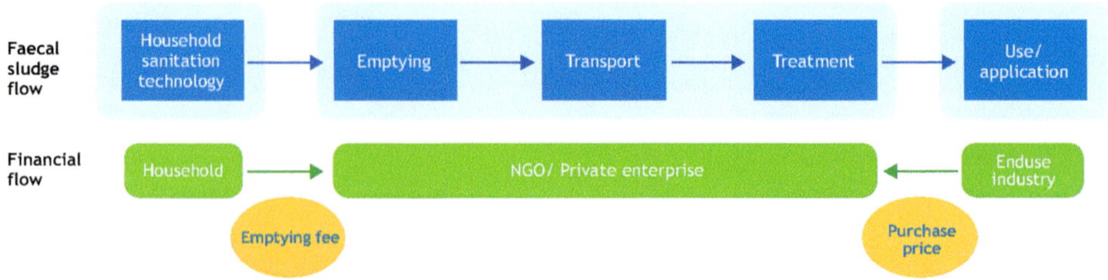

Figure 13.4 Model 2: Integrated c ollection, transport and treatment model.

This model could, however, serve as an entry point for the government authority to initiate budget support to not only strengthen the quality of service, but to reduce the need for discharge fees to cover operating costs, and thus reduce the amount of illegal discharge. Figure 13.4 presents a variation of this model, in which the operator responsible for treatment is not subject to the sludge or payment irregularities of the PE responsible for emptying.

The model depicted in Figure 13.4 appears similar to Figure 13.3, but the financial implications are significantly different. In Figure 13.4, a single private enterprise or non-governmental organisation (NGO) is responsible for the emptying, transport and treatment, thus eliminating the need for a discharge fee between the stakeholder responsible for C&T and the stakeholder responsible for treatment. There are several important financial and operational implications as a result of this difference which are explained below.

The private enterprise is responsible for collecting fees directly from the household-level toilet users. The enterprise receives no income from a discharge fee, but because the PE itself is not being charged a discharge fee, there is no need for cost recovery in the form of extra charges to the toilet user, and the toilet user may benefit from reduced emptying fees.

The market could respond in one of two ways; (i) with an efficient financial model including cross-subsidies between business activities, or by other independent C&T operators being driven out of business or to the margins of the market (e.g. in difficult, or hard to reach areas which are less profitable) or (ii) a non-optimised financial model could see the emergence of new, more competitive C&T operators who are able to undercut the multi-tasking enterprise, especially if the competing business saves costs by discharge without a permit, and if the legal framework does not enforce the proper payment and/or fines.

A variation of this model was documented in Bamako, Mali (Collignon, 2002; Bolomey *et al.*, 2003; Jeuland, 2004). There, IE Sema Saniya, an NGO owned and operated two vacuum trucks and a FSTP. With no discharge fee being charged, there was no incentive for illegal discharge, but the sustainability of the model has been called into question. The emptying fees required to cover the cost of transport and treatment were too high for many households and more cost recovery strategies were needed to ensure the financial sustainability of the system.

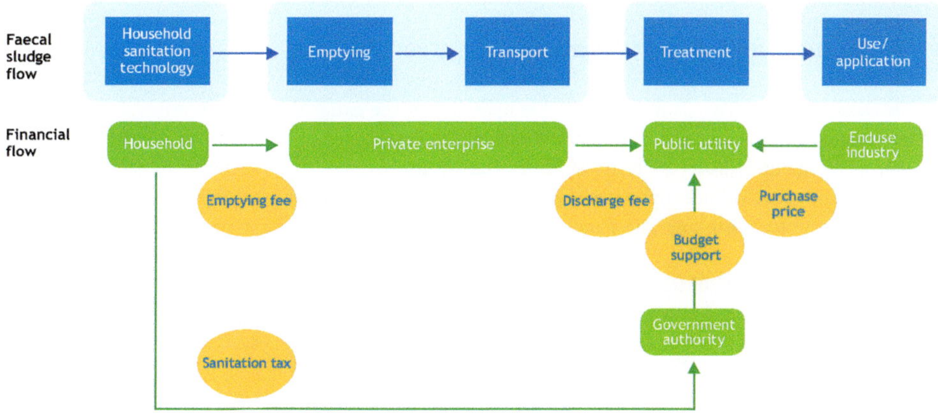

Figure 13.5 Model 3: Parallel tax and discharge fee model.

In the model presented in Figure 13.5, a sanitation tax is paid directly to the government authority by the toilet user, either through water, sewer, or property taxes. The utility is given budget support from the government authority that collects the sanitation tax. The utility therefore does not need to rely entirely on the discharge fee, and could lower it (in comparison to Model 1) thus reducing the total costs of the private enterprise. The discharge fee must therefore be high enough, such that operator can hold the PEs accountable for what they dump, but not so high that the toilet users are unable to afford the high emptying fees passed onto them by the C&T operators, or that the sludge is dumped illegally. This system is prone to corruption and under-servicing if the government authority is not competent or transparent in how it allocates it money. Furthermore, the financial balance is very much dependent on the consistent collection of the sanitation tax. Unstable land tenure, poor record keeping, corruption, transient populations and other features of fast-growing urban centres threaten the collection of a steady stream of user-based revenue. Fee collection is notoriously low in many government authorities and fluctuations in the sanitation fees can significantly affect the ability for the utility to make long-term O&M decisions if there are not reserves available from the authority to buffer the variation.

Management

Case Study 13.1 : Cambérène FSTP in Senegal

(Adapted from Mbéguéré *et al.*, 2010 and Dodane *et al.*, 2012)

In Dakar, Senegal, The Cambérène FSTP is operated by the national sanitation utility ONAS. The treatment facility includes settling/thickening tanks and unplanted drying beds, designed for 100 m^3/day of FS; about 41,500 people are serviced. The facilitiy receives sludge from septic tanks that are emptied by vacuum trucks operated by private collection and transport companies. The financial flow model at Cambérène follows the 'Parallel Tax and Discharge Fee' model described above (Figure 13.3).

Households pay 50 USD to private C&T companies to have 10 m^3 of sludge removed; this translates into approximately 5 USD/capita/year. Furthermore, households pay a sanitation tax to ONAS which amounts to about 2 USD/capita/year. The total payment per person, per year (7 USD) corresponds to about 2% of the average household budget of the Dakar population.

The C&T companies made large initial investments in their trucks which must be paid off over time, and this has been estimated as as a 0.3 USD/capita/year expense. The company must also pay a discharge fee to discharge the sludge at the FSTP: the fee amounts about to about 0.4 USD/capita/year. The remainder of the money earned goes towards O&M costs which include staff, fuel, overhead, repairs and maintenance to the truck; this total must be less than 4.3 USD/year in order for the company to make a profit.

ONAS has two main sources of revenue: the sanitation tax paid by households and the discharge fees paid by the C&T companies. To further generate income, and to improve nutrient cycling in the urban area, ONAS sells the dried FS to agricultural industries for use as a soil amendment. They generate about 250 USD/year (which, converted for comparability translates into about 0.007 USD/capita/year).

The daily operation and maintenance of the facility (i.e. electricity, salaries, etc.) costs about 1 USD/capita/year. The capital costs (i.e. the construction of the facility), annuali ed, were estimated to be 1.3 USD/capita/year (41,500 customers). A summary of the financial flows is shown in Figure 13.6.

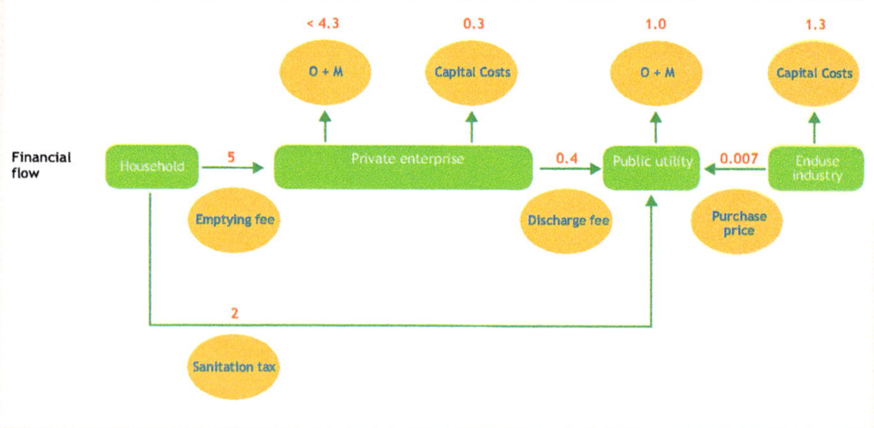

Figure 13.6 Financial flow in USD of the faecal sludge management system in Dakar, Senegal.

Management

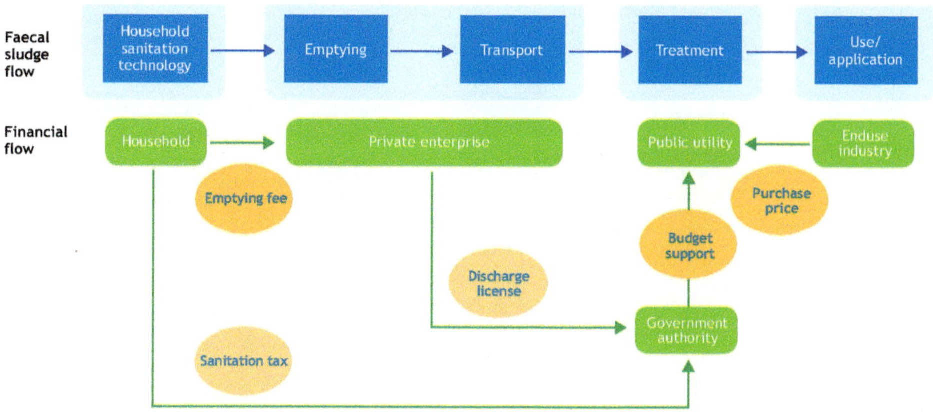

Figure 13.7 Model 4: Dual licensing and sanitation tax model.

In the dual licensing and sanitation tax model, as shown in Figure 13.7, the private entrepreneur who is responsible for C&T is not penalised with a discharge fee for each discharge at the FSTP, but instead is granted unlimited (or semi-limited) access to dump through a discharge license, thus reducing illegal discharge by those C&T operators who may not be able to afford the discharge fee.

Having to pay a discharge license, no matter how nominal, ensures that the government has more administrative control over the industry. Data on the number of operators, the revenue that is generated, the distances travelled etc. can be collected and used to advise policy. Furthermore, the discharge license means that the PE is recognised by the government, and theoretically, should have to pay fewer bribes, fees, or fines during the course of work. This model has been enacted in Kumasi, Ghana where the C&T businesses must obtain a discharge license which can be revoked if the emptier is found discharging anywhere but the official facility (Mensah, 2003; SANDEC, 2006). Discharge licenses have also been implemented in Nairobi's Kibera slum where they were sold yearly (Water and Sanitation Program Africa, 2005) and in Da Nang Vietnam, where they were sold monthly (Steiner *et al.*, 2003).

As explained in Chapter 4, the FS C&T industry has remained largely unrecognised. Its employees are ostracised and are often forced to work clandestinely or at night under threat of persecution or police scrutiny. It's informal nature means that it is beyond the realm of labour and health laws, so workers endure unsafe and humiliating conditions, without the basic rights afforded to other industries (Eales, 2005). Therefore, although obtaining discharge licenses may be costly and prone to corruption, licensing is one of the first steps towards formalising the industry, and potentially opening it to more transparent and effective policy interventions. Licensing is a mechanism that does not exclude the smallest operators (provided they can afford the one time fee, they are not penalised for frequent use of the FSTP), may help improve industry standards, while also improving working conditions for the labourers and service delivery for the toilet users.

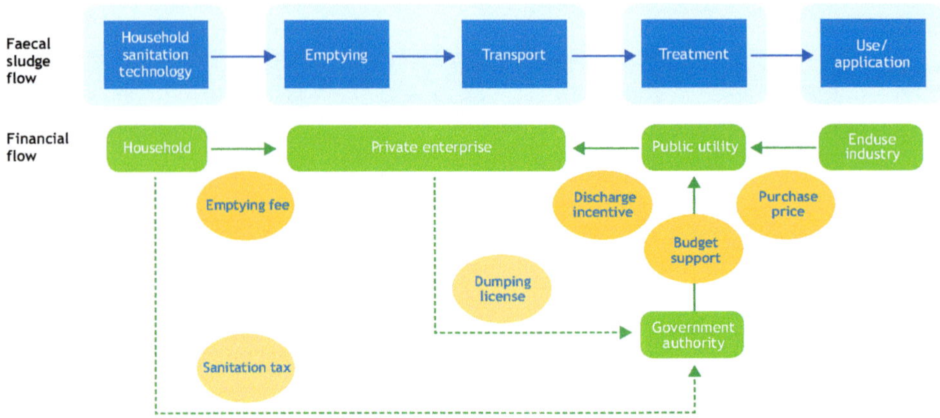

Figure 13.8 Model 5: Incentivised discharge model.

An important feature of the model shown in Figure 13.8 is the direction of the financial transfer from the public utility to the private entrepreneur. In this model, the FSTP operator pays the stakeholder responsible for C&T a discharge incentive to dump sludge at the FSTP. A financial model that includes discharge incentives could take a variety of forms. For this reason, the discharge license and sanitation tax flows in Figure 13.8 are left as dashed lines to indicate that they may or may not exist in this model, depending on the context.

As discussed previously, financial incentives can be used to encourage socially desirable behavior. In the case of discharge incentives, the payment is used to encourage sludge collection and reduce illegal discharge. These types of conditional cash transfers are still relatively new, and although results are promising in health and education programs, there is little data to support their use in sanitation programs (SANDEC, 2006).

This model is built on the theory that C&T stakeholders cannot afford the discharge fees charged by FSTP operators and so dump indiscriminately, causing damage to public and environmental health. Working under this scheme, the C&T operator would only have to recover a portion of the total operating costs from the emptying fee (the other portion would be made up by the discharge incentive). As a result, the collection service would be more affordable for poorer households, more sludge would be collected, less sludge would be discharged to the environment and the community as a whole would benefit.

Unfortunately, this scheme means that the FSTP operator would not receive revenue from discharge fees and yet would also be responsible for paying the discharge incentives. This model could only function with substantial government or donor support, which can be variable and inconsistent, leaving the FSTP operator with budget gaps. To prevent such shortcomings, sanitation taxes would likely have to be raised to cover the increased operating expenses of the treatment plant. The emptying fee could however be reduced, tightly regulated or done away with altogether. The toilet user would still be responsible for the sanitation tax, but would be relieved of the financial burden of paying for access to sanitation twice (i.e. sanitation tax and collection fee).

One concern with this model is the opportunity for C&T stakeholders to take advantage of the financial incentive, and rather than spending time and fuel to actually empty onsite systems, operators may attempt to receive the incentive for watered-down sludge or alternative liquids which could damage the treatment process and its financial viability. To control the type and quality of the sludge emptied at the FSTP, some type of quality assurance or quality control must be in place, such as a manifest program as described in Chapter 11.

A possible variation of the model presented in Figure 13.8 would be to include incentives for toilet owners who have their sludge removed by a certified service provider. This model would prevent homeowners from waiting until the onsite storage technology is overflowing, from dealing with an unlicensed C&T business, or from emptying it directly to the environment during the rainy season. No known examples of this variation have been put into practice. The logistics of administering such a program are complex as it would need to ensure the delivery and acceptance of reverse payments to households, and the subsequent fulfillment of the emptying service promised, would require widespread education and policy enforcement. A concise summary of the pros and cons for each of the models is presented in Table 13.3.

Figure 13.9 Slow moving city traffic can add significantly to the fuel and labour costs associated with collection and transport (photo: Linda Strande).

Table 13.3 Summary of pros and cons for each the financial models presented in this Chapter

Model	Pros	Cons
Model 1: Discrete C&T and Treatment Model	+ Households are free to choose the most competitive price on offer for emptying; + Timing of emptying is flexible and can be done when financially feasible + The household is not committed to a fixed sanitation tax	- The utility's operating expenses must be covered by the discharge fee
Model 2: Integrated C&T and Treatment Model	+ A single operator is able to optimi e the business model and improve efficiency; + Less potential for illegal discharge as the single entity will discharge at the self-run treatment works	- High fees may be passed onto the household
Model 3: Parallel Tax and Discharge Fee Model	+ Low-income households' that are not connected to the sewer may have lower C&T costs from cross-subsidies; + C&T operators may benefit from lower discharge fees + Collection and coverage increases	- C&T businesses may avoid discharge fees by discharge illegally
Model 4: Dual Licensing and Sanitation Tax Model	+ Industry regulation and legitimisation through licensing + Improvement in health and safety conditions; + Unlimited discharges minimises risk of illegal dumping	- The management of too many aspects of the service chain by one entity could prove difficult for a new business or NGO
Model 5: Incentivised Discharge Model	+ Emptying fees for households may be reduced; + Households that are difficult to access, or located far from the treatment plant, may become attractive to C&T operators because of incentives	- Incentives must be corruption-proof (e.g. not given for diluted sludge, seawater, etc.) - FSTP operator requires significant budget support to function budget support to function

13.4 FINANCIAL PERSPECTIVE OF A COLLECTION AND TRANSPORT ENTERPRISE

It is difficult to breakdown the allocation of costs and benefits within a FS system as each stakeholder views each financial transfer from their own, unique perspective. For example, an emptying fee is a cost for a household-level toilet user, while it is a benefit for a C&T operator. It is beyond the scope of this chapter to summarise all of the costs and benefits for each stakeholder operating within each type of model. Dodane *et al.* (2012) illustrate the distribution of costs and payments among household-level users, businesses, and the public utility in Dakar, Senegal and conclude that the FSM management system is 5 times less expensive than a sewer-based one. However, this study showed that 6% of the annualised cost of the FSM system is inequitably borne by household level users, and that the C&T companies are operating at no net annual profit. An analysis of C&T businesses provide an interesting case study because they serve as a simple, but useful way to illustrate how the various financial transfers described in this chapter affect operational sustainability.

Despite working at the social margins, the C&T business can be very competitive, forcing each entrepreneur to work at the edge of profitability. However, in spite of cutting costs wherever possible, C&T enterprises still cater to a client base that often finds their services too expensive. Furthermore, the business must pay fees to the utility for discharge, taxes to the government, as well as O&M costs to keep the equipment operational. The model that was presented in Figure 13.1 is the simplest example of the financial transactions that a C&T business is responsible for and yet, many of the actual payments (e.g. taxes, O&M) are not shown.

In order to demonstrate the variety and number of costs and payments associated with a small C&T enterprise (one of only several parts in an entire FSM system), an example is provided in Section 13.4.2. On completion of this example one should gain an understanding of the complexity and difficulty of designing, implementing, monitoring and optimising an entire FST system which includes all of the stakeholders and financial interactions will be obtained.

13.4.1 Future perspectives

Much of the financial sustainability of a C&T business depends on government policy and support. The supporting legal structures are essential to any financial policy designed to assist small business operators and household level users (see Chapter 12).

Short-term discharge incentives appear to be one of the most promising ways to strengthen the private sector, help clear the backlog of full pits and septic tanks, and generate steady-state conditions that can be further refined or manipulated through policy and/or financial mechanisms. Businesses need to develop a client base, optimise their routes and pay off their capital costs. Implementing discharge incentives for a short time (e.g. 5 years) could help to sustain small businesses and improve sanitation conditions drastically within a short period of time. Once businesses are established, incentives could be slowly reduced and eventually, discharge fees introduced. Donor-funded incentives could be a short-term, highly effective way of supporting small business generation while strategically addressing sanitation deficiencies. As is demonstrated in the example provided in Section 13.4.1, the removal or reversal of discharge fees could have had a profound impact on the sustainability of the C&T business and financial well-being of the owners.

Sanitation taxes, applied most equitably as a function of water usage, can help cover the cost of FSM. The money collected should be used to support the FSTP O&M, assist in regularly scheduling collection or maintenance of household sanitation technologies, offset the discharge fee or generate a fund for discharge incentives.

Licensing, in combination with genuine rights granted to licensees, and enforcement of fines when rights are abused (i.e. the withdrawal of the permit if the C&T operator is found to discharge illegally) would help to reduce corruption and illegal discharge. Different types of regulations and enforcement are discussed further in Chapter 12. Licensing is also the first step to formalisation of the sector, and would therefore open the businesses up to other policies and subsidies designed to support small businesses; perks which have historically been denied to informal workers.

More efficient trucks (i.e. newer, fuel-efficient vehicles), made available through lower import tariffs would significantly improve fuel consumption and help lower overall costs. More strategically located discharge/treatment facilities would reduce the travel distance, and importantly for the city, reduce time and fuel wasted idling in city traffic.

Discharging into transfer or relay stations, which are then emptied by larger vehicles, would allow small emptying businesses to spend more time emptying, and less time transporting (and in turn, earning more money) (Tilley *et al.*, 2008). If appropriate treatment and transport infrastructure exists, license holders could be permitted to dump into the sewerage system in order to reduce their travel time, and focus instead on emptying onsite technologies. This option is however, dependent on the proper design of the treatment technology to prevent overloading and blockages (refer to Chapter 5 for a summary of appropriate treatment technologies). Licensing revenues should be used to formalise sewer discharge stations and transfer stations.

A range of policies to support larger, multi-truck operators who can serve higher-paying, easier to reach clients as well as smaller- operators who can serve lower-paying, harder to reach clients, must be developed. As discussed in this chapter, there is no single model for efficient FSM, and experimentation and flexibility with novel financial mechanisms must be encouraged.

Areas for further research include understanding the financial flows and business models for existing and successful FSM enterprises. Since the sector is mostly informal, there is very little known in this area. There are currently very few examples of functioning FSM systems. Different business models must be tested and studied under different operating conditions to prove which will be the most robust and sustainable.

Finally, and perhaps most importantly, political will, (i.e. public support and acknowledgement of the FS industry), must be communicated from the highest levels down to traffic controllers. This will assist in reducing corruption, embarrassment and the current financial inefficiencies that exist in a business that is essential to the health and growth of the world's cities.

13.4.2 Case study example

Consider a small C&T business that is run by two brothers in West Africa. The dense urban area where they work includes about 250,000 residents and has a density of about 300 people/ha (UN-Habitat, 2003). By working 20 days a month, 12 months a year, and servicing 3 clients a day the brothers hope they can pay back their truck loan, cover their operating expenses, pay themselves a small salary and hopefully make a profit. The brothers each hope to earn 5 USD per day.

To determine if this is possible, use the information and formulae given below to calculate:

- the annual costs for operating the business by filling in a version of Table 13.4; and
- the minimum cost that they must charge households to cover their expenses.

Table 13.4 Table for summarising yearly operating and capital costs for a small C&T enterprise

Item	Yearly Costs (USD)	Percent of total cost (%)
Truck payments		
Discharge license		
Equipment		
Labour		
Fuel		
Discharge fees		
Maintenance		
Police		
Insurance		
Parking		
Taxes		
Administration		
Total		100

13.4.3 Problem information

Overalls, gloves, boots, shovels, and simple tools for breaking slabs and accessing pits will vary, but basic equipment will cost up to 100 USD/year (Water and Sanitation Program Africa, 2005).

The truck is the largest expenditure. The brothers decide on a used, 8 m^3 trucks that they can purchase for 20,000 USD (Steiner *et al.*, 2002). Because of the harsh working conditions, they expect the truck to last about 10 years before they have to replace it. In the dense urban areas the truck can travel at an average speed of 5 km/h, and it costs about 0.5 USD/km for fuel (assume an interest rate of 5% on their loan).

Equation 13.1:

Equivalent Annual Cost (EAC) = Capital Investment/Annuity factor

$$= \frac{\text{Capital Investment}}{\frac{1-(1+i)^{t}}{i}}$$

Where i is the interest rate and t is the repayment period

The discharge license has been set at 780 USD/year (for their large 8 m^3 vehicle) based on the Kenyan model (Water and Sanitation Program Africa, 2005).

When the truck arrives at the FSTP, it is charged 2 USD per full discharge (8 m^3)(Steiner *et al.*, 2003), but the operators usually charge the full price regardless of how much is discharged.

Table 13.5 Annual expenses (given in percent of the total, %) from a C&T enterprise operating in Bamako, Mali (adapted from Bolomey *et al.*, 2003; Jeuland, 2004)

Maintenance	Police	Salaries	Insurance	Parking	Tax	Admin.
20	10	15	2	1.5	2	15

To determine the daily transport distances, the following assumptions can be made:

- the area served is round, and that the average transport distance is half the radius;
- the FSTP is located in the centre of the area that they serve, and that the population density is homogenous; and
- the truck must return to the treatment plant after each household visit (i.e. the truck cannot empty more than one house with the same tank).

The remaining annual expenses can be calculated using the information given in Table 13.5.

In Table 13.5, "police" refers to the payment of 'fees' or 'taxes' to the police for transporting what is sometimes called 'dangerous matter' (Jeuland, 2004).

Based on this revised estimate, the average fee to the household would have to be about 22 USD, which is closer to the average rate and the brothers know that the willingness to pay of the toilet user is much less than they will actually be able to charge (Bolomey *et al.*, 2003). After completing their analysis, the brothers start to wonder how, if ever, their business could become profitable (i.e. how much they would have to charge their customers (question b)).

13.5 BIBLIOGRAPHY

Banerjee, A., Duflo, E., Glennerster, R., Kothari, D. (2010). Improving immunisation coverage in rural India: clustered randomised controlled evaluation of immunisation campaigns with and without incentives. British Medical Journal 340.

Bolomey, S., Koné. D. (2003). Amélioration de la Gestion des Boues de Vidange par le Renforcement du Secteur Privé: Cas de la Commune VI du District de Bamako. Dübendorf, Switzerland, EAWAG/SANDEC.

Collignon, B. (2002). Les enterprises de vidange mécanique des systümes d'aassainissement autonome dans les grandes villes africaines: Rapport de synthèse finale. PDM, PS-Eau, Hydroconseil, Chateauneuf de Gadagne, France.

Diener, S., Semiyaga, S., Niwagaba, C., Muspratt, A., Gning, J.B., Mbéguéré, M., Ennin, J.E., Zurbrugg, C., Strande, L. (2014). A value proposition: resource recovery from faecal sludge – can it be the driver for improved sanitation? Resources Conservation & Recycling (in press).

Dodane, P.H., Mbéguéré, M., Ousmane, S., Strande, L. (2012). Capital and Operating Costs of Full-Scale Faecal Sludge Management and Wastewater Treatment Systems in Dakar, Senegal. Environmental Science & Technology 46 (7), p.3705-3711.

Eales, K. (2005). Bringing pit emptying out of the darkness: A comparison of approaches in Durban, South Africa, and Kibera, Kenya. S. P. Series.

Eldridge, C., Palmer. N (2009). Performance-based payment: some reflections on the discourse, evidence and unanswered questions. Health Policy and Planning 24(3), p.160-166.

Gertler, P. J., Boyce, S. (2001). An Experiment in Incentive-Based Welfare: The impact of PROGRESA on Health in Mexico.

Management

Jeuland, M. (2004). Private Sector Management of Faecl Sludge: A model for the Future? Bamako, Mali, Swiss Federal Institute of Aquatic Science and Technology.

Kakwani, N., Soares. F., Son, H.H. (2005). Conditional Cash Transfers in African Countries. International Poverty Center, UNDP. Working Paper 9.

Klingel, F. (2001). Nam Inh Urban Development Project- Septage Management Study. Nam Dinh, Vietnam. Dübendorf, Switzerland, EAWAG and Colenco.

Mbéguéré, M., Gning, J.B., Dodane, P.H., Koné, D. (2010). Socio-economic profile and profitability of faecal sludge emptying companies. Resources, Conservation and Recycling 54 (12), p.1288-1295.

Mensah, K. (2003). Sanitation, Solid Waste Management and Storm Drainage Component. Medium term development plan for Kumasi. Kumasi, Ghana.

Robbins, D.M., Strande, L., Doczi, J. (2012). Sludge Management in Developing Countries: experiences from the Philippines. Water 21, Issue 4.

SANDEC (2006). Urban Excreta Management: Situation, Challenges, and Promising Solutions. 1st International Faecal Sludge Management Policy Symposium and Workshop, Dakar, Senegal.

Steiner, M., Montangero, A. (2002). Economic Aspects of Faecal Sludge Management- Estimated Collection, Haulage, Treatment and Disposal/Resuse Costs. Dübendorf, Switzerland, Swiss Federal Institute of Aquatic Science and Technology, 1st Draft.

Steiner, M., Montangero, A. (2003). Towards More Sustainable Faecal Sludge Management Through Innovative Financing: Selected Money Flow Options. Dübendorf, Switzerland, Swiss Federal Institute of Aquatic Science and Technology.

Water and Sanitation Program Africa (2005). Understanding Small Scale Providers of Sanitation Services: A Case Study of Kibera. Nairobi, Kenya, Water and Sanitation Program.

Tilley, E., Lüthi, C., Morel, A., Zurbrügg, C., Schertenleib, R. (2008). Compendium of sanitation systems and technologies. Swiss Federal Institute of Aquatic Science and Technology (EAWAG), Duebendorf, Switzerland.

United Nations Human Settlements Programme (UN-HABITAT) (2003). The challenge of slums : global report on human settlements, 2003. Earthscan Publications Ltd, London and Sterling, VA. 345 pp.

Water and Sanitation Program Africa (2005). Understanding Small Scale Providers of Sanitation Serivces: A Case Study of Kibera. Nairobi, Kenya, Water and Sanitation Program.

End of Chapter Study Questions

1. What are discharging incentives in FSM?

2. List three possible financial models for FSM and what the advantages and disadvantages of these models are.

3. Explain the pros and cons of the Dual Licensing and Sanitation Tax model.

Planning

Planning

Chapter 14

Assessment of the Initial Situation

Philippe Reymond

Learning Objectives

- Understand what is important to know at the beginning of the faecal sludge management planning process and identify what information needs to be collected.
- Be aware of the different methods and tools for collecting relevant data and know how to apply them.
- Know how to identify the shortcomings and challenges of an existing faecal sludge management system and be able to describe an enabling environment.

14.1 INTRODUCTION

The assessment of the initial situation, which is the first step in the planning process (Chapter 17), is crucial, as it provides the baseline information for decision-making. This chapter serves as a guideline for process leaders (Box 17.1) on which data to collect and how to carry out this assessment using a participatory approach.

The main goals of the assessment of the initial situation are to set the scene, understand the context, get to know the stakeholders and provide enough information to start elaborating the faecal sludge management (FSM) scenarios, including context-specific design parameters and therefore this stage is characterised mainly by data collection via different means. This data is collected step by step during the exploratory investigation, preliminary studies and feasibility study (see Table 17.1, FSM planning from A to Z), phases which relate in terms of participatory planning stages to the launch of the planning process, the detailed assessment of the current situation and the identification of service options respectively. Useful examples are provided by Dodane (2010) and Larvido and Dodane (2011) for Mahajanga, Madagascar and by Mikhael (2010, 2011) for Freetown, Sierra Leone.

This type of assessment gives a snapshot of the situation at the beginning of the project. It describes the existing service chain, starting with the types of latrines, the formal and informal sludge emptying sector, the organisation of the system and the links between the stakeholders. It also identifies the enabling environment (Section 17.2.1), government support, the legal and regulatory framework, institutional arrangements, skills and capacity, financial arrangements and socio-cultural acceptance.

Figure 14.1 Faecal sludge disposal into the ocean, Accra, Ghana (Google Earth, 2010).

If it does not already exist, an enabling environment must be built during the process if the project is to succeed (AECOM and SANDEC/EAWAG, 2010; Lüthi *et al.*, 2011a; Lüthi *et al.*, 2011b; Parkinson and Lüthi, 2013).

This chapter focuses on the information and data that must be collected and provides guidance on the collection procedure. The way to analyse this information and the various scenarios that can be encountered are described in related chapters, especially the chapters *Quantification, Characterisation and Treatment Objectives* (Chapter 2) regarding the design parameters, *Methods and Means for Collection and Transport* (Chapter 4) regarding the profile of manual and mechanical service providers, *Institutional Frameworks* (Chapter 12) regarding the laws, regulations, roles, responsibilities and institutional stakeholders, *Financial Transfers and Responsibilities* (Chapter 13) regarding financial flows and market studies, *Stakeholder Analysis* (Chapter 15) regarding the FSM stakeholders, and *Stakeholder Engagement* (Chapter 16) regarding the involvement tools. The chapter *Planning Integrated FSM Systems* (Chapter 17) helps to place the content of this chapter within the framework of the whole planning and engineering process and details the decision factors that must be investigated for technology selection. Figure 14.2 summarises how the different chapters contribute to the present one.

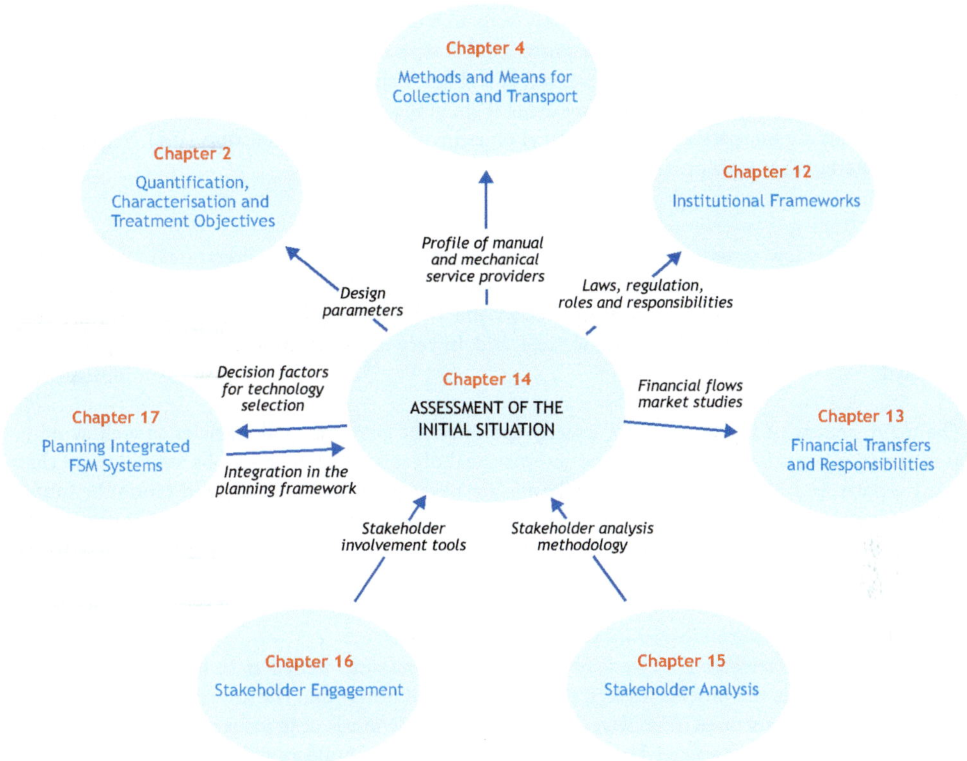

Figure 14.2 Links of the assessment of the initial situation to the other book chapters.

In addition, the human dimension of this assessment should not be overlooked. This is when the first contacts take place and trust starts to be built with the stakeholders. This is crucial for the rest of the project. The role of the local facilitator(s) is very important here, as they help to open doors and gain access to information. It should not be forgotten that data, if it exists, is not always readily available and that getting accurate information usually depends on the goodwill of local partners. Building trust relationships should be the first priority before rushing into data collection.

14.2 TOOLS AND METHODS FOR DATA COLLECTION

Collecting good quality, useful data is often not an easy process, especially in contexts where data is scarce, and either not collected or analysed properly, or, sometimes, hidden or manipulated for political or personal reasons. Governmental agencies usually have some reports, statistics and maps that can serve as a preliminary introduction. However, they should always be considered with care, and therefore the collection of primary data is recommended, if not essential. The best way to get a reasonably accurate estimation is to rely on several sources of information, which can be cross-checked and, if needed, complemented by further research.

The following paragraphs describe different ways to collect the necessary data for the assessment of the initial situation. It is important that the process leaders go out into the field and meet the stakeholders (Chapters 15 and 16), in order to obtain a first-hand grasp of the real situation. There are two main ways to get preliminary impressions: observation and discussion. The discussion is about finding out each FSM stakeholder's perspective and involving them in the project from the start. In that sense, data collection tools are inseparable from the stakeholder involvement tools (Section 16.4). Section 15.3 lists the different FSM stakeholders.

14.2.1 Literature review

The literature review consists of searching for data that already exists (grey literature, i.e. reports and maps, or white literature, i.e. publications). There is no need to 'reinvent the wheel', although existing documents should always be used with caution and the reliability of data assessed. Data quality (especially with statistics) is often questionable, and, in very dynamic contexts, may become quickly outdated.

The main sources of information are usually the different governmental agencies as well as non-governmental organisations (NGOs) and international organisations. It should be kept in mind that many reports, especially written by consultants, are never published officially and cannot be found on the Internet. Individual meetings with the various organisations and agencies are recommended, starting with a few key informants who are able to provide a rough idea of the information that is available.

14.2.2 Semi-structured interviews

Semi-structured interviews are one way to structure discussions aimed at collecting information. The interviewers are the process leaders, usually with facilitator(s), and the interviewees are the FSM stakeholders. Semi-structured interviews can be held with individuals or in focus groups (Section 16.4). They require time and experienced interviewers, but they help to build a solid basis for further work.

Semi-structured interviews are conducted with a fairly open framework which allows for focused, two-way communication (FAO, 1990)[1]. They can be used both to give and to receive information. Not all the questions are designed and phrased ahead of time, hence the 'semi-structured' nature of these interviews. The majority of questions are created during the interview, allowing both the interviewer and the person being interviewed the flexibility to probe for details or discuss issues. This freedom helps interviewers to tailor their questions to the interview context/situation, and to the people they are interviewing. Often the information obtained from semi-structured interviews will provide not just answers, but the reasons for the answers. It also helps process leaders and field staff to become acquainted with community members. Indeed, data collection is not the only purpose of the interviews; it is also an opportunity for discussion, exchange and trust-building.

Semi-structured interviews should be prepared in advance, with key questions listed in an interview guide. Box 14.1 provides basic interview guides for FSM key stakeholders.

1 For more information on semi-structured interviews:
CLUES Toolbox: Tool T2 – Interview Methods and Questionnaire Examples (www.sandec.ch/clues)
SSWM Toolbox: www.sswm.info/category/planning-process-tools/exploring/exploring-tools/preliminary-assessment-current-status/semi

Box 14.1: Interview guides for some key faecal sludge management (FSM) stakeholders
(Adapted from Koanda, 2007a; Reymond, 2008)

FSM stakeholders are described in Section 15.3 and 15.4. Guides for semi-structured interviews with municipal authorities/the mayor, municipal technical services and mechanical operators are provided here below. As 'guides', their primary aim is to help the interviewer to remember the important discussion topics.

Interview guide for the municipal authorities/the mayor

1 Importance of sanitation for the municipal authorities (sanitation in general, including wastewater and solid waste)

2 Importance of FSM for the municipal authorities

3 Role of the municipality in sanitation and FSM

4 Financial arrangements
- Finance arrangements for water and sanitation
- Taxes, grants and subsidies
- Management of this budget for water and sanitation

5 Legal and regulatory framework
- Laws and regulations
- Municipal decisions
- Means of enforcement
- If no regulations, is there an opportunity to publish municipal decrees?

6 Current practices
- Existing infrastructure: sewer networks, disposal sites, treatment units
- Sludge pumping trucks: number, property, management
- Public latrines: number, volume, management
- Roles and responsibilities (who is in charge of what – e.g. sewer maintenance, FS collection, solid waste collection)

7 Institutional setup
- Other governmental agencies involved in sanitation – links and relationship with the municipality

8 Socio-cultural acceptance
- Perceptions of the population towards current situation and existing initiatives
- Enduse and resource recovery practices and potential

9 Proposals for improvement, needs of the municipality

Interview guide for the municipal technical services

1 Management of sanitation (wastewater, faecal sludge (FS) solid waste)
- Roles and responsibilities (who is in charge of what – e.g. sewer maintenance, FS collection, solid waste collection)
- Number of staff
- Operating mode

2 Organisation of FSM
- FS collection: practices, number of trucks, staff
- Quantity of collected sludge; records?
- FS disposal sites
- Existing and potential for enduse and resource recovery
- Public places (schools, markets, public latrines, mosques, temples, etc.)
- Number, volume and management of public latrines

3 Organisation of solid waste management
- Collection
- Public infrastructure (market, slaughterhouse)
- Community level (markets, abattoirs, schools, public places)
- Volumes
- Disposal, enduse, resource recovery (soil amendment, fuel, etc)

4 Finance (available resources)
- Municipal budget
- Taxes
- External support

5 Legal and regulatory framework
- Laws and regulations
- Municipal decisions
- Means of enforcement

6 Proposals for improvement, needs

7 Future possibilities in the pipeline

Interview guide for mechanical FS operators

1 General description of the company
- Equipment
- Staff
- Tariffs
- Relationship with the municipal authorities (formal?)
- Juridical status
- Taxes

2 Quantity of sludge collected
- Capacity of the truck(s)
- Number of rotations per truck per day/month/year
- Seasonal variation or any other significant variations in emptying services over time?
- Availability of records/accounting

3 Types of emptied latrines

4 Emptying frequency

5 Emptying of public latrines?
- Special mandate from the municipality?
- If public and private latrines, organisation in terms of timing: sludge removal of public latrines all at same time?

6 Disposal/enduse
- Disposal sites
- Discharging in agricultural fields?
- Collection of dried sludge at the disposal sites?
- Enduse practices?

7 Partnerships, customers

8 Proposals for improvement

14.2.3 Household-level surveys

Surveys or questionnaires are a way of collecting information systematically, so that data collected from different sources can be easily compared and analysed quantitatively, e.g. using statistics. In FSM, they are especially used to collect data at the household level in order to assess the practices, perceptions and sanitation status (Section 14.3.4). Data collected at this level allows to quantification and characterisation of the FS to be treated (Chapter 2).

Before preparing a survey, it is important to know exactly which data is needed and what it will be used for. The following points are important to be kept in mind (adapted from Tayler-Powell, 1998):

- What is the purpose of the data to be collected and its expected use (e.g. frequency, percentage)?
- Is the information available elsewhere?
- Stick to only the necessary questions, so as not to overburden the persons being surveyed, except for a few contact questions at the beginning to put the interviewee at ease.
- Try to view the questions through the respondent's eyes wording is importantas as is understanding and utilising the social norms, the specific vocabulary, and being aware of context-sensitiveness [2].
- The response or information obtained is only as good as the question is.

2 For example, a question like 'Do you discharge sludge directly on agricultural fields?' may threaten a truck operator, who is usually aware of the non-conformity – or even illegality - of such a practice; he may then answer 'no', even if he does. Thus, the question should rather be formulated as: 'Some farmers are known to ask for sludge on their fields. Have they ever contacted you, and how?'

Finally, when carrying out a survey, representativeness of the sample is key. Indeed, households may be very different in nature, e.g. in terms of income, cultural background, tenure (owner-occupier versus tenants), and informal versus formal developments. Detailed information on how to build a representative survey can be found in sociological literature (Groves *et al.*, 2009).

Case Study 14.1 : Key question areas for the household-level surveys

(Adapted from Koanda, 2007b and Reymond and Ulrich, 2011)

Household-level surveys must be comprehensive but not burden the interviewee, who may then lose focus. They must reflect the integrated approach to sanitation and highlight the practices, constraints and needs of the population.

Figure 14.3 Householder survey, Nile delta, Egypt (photo: Colin Demars).

The following aspects need to be part of the household-level survey in a FSM planning process:

- characterisation of the interviewee: status, family, cultural background, household size
- water supply: water sources, water quality, service quality, water consumption, costs
- hygiene and sanitation:
  - type of on-site sanitation technology (or open defecation), number of users
  - type of emptying services ('what happens when the pit is full') – if no sewers: mechanical/manual, public/private, frequency (winter/summer or dry/rainy season), cost, perception of cost and service, willingness to pay for improved services
  - if sewer network: type of sewers, problems encountered, discharge point
  - greywater management
  - solid waste management: disposal, service, cost
  - stormwater management
  - in rural areas: animal manure management – disposal/enduse practices
- institutional/organisational aspects: who is responsible for each service, positive/negative aspects
- environmental awareness: perception of cleanliness and health impacts, willingness to improve
- communications channels: main information sources, information on consumption habits

14.2.4 Qualitative field observations

While field visits are a powerful tool to expose all the stakeholders to reality, they are also a good way for the process leaders to understand the reality better, to cross-check the available information by observing and discussing with people, and to build trust onsite with the main stakeholders (Figure 14.4). They provide an introduction to the existing sanitation services and an initial understanding of the conditions from the perspective of local residents.

Quantitative household-level surveys are essential for good quantitative data, but freer observation is also very important. This can include site visits, possibly including transect walks, one-to-one semi-structured (narrative) interviews with householders, and focus groups with community members, if necessary specifically focusing on particular vulnerable groups (Section 16.4). Interviewing people leads to information on their perceptions of reality and on what they think they do, but when one wants to know what people really do, one has to observe them.

For observation, transect walks are recommended. Routes are chosen so that they cover the greatest diversity regarding the water, sanitation and agricultural issues (Figure 14.5).

Figure 14.4 Accompanying a faecal sludge service provider at work, Nile Delta, Egypt (photo: Philippe Reymond).

Figure 14.5 Transect walk in Nakuru, Kenya, including discussions with households (photo: Philippe Reymond).

14.2.5 Mapping

Mapping is essential for a clear and extensive analysis of the existing situation, especially when it comes to understanding the city structure (neighbourhoods with different levels of revenues, main axes - Section 14.3.8) and identifying treatment sites (Section 14.4). Mapping has become much easier in recent years with the democratisation of satellite images (e.g. Google Earth) and geographical information systems (GIS) (WSUP, 2011).

A participatory mapping exercise is also recommended, as it is a good way to involve selected stakeholders (Section 16.4). Particularly important is the identification of key elements, such as existing disposal sites or obstacles for emptying trucks (e.g. road segments prone to traffic jams and poor quality roads).

14.2.6 Laboratory analyses

In FSM, where a comprehensive database on FS characteristics does not yet exist, it is usually necessary to carry out sampling campaigns and analyses in order to be able to characterise the sludge on a site-specific basis. Sludge characteristics vary significantly between and even within cities, and it is important to obtain first-hand data (Figure 14.6). Parameters to be measured and types of sampling campaigns are described in Chapter 2.

Figure 14.6 Septage sampling and analysis with portable laboratory equipment in the Nile delta, Egypt (photo: SANDEC).

Figure 14.7 How to read a SWOT analysis matrix (Schall, 2004).

14.2.7 Strengths, weaknesses, opportunities and threats analysis

When carrying out the initial assessment, it is important to clearly determine what are the Strengths, Weaknesses, Opportunities and Threats (SWOT) of the environment in which the FSM system has to be developed, especially the organisational and institutional framework (Chapter 12), as well as the key stakeholders (Chapter 15). The SWOT matrix shows the positive and negative factors that have to be dealt with (Figure 14.7); setting them out clearly in this way makes it possible to take action in order to maximise the potential of the strengths and opportunities while minimising the impact of the weaknesses and threats. The factors can be categorised according to the components of the enabling environment (Section 17.2.1).

Box 14.2 provides an example of a SWOT Matrix.

Box 14.2: Example of a SWOT matrix

A SWOT analysis helps to visualise the positive and negative factors that may influence the FSM project. Examples of questions to be answered are. Are the existing laws enabling (strength/weakness)? Which skills are currently missing (weaknesses)? Which organisation may help implement the project (opportunity)? How organised is the FS private sector (strength/weakness)? Which stakeholder may support/hinder the process (opportunity/threat)?

Figure 14.8 shows what a SWOT matrix could look like for a typical secondary city in a low-income region. This is a simple example for the sake of clarity. However, a SWOT analysis is usually much more exhaustive, and hence it has added value to further categorise the factors, e.g. through the elements of the enabling environment (Section 17.2.1). Further potential challenges linked to stakeholders can be found in Section 15.4 and Chapter 12, Institutional Frameworks. A brief review of challenges and strengths in South-East Asian countries is provided in AECOM and SANDEC/EAWAG (2010).

Figure 14.8 Example of a SWOT matrix.

14.3 DATA TO BE COLLECTED

The following paragraphs detail the data to be collected during the assessment of the initial situation. There is not set way in which to collect the data. Data should be collected from different angles, with a combination of the methods described above, in order to cross-check the various sources.

14.3.1 General context

Understanding the general context helps to obtain a global picture of the situation and to understand the core constraints and potentials of a city. The most important data to be collected is:

- population and demography: number of inhabitants, number of people per household, population density and growth rate, type of housing;
- water and hygiene: drinking water coverage and infrastructure, drinking water sources, types of supply (e.g. networks, taps in houses, fountains, trucks), operators (public/private), prevalence of diseases related to faecal matter;
- physical characteristics: geomorphology, hydrologic basins, areas prone to flooding, type(s) of soil, groundwater table;
- climatic data (Section 14.3.7);
- storm water management;
- main elements of the city structure (as described in Section 14.3.8); and
- local economy: main economic activities in the city, main sources of household revenue, average income.

14.3.2 Sanitation sector

The sanitation sector is the backbone on which FSM has to be built. Understanding the sanitation sector includes obtaining information on:

- sanitation stakeholders and their role (which is presented in Chapter 15);
- collection/treatment/disposal facilities: sewered/unsewered areas, emptying modes (manual/mechanical), organisation (public/private), disposal sites, tariffs, solid waste management, enduses and resource recovery initiatives. (Table 14.2 summarises the relevant information on existing services);

Table 14.2 Relevant Information about existing services (Parkinson *et al.*, 2008)

Latrines and onsite treatment	
Water availability	Information on existing water supply services (including daily consumption per household) can be used to estimate daily wastewater production
Sanitation facilities	Current levels of service (household and shared facilities) including approximate household coverage and number and location of communal or public toilets
Onsite treatment	Types of onsite sanitation system serving households with households connections
Waste collection and conveyance	
Existing sewerage infrastructure	Coverage of sewerage and proportion of household with household connections
Faecal sludge and septage collection services	Coverage and frequency of servicing
Offsite wastewater treatment and reuse	
Wastewater treatment	Location and types of wastewater treatment infrastructure (if any exists)
Discharge or enduse	Location where wastewater and faecal sludge is disposed or endused

- analysis of the institutional framework (Chapter 12). Chowdhry and Koné, (2012) and AECOM and SANDEC/EAWAG, (2010) provide an overview of institutional frameworks in several countries in Asia and Africa;
- analysis of the legal and regulatory framework (Section 14.3.5); and
- financial analysis of the existing situation, i.e. the financial flows between the existing FSM stakeholders (Chapter 13).

14.3.3 Profile of manual and mechanical service providers

Private manual and mechanical service providers are the core of the initial FSM business. They address the needs of the population and have invaluable information about the types of latrines, the characteristics and quantities of sludge as well as the respective seasonal variability and the problems they encounter for collection, transporting and delivering. The interviews should also help in identifying existing and possible discharge sites and transport obstacles in the city, e.g. traffic jam zones (section 14.3.8). In addition it is recommended to team up with service providers for sampling during the FS characterisation and quantification studies, be it latrines or at the outlet of pumping trucks.

The following information should also be collected:

- socio-professional profile;
- frequency of the activity (especially for manual service providers): main/regular/occasional, and why;
- tariffs and tariff breakdown;
- problems encountered, e.g. accessibility, social pressure, health;
- staff and equipment; and
- methods employed.

The profile of FS service providers has been established, for example, in Dakar, Senegal (Mbéguéré *et al.*, 2009), Freetown, Sierra Leone (Mikhael, 2011), and Mahajanga, Madagascar (Larvido and Dodane, 2011). A more global financial analysis of emptying and transportation services in Africa and Asia is to be found in Chowdhry and Koné (2012).

14.3.4 Practices at household level

The household level is where FS is produced and stored and, as such, is a major factor in determining the quality and quantity of the sludge to be treated. Household surveys and interviews with the authorities and manual and mechanical operators should provide the necessary relevant information about the current status (Section 14.2.3 and Boxes 14.1 and 14.2). This includes:

- latrine types;
- emptying mode and means: manual/mechanical, equipment, staff, practices;
- emptying frequency according to each emptying mode;
- seasonal variability;
- number of operators per emptying mode;
- proportion of manual and mechanical emptying
- tariffs; and
- perceptions: which drivers influence the choice of an emptying mode, assessment of tariffs, capacity to pay, willingness to pay for improved services, proposed tariffs.

It should be noted that households usually do not have a clear vision of the volumes of sludge removed.

It may also be worthwhile to ask similar questions to people responsible for non-domestic buildings, such as hotels, restaurants, schools, public toilets and religious buildings. This usually represents a regular and lucrative segment for private service providers.

14.3.5 Legal and regulatory framework

At the national level, there are in most cases general laws and regulations about water, the environment and health. However, it is still rare to find texts specific to FSM, be it at the national or local level (Chapter 12).

Figure 14.9 Faecal sludge manual emptying and burying within the courtyard (photo: Linda Strande).

The information to be collected includes:

- laws and regulations;
- legal structures in charge of application; and
- enforcement.

As laws and regulations are often stringent but not enforced in practice (either because of lack of will, or because they are inappropriate), it is important to take a reality check (Lüthi *et al.*, 2011). One should understand what is tolerated and what is not. The relevant stakeholders should be consulted to determine how reality compares to the written procedures. Building inspectors, plumbers, contractors, municipal engineers and planners, and officials from the relevant ministries (e.g. environment, housing, construction, health, etc.) will all have invaluable information about what they would accept and approve in practice. It is advisable to show the preliminary assessment to the relevant decision makers so that it can be corrected and amended.

14.3.6 Estimation of design parameters

To design a FSTP properly, sludge quantities and characteristics should be estimated on a case-by-case basis, as they vary significantly between cities. Methods to quantify and characterise sludge are explained in Chapter 2 (Section 2.2 and 2.3) and the criteria for the selection of appropriate options are given in Section 17.4 and the related technology selection scheme (Figure 17.10).

All the data related to the formulas, parameters and criteria should be collected. This is a complex process, where several data collection methods are used. The main three are surveys of households and service providers, collection of climatic data, and sampling campaigns.

14.3.7 Climatic data

Climate is a key factor for the selection of treatment options (Section 17.4), especially the amount of rain and its distribution over time. It affects a FSTP in two ways:

- directly, as it affects sludge dewatering; and
- indirectly, as it affects the filling rate of latrines, the emptying frequency and, thus, the quantity and characteristics of the sludge to be treated in the FSTP. Such patterns can be estimated through semi-structured interviews with the manual and mechanical operators, and with households.

The main climatic data to be collected is:

- temperature over time;
- quantity of precipitation, maximum, minimum and distribution over time; frequency of rain episodes; seasonality (e.g. dry and rainy season); and
- evaporation rates, which allow a hydric balance between precipitation and infiltration, and gives an idea of the length of time necessary to dry the sludge. Runoff water can also be added to the balance.

Ideally, daily climatic data should be obtained for a period of 10-20 years for the study area, in order to best understand the variations. If evaporation is not measured, it can be deduced approximately from the temperature, wind and humidity data. Note that there has been very little investigation into the water retention capacity of sludge so far, implying difficulties in forecasting dewaterability.

14.3.8 Spatial data and city structure

It is crucial to understand how a city is organised and around which features. Factors like population density, socio-economic stratification, types of housing, topography, accessibility, traffic, the presence of existing sewer lines and the quality of service provision often influence the sludge emptying patterns (Case Study 14.2). These also have an influence on the selection of locations for treatment sites and transfer stations (Section 14.4). Access, environmental conditions of the potential sites (e.g. prone to flooding, type of soil, groundwater table) and land tenure are all major spatial decision factors.

Understanding the city may also have implications in terms of stakeholder analysis and engagement. Different sub-units may have different private entrepreneurs, be they mechanical or manual service providers, with their specific practices and disposal sites. Different administrative units will have different leaders, whether they are governmental or traditional.

Existing city maps and land registers can provide a considerable amount of relevant information, but they rapidly become out of date as cities expand and new settlements spring up. A common problem is that unplanned informal settlements are literally not on the map. It may therefore be necessary to prepare some simple but accurate up-to-date maps to ensure that these areas are not neglected in the planning of service improvements (WSP, 2008).

Some key questions about the spatial analysis are (modified from WSP, 2008):

- What sanitation infrastructure and services are in place, and how effective are they?
- Where are the sanitation problems most acute?
- Where is there a need for new infrastructure or services, and where is there a need for upgrading?
- Which areas should be prioritised for improvement?
- Where are the potential sites for FSTPs?
- Which areas are inaccessible for mechanical emptying?
- Where are the potential interfaces between these inaccessible areas and the city-level services?

The main outcome should be a clear understanding of the problems to be addressed, both in terms of location and type, at the household, neighbourhood and city levels.

14.3.9 Enduse practices and market studies

Chapter 10 details the potential endproducts and possibilities for resource recovery. The opportunities in a specific context should be investigated from the start, as this may influence the selection of technical options, as shown in the Technology Selection Scheme in Chapter 17.

The best method of finding out how the sludge is currently treated, disposed of or used is through interviews with the different FSM stakeholders. Observation of the disposal, treatment or enduse practices and visits to the corresponding sites are essential (Klingel *et al.*, 2002). The interest of potential endusers should be assessed; and if there is interest, the feasibility should be checked. This encompasses basic market studies, assessment of the willingness to pay, and scenarios for the supply chain. This is because bringing a product to a customer usually has a cost, which potential endusers may not be able to cover.

Figure 14.10 A farmer collecting dried faecal sludge at an informal disposal site in Togo (photo: Philippe Reymond).

Case Study 14.2: Urban complexity

(Adapted from Parkinson *et al.*, 2011)

It is important to think of the city as a patchwork of different domains and physical environments, each of which presents their own challenges and opportunities. A city can be divided into four typical urban settings: (i) informal settlements; (ii) peri-urban interface; (iii) planned urban development areas; (iv) inner-city middle- and high-income settlements (Figure 14.11). Each setting features different physical, spatial, demographic and socio-economic factors, leading to different dynamics and highlighting the fact that a range of sanitation technologies and management schemes are required to solve large-scale urban sanitation deficiencies. Integrating all these different areas and creating interfaces where needed is key in FSM planning.

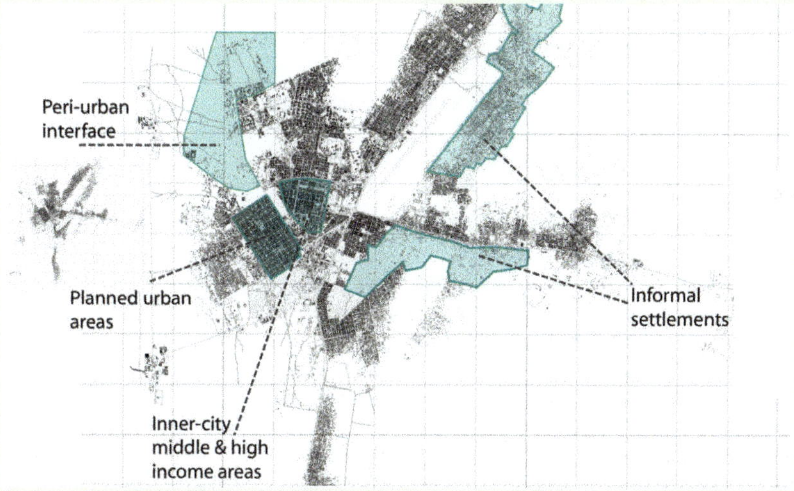

Figure 14.11 Identification of different urban settings (figure: Christoph Lüthi).

Figure 14.12 Two different urban settings (left: Kibera, Nairobi, Kenya; right: Nouakchott, Mauritania).

Planning

Box 14.3 provides an interview guide for farmers, which aims to assess the practices and needs of sludge use, their willingness to pay as well as the characteristics of similar products on the market.

Box 14.3: Interview guide for farmers

1 Localisation of the farms

2 Types of crops

3 Fertilisers
- Use of manure (cow, sheep, goat, chicken), sludge, compost, chemicals?
- Efficiency and price of the different fertilisers

4 Disposal / enduse
- Direct discharge on fields
- Transport to disposal sites
- Post-treatment before enduse (e.g. storage, composting)?

5 Health risk perception related to the use of FS

6 Periods when sludge is needed

Jan.	Feb.	March	April	May	June	July	Aug.	Sept.	Oct.	Nov.	Dec.
❐	❐	❐	❐	❐	❐	❐	❐	❐	❐	❐	❐

Cross the months where there is a need (X)

7 Periods where the other types of fertilisers are lacking

Jan.	Feb.	March	April	May	June	July	Aug.	Sept.	Oct.	Nov.	Dec.
❐	❐	❐	❐	❐	❐	❐	❐	❐	❐	❐	❐

Cross the months where there is a need (X)

8 Conditions for acquisition of treated FS
- price

14.4 CHARACTERISATION, EVALUATION AND SELECTION OF TREATMENT SITES

A well-thought out selection of treatment sites is crucial. There are examples, such as Bamako, where a FSTP was constructed but never used because the site was inappropriate. The selection of sites should be carried out according to city size, city configuration, number of available sites and the spatial distribution of the emptying companies, and several treatment sites should be considered. The optimum plant size has to be determined on a case-by-case basis as it depends on the local context (e.g., labour cost, land price, treatment plant scale, haulage distance, and site conditions, AECOM and SANDEC/EAWAG, 2010).

Existing disposal sites and potential treatment sites should be identified at the beginning of the planning process (Table 17.1), within the framework of the interviews and field visits during the detailed assessment of the initial situation. The evaluation of the identified sites should be carried out before the selection of technical options, as their characteristics may influence the selection.

Figure 14.13 Faecal sludge trucks stuck in traffic, Dakar, Senegal (photo: Linda Strande).

It is fundamental to involve the private collection and transport service providers in the selection process, as they are most affected. Their practices, constraints and needs should be understood, especially:

- their routes and disposal sites;
- the problems they face on the streets (traffic, police fees);
- the average distance and duration of the trips; and
- the money they gain per trip, in total and after deducting the fuel cost and maintenance costs of the trucks.

They will be able to say whether it is *practically* and *financially* possible for them to drive and deliver to the sites listed (see also Section 14.3.3 and Chapter 4).

14.4.1 Identification of treatment sites

The identification of existing sites, former sites and potential sites is carried out through discussions with the key stakeholders. The following stakeholders may be particularly helpful:

Manual and mechanical FS operators: They have knowledge of the discharge sites. It is also important to ask about former sites, or direct delivering to farming areas. It should be borne in mind that they may be reluctant to talk about sites that are illegal.

Endusers: Firstly, areas where sludge is used should be identified. Then, for example, if it is agriculture, farmers can provide information on where they find the sludge. They can also give information about former dumping sites, or temporary sites. It is interesting to cross-check this with the information from emptying service providers. This also gives indications on how the enduse market could be structured.

Table 14.3 Criteria for site evaluation with *sine qua non* (essential) conditions

Criteria	*Sine qua non* conditions
1. Average transport distance for mechanical service providers	Acceptability and affordability for service providers, as defined during interviews
2. Accessibility	Ease of access
3. Surface area	Surface area > 0.3 ha
4. Land ownership and price	Guarantee to be able to buy, at a reasonable price
5. Neighbourhood/potential for urbanisation	Risk of future access due to urbanisation
6. Topography	No risk of flooding
7. Soil type	Free soil (unconsolidated)
8. Groundwater table	> 2 m. deep
9. Opportunities for disposal of treated effluent and sludge	Must have disposal and enduse possibilities

Municipal authorities: The state may own available land. This would be a good option, as the municipality would then immediately be more involved.

Traditional authorities: Very often, land is still in the hands of traditional cultural leaders. They may be willing to provide land for public interest.

Politicians, landowners, town planners, residents, operators and users are all likely to have differing priorities and requirements as to where the infrastructure is located. Decisions may be heavily biased. Political pressures or available space may override what is considered appropriate for the user and host community (Scott, 2013). Siting infrastructure in the wrong location is likely to impact adversely on the long-term sustainability of the service.

It is common for cities not to have an updated land registry. Particular emphasis should be placed on finding out who owns the identified sites. GPS and Google Earth (see also Section 14.2.5) can be valuable tools to reference and assess the areas surrounding the sites and they can replace missing or outdated maps.

14.4.2 Characterisation and evaluation criteria

Nine criteria are proposed in Table 14.3 to characterise and evaluate potential sites, providing a good basis for decision-making. Some circumstances can lead to the immediate exclusion of a site. These are expressed as *sine qua non* conditions; if any of them is not valid, the site is considered as not appropriate.

Additionally the following information should be collected for each existing site:

- when the site is used (seasonality?);
- frequency of use; and
- city neighbourhoods that are served by this site.

Distance from emptying to delivering and accessibility of the site are major issues. A site that is too far away or has poor accessibility may also result in FS operators reverting to the former unsafe disposal sites. Collection service providers and vacuum truck drivers are always aware of haulage time and cost. The haulage of relatively small FS volumes (5-10 m^3 per truck) on congested roads over long distances in large urban agglomerations is financially unfeasible. A site that is too far away implies fewer trips per day, less revenue and more fuel costs for the FS operators. Very often they will add these costs to the emptying fees, in order to ultimately reach the same revenue. A price hike may then discourage households from using this service and cause them to turn to informal, unhygienic practices.

The surface area needed for a FSTP is determined by its technical design. However, if there is no single site large enough but instead there are several mid-sized sites, splitting the treatment units should be considered.

It is important that the site be bought by the institution in charge of the FSM system. Renting a site for treatment units is not a good option, as it will always be under threat from closure without notice. Land price is often another big issue. This should be kept in mind when preparing the budget for the FSTP.

The immediate environment of the site (ground and surroundings) is also of importance and the following should be taken into consideration:

Neighbourhood - nuisance: A FSTP can generate nuisance, especially bad odours. For this reason, it should be located at an appropriate distance from residential areas. It is also important to consider how the city will develop in the future.

Neighbourhood - synergy: If the site is surrounded by farming areas, the treated effluent may be directly used for irrigation, with the additional value of nutrient recycling. This would also greatly facilitate the enduse of sludge, if farmers are interested.

Topography: The FSTP should neither be threatened by flooding nor by erosion.

Soil type: This particularly affects the costs of excavation. In areas where mechanical means are scarce and most work is done manually, lateritic and other hard soils should be avoided. It is rare to find pedological maps; the soil characteristics should be assessed onsite, with the help of local residents.

Groundwater table: A high groundwater table may jeopardise the lifetime of the concrete and infrastructure. To assess the groundwater table, it is a advisable to look in any nearby wells, or to ask neighbours.

Disposal of the treated effluent also has to be planned. Even if treated, the effluent should not be disposed of directly into water. As mentioned above, it can be used for irrigation, or, if not possible, infiltrated in a leaching field. If wells are present within 100 meters downstream of the FSTP, signs should make it clear that the water is not of drinking water quality (i.e. non-potable).

14.4.3 Number of sites

The average haulage distance from the houses where FS is collected to the FSTP and the actual size of the plant are very decisive variables for the total cost of the disposal system as well as for its efficiency and sustainability (Strauss and Montangero, 2003). Given the difficulty of collecting FS and hauling it across cities to designated disposal and treatment sites, medium-scale FSTP in easily accessible locations may significantly reduce collection and haulage costs (Figure 14.14). Capital, operation and maintenance costs decrease with increasing plant size. However, since larger treatment plants require longer haulage distances between pits and disposal sites, costs escalate for collection a company, which in turn increases the risk of indiscriminate and illegal disposal (Chowdhry and Koné, 2012).

FS treatment can be optimised through levels of decentralisation, as most FSTPs are made up of relatively low-cost and modular treatment technologies. The selection of several sites could be a better match for the logistics of collection and for transport companies, and could lead to lower prices for emptying services.

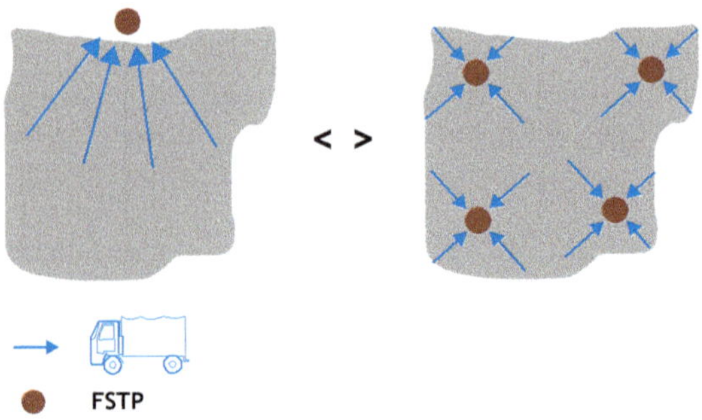

Figure 14.14 Centralised versus semi-centralised approach for site selection (Strauss and Montangero, 2003).

Implementation of several FSTPs implies a more detailed analysis of 'FS basins' (all the points in the city from where the FS is brought to the same area by FS operators; in analogy with hydrological basins) and the quantities that would arrive in each of the plants, in order to reduce the risk of over- or under-design. It would be problematic if, in the case of two FSTPs of the same size, most trucks went to one and not to the other because an important parameter has been missed during the assessment of the initial situation.

14.4.4 Sludge from manual emptying

The issues discussed above are mainly related to mechanical emptying. Manual emptying represents additional challenges, as manual service providers cannot transport the sludge far from the pits. Manual operators mainly work in areas where mechanical emptying is considered too expensive, or where pumping trucks cannot access (see also Chapter 4). It is very often not possible for them to dispose of the sludge safely in the neighbourhood where they work and it is thus important to link them with mechanical service providers.

A solution is to build transfer stations (or underground holding tanks, Tilley *et al.*, 2014) which are accessible to pumping trucks and which are close to where manual service providers work. Light transport gear should be provided so that they can easily bring the sludge to these collecting points. These holding tanks would then be managed as any onsite facility by mechanical operators. Emptying collecting points like these are best financed by the community or the municipal authorities (Figure 14.15).

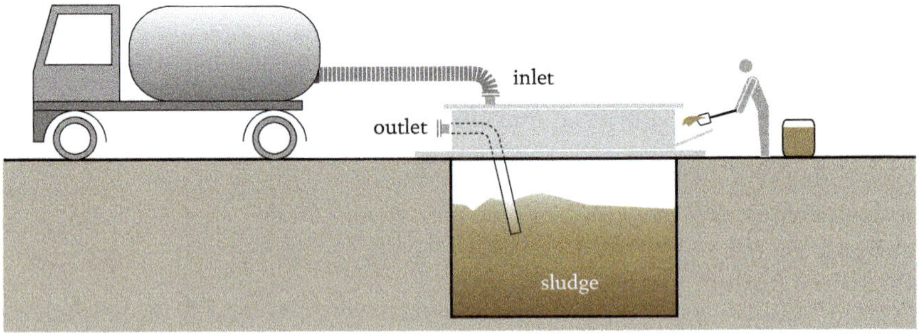

Figure 14.15 Schematic of a transfer station (Tilley *et al.*, 2014).

Planning

14.5 BIBLIOGRAPHY

AECOM and SANDEC/EAWAG (2010). A Rapid Assessment of Septage Management in Asia: Policies and Practices in India, Indonesia,Malaysia, the Philippines, Sri Lanka, Thailand, and Vietnam. USAID. Bangkok, Thailand.

Chowdhry, S., Koné. D. (2012). Business Analysis of Fecal Sludge Management: Emptying and Transportation Services in Africa and Asia. Bill & Melinda Gates Foundation.

Dodane, P. H. (2010). Assainissement de la Ville de Mahajanga: Situation Existante, Zonage et Priorités d'Action. (Rapport Final) France.

FAO (1990). The community's toolbox: The idea, methods and tools for participatory assessment, monitoring and evaluation in community forestry. Rome.

Groves, R.M., Fowler Jr, F.J., Couper, M.P., Lepkowski, J. M., Singer, E., Tourangeau, R. (2009). Survey methodology. Wiley.

Klingel F., Montangero, A. (2002). Faecal Sludge Management in Developing Countries - A Planning Manual / Gestion des boues de vidange dans les pays en voie de développement - Un manuel de planification. Dübendorf, EAWAG/SANDEC.

Koanda, H. (2007a). Fiches d'entretien. N'Kotchoyem. Sokodé, Togo, CREPA.

Koanda, H. (2007b). Projet AECM - Questionnaire d'enquêtes ménages pour l'élaboration de l'état des lieux sur les services urbains de base, CREPA.

Larvido, A., Dodane, P.H. (2011). Caractérisation du Secteur Informel de la Vidange des Latrines dans la Ville de Mahajanga, Madagascar. France.

Lüthi, C., Morel, A., Tilley, E., Ulrich, L. (2011a). Community-Led Urban Environmental Sanitation Planning: CLUES. Complete Guidelines for Decision-Makers with 30 Tools. Dübendorf, Switzerland, Swiss Federal Institute of Aquatic Science and Technology (EAWAG).

Lüthi, C., Panesar, A., Schütze, T., Norström, A., McConville, J., Parkinson, J., Ingle, R. (2011b). Sustainable Sanitation in Cities - a Framework for Action. Sustainable Sanitation Alliance (SuSanA) & International Forum on Urbanism (IFoU), Papiroz Publishing House, The Netherlands. Available from www.susana.org.

Mbéguéré, M., Gning. J.B. (2009). Socio-Economic Profile of Domestic Faecal Sludge Emptying Companies. SANDEC News. Dübendorf, Switzerland, EAWAG.

Mikhael, G. (2010). Demand Assessment for Sanitary Facilities and Services. Sanitation Market Assessment, Freetown, Sierra Leone. Report GOAL Sierra Leone. Vol. I.

Mikhael, G. (2011). Assessment of Faecal Sludge Emptying Services. Sanitation Market Assessment, Freetown, Sierra Leone. Report GOAL Sierra Leone Vol. II.

Parkinson, J., Lüthi, C. (2013). Sanitation21 – a planning framework for improving city-wide sanitation services. London, UK.

Parkinson, J., J. McConville, Norström, A., Lüthi, C. (2011). Framework for Dealing with Urban Complexity. In: Sustainable Sanitation in Cities - A Framework for Action. SuSanA and IFoU. The Netherlands, Papiroz Publishing House.

Parkinson, J., Tayler, K. (2008). A Guide to Decisionmaking: Technology Options for Urban Sanitation in India.

Reymond, P. (2008). Elaboration d'une méthodologie permettant de déterminer une option durable pour le traitement des boues de vidange dans une ville moyenne d'Afrique subsaharienne - Application à la ville de Sokodé, au Togo. MSc. Thesis, EPFL, EAWAG/SANDEC.

Reymond, P., Ulrich, L. (2011). Questionnaire for Household Survey. ESRISS Field Material. Cairo, EAWAG/ SANDEC.

Schall, N. (2004). Practicioner's guide: Strengths, Weaknesses, Opportunities & Threats (SWOT). Methodfinder.

Scott, P. (2013). Dealing with land tenure and tenancy challenges in water and sanitation services delivery. Topic Brief n°6. WSUP.

Strauss, M., Montangero, A. (2003). FS Management -Review of Practices, Problems and Initiatives. Capacity Building for Effective Decentralised Wastewater Management. Dübendorf, EAWAG/SANDEC.

Tayler-Powell, E. (1998). Questionnaire Design: Asking questions with a purpose.

Tilley, E., Lüthi, C., Morel, A., Zurbrügg, C., Schertenleib, R. (2014). Compendium of Sanitation Systems and Technologies. Swiss Federal Institute of Aquatic Science and Technology (EAWAG) and WSSCC. Dübendorf, Switzerland. 2nd revised edition. Available from www. sandec.ch. WSP (2008). Technology Options for Urban Sanitation in India. A guide to Decision-Making, WSP New Delhi.

WSUP (2011). GIS & mapping tools for water and sanitation infrastructure. Practice Note.

End of Chapter Study Questions

1. What types of data need to be collected during the initial assessment phase for FSM?
2. List the climatic data that should be collected in the initial assessment of FSM and explain the importance of this information.
3. When interviewing mechanical FS operators, what key information needs to be obtained from them?

Chapter 15

Stakeholder Analysis

Philippe Reymond

Learning Objectives

- Understand why stakeholder analysis is important for faecal sludge management project design.
- Be able to perform a stakeholder analysis for faecal sludge management projects and identify and characterise key stakeholders and relationships between them.
- Understand the main interests and constraints of stakeholders.
- Understand how key stakeholder selection evolves through the planning process and link the iterative approach with the planning framework in Chapter 17.
- Be able to determine those who need empowerment, motivation, incentives, capacity-building and/or information.

15.1 INTRODUCTION

Managing faecal sludge (FS) at the city level in an efficient and sustainable way requires the involvement and support of all the concerned stakeholders, i.e. the "key stakeholders". With "stakeholder" is meant "*any group, organisation or individual that can influence or be influenced by the project*". In short, 'people who matter'. Neglecting the needs, priorities and interests of people as well as their cultural and economic reality is one of the major causes of failure for water and sanitation programs in low- and middle-income countries.

In order to be able to understand and engage stakeholders, stakeholder analyses should be performed. Stakeholder analysis is the process of identifying and characterising stakeholders, investigating the relationships between them, and planning for their participation. It is a vital tool for understanding the social and institutional context of a project or policy. Its findings can provide early and essential information about who will be affected by the project and who could influence the project (positively or negatively); which individuals, groups, or agencies need to be involved in the project and how; and whose capacity needs to be built to enable them to participate (Rietbergen-McCracken and Narayan, 1998; Koanda, 2006). It is an iterative process, which is structured in this chapter in five main steps (Section 15.5). The challenge is to ensure a continuous monitoring of stakeholders in order to adjust the role of each of them in the process and to capture the dynamic nature of their needs, priorities and interests. Stakeholder analysis is thus a cross-cutting task that recurs throughout the whole FS management (FSM) planning process.

Koanda (2006) showed that stakeholder analysis is an appropriate tool for FSM planning. It provides a foundation and structure for the participatory planning, implementation, and monitoring of the project. As such, this chapter is closely linked to Chapter 16 *Stakeholder Engagement*, Chapter 12 *Institutional Frameworks*, and Chapter 17 *Planning Integrated FSM systems*. Section 15.5 of this chapter provides a guide on how to identify and prioritise key stakeholders in FSM activities while Chapter 16 provides information on how to engage them and allocate their roles and responsibilities appropriately and Chapter 12 on how to organise them in a sustainable management scheme. The stakeholder analysis approach proposed in this book is fully contextualised in FSM planning framework (Table 17.1), where its five steps are included as specific activities in the planning process. The main goal of each of these activities is to structure the information gathered and to determine the involvement strategy for the next phase.

Stakeholder analysis should be carried out by the process leaders (Box 17.1 in Chapter 17) and their facilitator(s) and stakeholders themselves can also be involved in the analysis. This should be started at the very beginning of the project. The main tools to carry out the analysis are informal and semi-structured interviews, focus groups, as well as field visits (Sections 14.2, 14.3 and 16.4).

Stakeholder analysis is mainly about understanding people and their feelings. Close relationships should be developed between the process leaders, the facilitators and the stakeholders. Trust-building is one key element of the stakeholder analysis, as it is for the whole planning process (Figure 15.1).

After a short overview of the stakeholder analysis process and the approach proposed in this book this chapter describes how to identify FSM stakeholders and then describes how to characterise them, giving practical insight into their typical interests, constraints and needs. Finally, the different steps of the stakeholder analysis through the planning process are explained and illustrated by a step-by-step case study.

Figure 15.1 Field visit with key sanitation stakeholders in Sokodé, Togo (photo: Philippe Reymond).

15.2 STAKEHOLDER ANALYSIS: WHY AND HOW

Stakeholder analysis has become increasingly popular with a wide range of organisations in many different fields, and it is now used by policy makers, regulators, governmental and non-governmental organisations (NGOs), businesses and the media (Friedman and Miles, 2006). Key questions that stakeholder analysis helps to answer are, for example: how can the relative interests and influence of different stakeholders be taken into account? And how can diverse stakeholders be adequately represented? As such, stakeholder analysis is seen as an approach that can empower marginal stakeholders to influence decision-making processes (Reed *et al.*, 2009). It is also used to work more effectively with stakeholders, facilitate transparent implementation of decisions or objectives, understand the policy context and assess the feasibility of future policy options (Brugha and Varvasovsky, 2000).

In FSM, the stakeholder analysis process is particularly important in order to:

- identify who to involve and at which level of participation, at the different stages of the planning and implementation process (see also Section 16.6);
- understand who has what interest and who is influential in supporting or in blocking/delaying/rejecting the project;
- identify conflicts of interests between stakeholders;
- identify relations between stakeholders that should be improved and strengthened;
- structure the knowledge about project stakeholders and share it with others;
- understand how to deal with the different people; for example, it should be clear who needs to be empowered, who needs to be informed and who should be dealt with in a particularly careful way (potential threats); and
- in partnership with governments and implementing agencies, assess how best to harness the positive aspects of the informal sector, minimise the negative aspects, and look for genuinely effective ways of creating effective links between the formal and the informal (Cacouris, 2012).

The stakeholder analysis method proposed in this book follows an approach of 'analytical categorisation' based on levels of interest and influence . Several 'attributes' or 'categorisation factors' help to identify who is important and/or influential and why. In order to structure the dynamic process in better defined activities, five formal steps are proposed, which follow the planning process illustrated in the FSM planning framework (Table 17.1):

STEP 1 Identification and preliminary characterisation of the stakeholders
STEP 2 Characterisation and selection of key stakeholders
STEP 3 Reassessment of key stakeholders according to validated options
STEP 4 Reassessment according to the Action Plan
STEP 5 Reassessment before the inauguration of the FSTP

Stakeholder analysis is a powerful tool to understand how people think and act. The information gathered should however be dealt with carefully, as it often involves sensitive information. Many interests are covert and agendas are partially hidden (ODA, 1995).

Findings from a stakeholder analysis are best recorded in tables and matrix diagrams (see Section 15.4 and Case Study 15.1), and the risks and assumptions arising from the analysis should be included in the logical framework of the project (ODA, 1995). These records will be revised throughout the whole process.

Figure 15.2 Focusing on stakeholders and market demand, Dakar, Senegal (photo: Linda Strande).

15.3 IDENTIFICATION OF STAKEHOLDERS

Stakeholder identification is one of the first tasks when starting a new project (Figure 15.2). Collaboration with local facilitators is essential to get the situation under control quickly. Identifying stakeholders is an iterative process, during which additional stakeholders are added as the analysis develops, for example, using expert opinion, focus groups, semi-structured interviews (see Section 14.2), snow-ball sampling (i.e. 'people who know other people, etc.'), or a combination of these (Reed *et al.*, 2009)[1]. It is all about contacting resource persons, who know the situation well and have access to the most important and influential stakeholders. Very often in low- and middle-income country contexts, a process leader must be introduced by a third local party in order to be able to get started and work efficiently from the beginning.

The more people that are met, the less likely it will be that any of the important stakeholder groups are missed. At each meeting, stakeholder identification can be done through a brainstorming process to collect an exhaustive list of people, groups or institutions (NETSSAF, 2008). Stakeholder mapping can be used as a tool to visualise the different stakeholders and their relationships.

In some countries, stakeholders to be involved in FSM have been defined in a national sanitation strategy. Such strategies may also mention who, out of the public and private sectors, is responsible for wastewater and excreta management, construction of latrines and sludge emptying.

15.3.1 Faecal sludge management stakeholders

In general, stakeholders who should be involved in a FSM planning process can be classified in eight categories, as detailed below. These stakeholders are further described in Table 15.3.

1 Reed *et al.* (2009) amalgamated in a table the different stakeholder analysis methods, including the resources required, the level of stakeholder participation, and the strengths and weaknesses of each method.

Municipal authorities

- mayor;
- municipal technical services (environment, sanitation, hygiene and public health); and
- municipal police.

Regional and national authorities

- different Regional Directorates (RD) e.g. Sanitation, Health, Hydraulics, Water Company, Public Works, Statistics, Urbanism & Habitat, Local Development, Agriculture.

Utilities

- public, semi-private (parastatal) or private (commercialised).

Traditional authorities and influential leaders

- ethnic leaders;
- neighbourhood leaders; and
- religious leaders.

Small-scale FS businesses

- mechanical service providers; FSM business owners, FSM business owner associations or interest groups; and
- manual service providers; pit emptier associations or interest groups.

There are numerous cases where someone owns and operates one truck (Chowdhry and Koné, 2012). There are others where FSM business owners and workers are not the same and it should therefore not be assumed that owners and operators have the same interests/influence (see also Section 14.3.3).

Organisations active in sanitation

- community-based organisations (CBOs);
- local or international NGOs with sanitation activities (including latrine construction and solid waste management);
- universities and research centres; and
- donor agencies.

Potential endusers

- farmers, farmer associations and institutions helping farmers;
- breeders, breeder associations and institutions helping breeders; and
- fuel consumers, such as companies needing combustible matter or biogas.

Households

- users; and
- owners (landlords in the case of tenant housing).

It is important to distinguish between users and owners here. For owner occupiers they are one and the same, however in the case of rental properties this distinction is critical. Tenants often pay for emptying services, not landlords (Figure 15.3) (Scott, 2011). In the case of public latrines for example, it makes sense to also consider user associations.

In all cases, the two following questions should be answered (ODA, 1995):

1. Have all the potential supporters and opponents of the project been identified?
2. Have vulnerable groups with an interest in the project been identified?

Figure 15.3 A few faecal sludge stakeholders: head of a household, househeld service provider (photo: Philippe Reymond).

Of course, stakeholders vary in each context. The institutions, modes of organisation, environment and culture vary from one region to the other, including the attitude towards human excreta. This list can be used as a guideline, but any case should be investigated and looked at on an individual basis.

The institutional setup and existing mode of organisation (see Chapter 12) are the skeleton on which the planner has to build and they have an important influence on the particular stakeholder configuration. In cities where FSM is not organised, the sector is mostly private and informal. In other cases, the State may delegate the management of the sector to utilities, be they public, semi-private or private.

15.3.2 Differences between large and medium-sized cities

Scale has an impact on the type and number of stakeholders and the way in which they are engaged. Large cities (i.e. main cities, characterised by heterogeneous neighbourhoods and a certain standard of housing and income, versus medium-sized or secondary cities with a more homogenous structure) generally present the following features:

More stakeholders: In large cities, there are more stakeholders in each category, especially mechanical and manual service providers, NGOs, farmers, traditional leaders, and politicians. While all service providers can be met individually in a medium-sized city, they may need to be organised into associations with representatives in larger cities (as already done in Dakar, Ouagadougou and Kampala).

Several cities in one: In large cities, different city parts or neighbourhoods may be compared to several medium-sized cities, each with their own private entrepreneurs, traditional leaders (and maybe also political leaders), and disposal sites. For the stakeholder analysis, it may be relevant to consider characteristic city parts separately, in addition to the city level itself.

More endusers, distributed differently: Different farming patterns and the presence of industries may offer enduse opportunities that are not available in secondary towns, like sludge enduse as a combustion fuel.

15.4 CHARACTERISATION OF STAKEHOLDERS

Stakeholder characterisation provides the necessary information on how to best involve each stakeholder and, at the end of the process, how to best attribute roles and responsibilities. It also paves the way for the key stakeholders' selection.

15.4.1 Information to be collected

Stakeholders should be characterised according to the following attributes (Koanda, 2006):

Main interests: Consultation with the stakeholders should be carried in order to determine how each interest can be taken into account in the future FSM system.

Strengths: Establish what the process leaders can count on.

Weaknesses: Establish where information, empowerment and capacity-building are needed.

Opportunities and threats: Characterise the potential positive (and negative) perspectives for the project.

Relationships between stakeholders: This includes for example, hierarchy, friendship, competition and professional links. Good, or bad, relationships may determine which working group can be built and where the best alliances to push the project forward lie. Trust and diplomacy are very important.

Impacts: The type of impact that the project has on a stakeholder also determines the measures needed to maximise positive impacts and mitigate negative impacts.

Involvement needs (including training needs): The action required results mainly from identified interests, weaknesses and potentials.

Chapter 14 (Assessment of the Initial Situation) highlights different tools and methods for data collection. Analysis of strengths, weaknesses, opportunities and threats are part of the wider SWOT analysis (Strengths, Weaknesses, Opportunities, and Threats). Chapter 16 gives guidance on how to translate stakeholders' characteristics into an involvement strategy.

The information collected can be amalgamated in a stakeholder table, as shown in Table 15.1.

Table 15.1 Example of a stakeholder table to summarise the stakeholders' characteristics

Stakeholders	Interests	Strengths	Weaknesses	Opportunities/ threats	Relationships	Impacts	Involvement needs
Stakeholder a							
Stakeholder b							
Stakeholder c							
...							

Planning

The relationships between stakeholders can be represented in a diagram of relationships. Such exercises in *stakeholder mapping* are particularly useful for stakeholders to visualise the situation in the case of participatory stakeholder analysis.

Relationships with and between stakeholders evolve through the process. In the beginning, mainly general groups or positions are considered (e.g. mechanical service providers and the municipal authorities). But as the project develops and close relationships emerge between the process leaders and stakeholders, there is a shift towards specific individuals (e.g. specific service providers, farmer leaders or influential politicians). Until service combinations are chosen and validated, relationships can mostly be described as informal between stakeholders. The process relies on discussions, interviews and meetings. However, once an Action Plan has been defined, many relationships become formal and contractual with specific individuals or companies.

15.4.2 Influence and interest

It is important to differentiate between two different types of opportunities and threats: the influence over the project and the interest in the project (adapted from ODA, 1995). These two key concepts can be defined as follows:

Influence is the power that stakeholders have on the project i.e. to control which decisions are made, facilitate their implementation, or affect the project negatively. Table 15.2 gives insight into which factors can confer influence.

Interest characterises stakeholders whose needs, constraints and problems are a priority in the strategy, e.g. sludge service providers, endusers, households, and sanitation authorities.

This distinction is particularly important for minorities and low-income groups, like manual service providers, low-income households and farmers, which are often not given a voice. It may require special efforts to enable these stakeholders to become active participants to ensure that their needs will be met (see 'empowerment' – Chapter 16). For the success of an initiative, it is important to know whether (and how) a stakeholder can take action and how he/she can be involved.

Table 15.2 Variables affecting stakeholders relative influence (adapted from ODA, 1995)

INFLUENCE FACTORS	
Within and between formal organisations	**For informal groups**
Hierarchy (command and control, budget holders)	Social, economic and political status
Leadership (formal and informal, charisma, political, familial)	Degree of organisation, consensus and leadership in the group
Control of strategic resources for the project	Degree of control of strategic resources significant for the project
Possession of specialist knowledge (e.g. engineering staff)	Informal influence through links with other stakeholders
Negotiating position (strength in relation to other stakeholders in the project) - personal connections to ruling politicians	Degree of dependence on other stakeholders

	Low influence	High influence
Low interest	Stakeholders are unlikely to be closely involved in the project and require not more than information-sharing aimed at the 'general public' *Information*	Stakeholders may oppose the intervention; therefore, they should be kept informed and their views acknowledged to avoid disruption or conflict *Consultation - Information*
High interest	Stakeholders require special effort to ensure that their needs are met and their participation is meaningful *Consultation - Empowerment*	Stakeholders should be closely involved to ensure their support for the project *Consultation - Collaboration Empowerment / Delegation*

Figure 15.4 Use of the influence-interest matrix to identify involvement needs and participation levels (adapted from Rietbergen *et al.*, 1998).

The next section provides criteria that help to categorise stakeholders according to their influence and interest. Once the stakeholders are characterised, the process leaders can amalgamate the outcomes into an influence-interest matrix (Figure 15.4). The influence-interest matrix serves as a decision-making tool for how to deal with the respective stakeholders and identify participation levels, as explained further in Section 16.3 (adapted from (ODA, 1995; Rietbergen *et al.*, 1998; IIED, 2005). Combined with the stakeholder selection table, it is a baseline document that helps to communicate the situation to external persons and which can be easily updated during further steps of the process.

When analysing influence and interest, it is necessary to understand to what extent a stakeholder is influential or interested, and, eventually, what impact(s) the stakeholder can have on the project, what impact the project can have on him/her ('their stake') and how he/she can be involved (see Chapter 16) with reference to the above-mentioned opportunities and threats linked with each stakeholder).

15.4.3 Selection criteria for key stakeholders

Key stakeholders in a FSM project are those whose interest and influence are most at stake. Six criteria or 'attributes' are proposed to select them. As soon as a stakeholder matches one of these criteria, he/she should be considered as a key stakeholder:

C1 Activity linked with FS management
C2 Political power
C3 Potential support or threat
C4 Ability to get funding
C5 Ownership of a potential treatment site
C6 Potential user of a treatment endproduct

These attributes refer either to interest, influence or both, and can be classified accordingly, as shown in Figure 15.5. In this way, populating the influence-interest matrix is simplified. For example, a stakeholder who has an activity in FSM (C1) will be considered as having an interest; a stakeholder that has an activity in FSM (C1) and ability to get funding (C4) has an interest and is influential. This process is illustrated in Case Study 15.2.

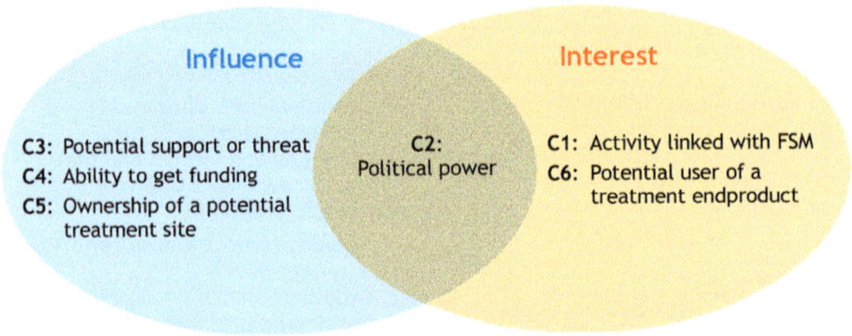

Figure 15.5 Classification of key stakeholder identification criteria according to influence and interest (figure: Philippe Reymond).

15.4.4 Amalgamation of FSM stakeholders' main characteristics and involvement needs

Table 15.3 illustrates the FSM stakeholders' typical interests and needs, opportunities for the project and actions to be undertaken in terms of involvement. Section 16.3 gives further information on how to develop a stakeholder strategy based on the stakeholder analysis, how to determine participation levels (Section 16.4) and how to determine the most appropriate involvement tools (Section 16.6).

15.4.5 Practical problems faced by faecal sludge management stakeholders

Stakeholders may experience practical problems during the planning/implementation process. They can be financial, educational, cultural or personal. The stakeholder analysis helps understand these problems, through the interviews and meetings it involves and it is part of the SWOT analysis (Section 14.2.7), as these factors can be considered as weaknesses in the current situation or threats for the process. Usually, these problems can be prevented or dealt with through information, capacity building/reinforcement, and, last but not least, diplomacy (see 'Involvement Tools', Section 16.4).

The following problems are commonly faced by stakeholders and some ways to deal with them are provided:

Lack of agency to participate:

Some important stakeholders lack influence and recognition: This can happen, for example, with manual service providers and farmers. Such groups need to be empowered, for example through an organisation such as a group or syndicate, which will enable their voice to have an appropriate influence in the planning and operation of FSM (see Chapter 16 for the selection of appropriate involvement tools). Informing the population of the importance of the service that these groups deliver and that these services may be improved with better FSM will also help to improve their status.

Some individuals cannot read, write or speak the official language: Illiteracy (either total or in the common local or official language) is disempowering. This may be particularly relevant to the lower-income groups and engagement and communication need to be adapted appropriately. Information must be adapted to the target audience. In all cases, in oral cultures, emphasis should be placed on illustrated methods of communication.

Lack of money: Many stakeholders may have very little income and project planning events may involve costs for them and be perceived as lost time for their usual business. Sometimes it is worth considering paying transport costs and food when gathering people together for meetings. Otherwise, attendance may be low, especially when considering the lowest-income groups.

Table 15.3 Typical characteristics of the main stakeholders and actions to be undertaken (adapted from Koanda, 2006)

Stakeholder categories	Main interests	Opportunities	Involvement needs and required actions
Municipal authorities	• Public health • Cleanliness of the city • Collection and management of sanitation fees	• Power for enforcement through regulatory framework and police • Management of treatment units • Link with other stakeholders, existing contracts and authorisations • Development of social services	• Sensitisation, need for capacity building, collaboration • Institutional and regulatory frameworks often need to be developed and their application enforced • Often lack financial, human resources and land • Involve them in the financing scheme
Regional and national authorities	• Respect for laws and regulations • Capacity building • Master plans	• Collaboration between agencies, development of synergies • Support for baseline data	• Sensitisation, information
Utilities	• Sufficient revenues • Municipal, regional or national priorities	• Collection, transport and treatment under the same umbrella • Cross-subsidy to allow social service	• Collaboration, sensitisation • Ensure that they act as 'public services' reaching low-income areas and not only upper-class neighbourhoods
Traditional authorities	• Public health	• Support and land property	• Consultation, information, sensitisation
Small-scale FS businesses • **Mechanical service providers**	• Sufficient revenues • Disposal sites close to working area • Clarification of legal status, better image	• Increase in quality of service • Lower emptying price • Collaboration with manual service providers	• Organise in association (empowerment) • Organise the market • Control the respect for rules • Contracts/licenses should be issued by municipal authorities
• **Manual service providers**	• Sufficient revenues • Gain status, social recognition • Reduce risk at the workplace	• Improvement of working conditions	• Organise in association (empowerment) • Empowerment ('give them a voice') and capacity building • Organise a service of collection and transport or transfer of sludge
Organisations active in sanitation	• Wellbeing of citizens • Clean environment • Capacity building • Visibility	• Experience in sanitation advocacy • Existing structures, human resources and competencies • Contact with households • Capacity to obtain funding	• Some organisation can be of great help (facilitation, experience, and international funding). • Their relationship with the authorities should be investigated
Potential endusers	• Affordable and safe products • Yield increase	• Increase WWTP's revenue through selling of endproducts	• Create enduser groups (empowerment) • Market study, and willingness and capacity to pay
Households (users and owners)	• Affordability of collection service • Clean environment	• Pressure on municipal authorities and service providers • Pay more for a better service • Better management of onsite systems	• Information, sensitisation for behaviour change, especially management of onsite systems • Assessment of willingness and capacity to pay • Advice for latrine construction

Constraints in the sludge emptying business:

Costs of sludge transport: This is a key issue for manual and mechanical service providers. The further they have to travel to discharge, the less trips they can carry out every day and, for the mechanical service providers, the higher the fuel expenses that they have. These costs are often transferred to the household, making the service unaffordable for many. This issue should be thoroughly discussed and understood before choosing treatment sites (see Section 14.4).

Lack of available land for FSM activities: This is typical where a local administration has been superimposed on the traditional land management systems and FSM activities are not mainstreamed into municipal service delivery. There are often multiple claims to land through the official and traditional land delivery systems. Resolving land conflicts can be a lengthy and often politically charged process which often gives informal or traditional land owners a crucial influence on the site selection for a FSTP.

Lack of resources/capacities:

Lack of management capacities: This is very often the case within the municipal entities. Close involvement in the FSM planning process and capacity building are beneficial, as well as exchanges with successful municipalities in the region.

Lack of human resources: Technical services are frequently understaffed, which is also a consequence of weak priority definition. Synergies with other institutions, organisations or private companies could be created during the participatory process and could relieve the technical services (e.g. delegation, and public-private partnership). The project could also employ and finance additional human resources within the municipality.

Laws incomplete and/or not enforced: Very often, the needed legal framework does not exist or is not enforced, as a result of lack of political will. Planners should help the municipal authorities to build their own legal framework, for example, through decrees issued by the mayor (see Chapter 12). If these measures are efficient, they may be taken up later at a regional or even national level.

Poor tax recovery: This can be a result of the previous point. Very often, the administration does not have the power to collect fees and, simultaneously, households are not willing to pay. This can be understood if the municipality is not providing the services that the fee is supposed to cover. In this case, transparency should be increased and the population informed.

Tensions between stakeholders:

Power games/competition: A lack of coordination and collaboration is often observed within institutions (administrative units, NGOs), between institutions and in the private sector (e.g. utilities vs. the informal sector). In some societies, information is considered to be power and there is reluctance to share it. The lack of information sharing is also a symptom of conflict of interest, overlap of institutional mandates and/or lack of an institutional home. The best way to deal with it is to pass on information and to show that working together and sharing information will be beneficial for everybody (see Chapter 16).

Lack of communication and coordination within and between agencies: This is often linked to the previous point – i.e. nobody knows what the others are doing or they take action independently, in the hope of work, prestige and/or funds. The solution to this issue is the same as pointed out above (passing on information and showing that working together and sharing information is beneficial for everybody).

Tensions between formal and informal sectors: Public water and sanitation utilities with monopolistic services are often unsupportive of small-scale entrepreneurs (Lüthi *et al.*, 2011). They may put pressure on the informal stakeholders (even when they themselves cannot provide a satisfactory service). Utilities do not obey the same logic as informal private entrepreneurs. If both categories are present, care must be taken to match their respective interests.

Awareness and behaviour:

Lack of awareness: Many stakeholders are unaware of the health and environmental burden due to lack of FSM. A lot of effort should be put into informing them and making sure that they understand the implication of each decision, in order not to have any unpleasant surprises during the implementation. Capacity building and reinforcement mainly aim to help stakeholders to take informed decisions (Chapter 15). Later on, during the implementation phase, they aim more to teach stakeholders how to deal with their respective roles and responsibilities.

15.5 IN PRACTICE: ITERATIVE SELECTION OF KEY STAKEHOLDERS

As the FSM planning process develops, knowledge of the initial situation deepens, data is gathered and more people are met, the way to proceed becomes clearer. Decisions are taken, which may have an impact on who is involved and how to move forward. Key stakeholders selected at the beginning may no longer be important, or, on the contrary, may gain importance or influence, and new stakeholders may appear. Consequently, it is fundamental to constantly observe the situation and to adapt to it. Stakeholder analysis is not only a task undertaken during the assessment of the initial situation (see Chapter 14), but it is an iterative activity throughout the whole planning process (see the *Planning Framework*, activities A, B, G, O, R and W).

For the purposes of clarity, five formal steps are proposed, which follow the planning process illustrated in the FSM planning framework (Table 17.1), and more specifically, the planning phases (see Section 17.4). These steps are considered to be primary activities of the planning process:

STEP 1: Identification and preliminary characterisation of the stakeholders (Activities A & B in the FSM planning framework)
STEP 2: Characterisation and selection of key stakeholders (Activity G in the FSM planning framework)
STEP 3: Reassessment of key stakeholders according to the validated options (Activity O in the FSM planning framework)
STEP 4: Reassessment according to the Action Plan (Activity R in the FSM planning framework)
STEP 5: Reassessment before the inauguration of the FSTP (Activity W in the FSM planning framework)

The stakeholders are continously reassessed as a function of their interest and influence with the help of the selection criteria. The main goal is to make informed decisions on how to best involve the different stakeholders in the process. The role of the process leader and his facilitator(s) is crucial. A close relationship to local stakeholders and soft skills are needed to 'feel the pulse of what is going on'.

15.5.1 STEP 1: Identification and preliminary characterisation of the stakeholders

At the beginning of the planning process, during the preparatory phase, the process leaders together with the local facilitator(s) carry out a preliminary assessment of the initial situation and a first inventory of stakeholders (Activity A). Then, at the beginning of the preliminary studies, they extend the first contact into a formal identification and preliminary characterisation of the latter (Activity B), before the official launching of the project. This makes it possible to get a first idea of who is there and who has to be involved, which will be used as a basis to send out invitations for the launching workshop. Great care should be taken not to miss out any influential person at this stage; otherwise the project could get off on the wrong foot.

A preliminary stakeholder table and a first diagram of relationships can be drawn, as illustrated in the Case Study 15.1. It is important to immediately begin to consider the relationships between the stakeholders. These relationships will become clearer throughout the process, and will be best understood through informal discussions.

The two outputs of Step 1 of the stakeholder analysis are:

- a draft of the stakeholder table; and
- a diagram of the relationships.

The main goal in this step is to find out who the stakeholders are and how best to involve them during the preliminary studies, for example, who to invite to the launching workshop and who to interview.

Case Study 15.1: Stakeholder analysis in a medium-sized West African city – Part I
(Adapted from Reymond, 2008)

STEP I – Identification and preliminary characterisation of the stakeholders and their relationships
(Activities A & B in the FSM planning framework, Table 17.1)

In this theoretical example, consultants have the task of designing a new FSM system for a medium-sized secondary city in West Africa. During the first few weeks in the field, they identify the FS stakeholders and make a preliminary characterisation. Three mechanical emptying service providers are working in the city, permanently or temporarily, two of them being private (mechanical service providers 1/2), and the third working as an NGO (NGO1). Sanitation is managed by the municipal authorities, which follow the rules of various Regional Directions (e.g. Public Health, Urban Planning, etc.). The latter have no political power but may threaten the project. In parallel, the city is ruled by traditional leaders, which own most of the land. Three other NGOs (NGO 2/3/4) are active in sanitation, especially in solid waste management. Farmers and cattle breeders who are potentially interested in the endproducts of the FSTP are present both in and outside the city. This relationship is shown in Figure 15.6.

NGO1 receives funds from abroad and owns a potential treatment site. Moreover, its leader is quite influential in the city. NGO2 is an international NGO, with important financial resources and influence on the municipality. NGO3 owns a potential site. NGO4 provides a small-scale solid waste collection service.

Households are the main users of the future system and have the biggest 'stake' of all the stakeholders. It is extremely important to understand their current practices, main constraints and needs.

The results of this first phase are summarised in a stakeholder table (see Section 15.4.1) and in a diagram of relationships (Figure 15.4). Soon after that, the launching workshop of the planning process is organised. Knowledge about the stakeholders will then be increased during the assessment of the initial situation.

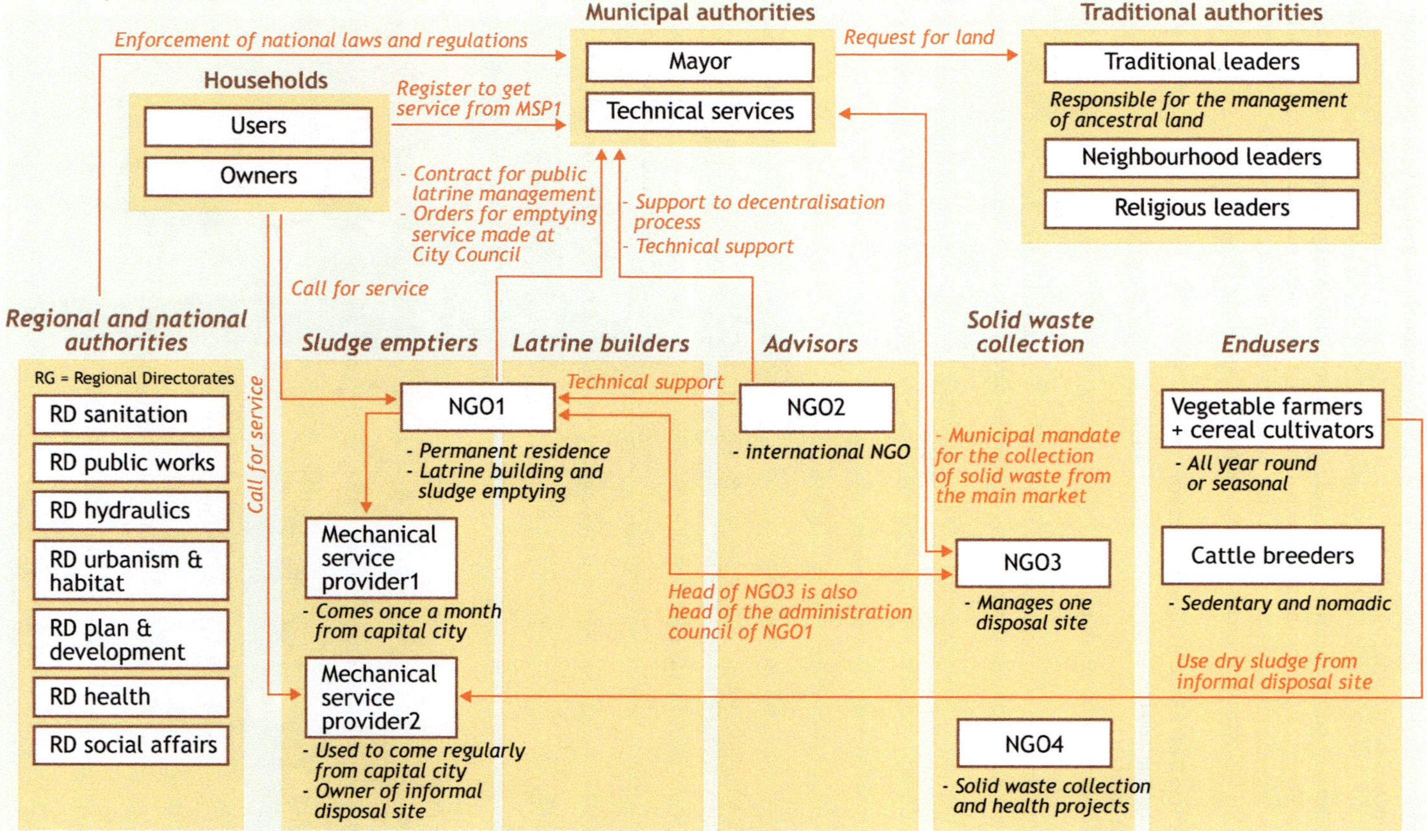

Figure 15.6 Example of a diagram of relationships between faecal sludge management stakeholders.

15.5.2 STEP 2: Characterisation and selection of the key stakeholders

At the end of the preliminary studies, i.e. the assessment of the initial situation, the characterisation of the stakeholders is refined (Activity G). At this stage, the sanitation practices and needs, potential organisational modes and potential sites for treatment are identified. The vague groups of people at the beginning have become people that the process leaders now know individually. A detailed stakeholder table can be developed. Based on the accumulated knowledge, the influence and interests of each stakeholder can be assessed and the key stakeholders identified, based on the criteria presented below. This results in the first influence-interest matrix, as shown in Case Study 15.2.

The outputs of Step 2 of the stakeholder analysis are:

- a detailed stakeholder table;
- an influence-interest matrix; and
- an updated diagram of relationships.

The main goal in this step is to find out how best to involve the stakeholders during the feasibility study. It is especially important to determine who to involve for the detailed evaluation of the options and, at the end of the phase, for the validation of the selected scenario. Forgetting a key stakeholder in the validation of options may have severe negative impacts later on in the planning or implementation process.

Figure 15.7 Faecal sludge truck driver in Togo (photo: Philippe Reymond).

Case study 15.2: Stakeholder analysis in a medium-sized West African city – part 2

(Adapted from Reymond, 2008)

STEP 2 – Characterisation and selection of the key stakeholders

(Activity G in the FSM planning framework, Table 17.1)

At the end of the preliminary studies, the process leader has a greater knowledge of the different stakeholders, which enables the preparation of a detailed stakeholder table and the selection of key stakeholders according to the proposed criteria. The process is illustrated in Table 15.4 with the information provided in Step 1 of the case study. Based on this information, the corresponding influence-interest matrix can be completed (Figure 15.8).

Table 15.4 Matching stakeholders with selection criteria in a stakeholder table

	Criteria					
Stakeholder	C1 Activity FSM	C2 Political power	C3 Support threath	C4 Funding	C5 Ownership site	C6 Enduse
Municipal authorities		▪	▪	▪		
Regional Directorates			▪			
Traditional authorities			▪		▪	
Mechanical service provider 1	▪					
Mechanical service provider 2	▪				▪	
NGO1	▪		▪	▪	▪	
NGO2			▪	▪		
NGO3			▪			▪
NGO4			▪		▪	▪
Farmers						▪
Cattle breeders						▪
Households	▪		▪			

	Low influence	*High influence*
Low interest		Regional Directorates Traditional authorities NGO2
High interest	Mechanical service provider 1 Farmers Cattle breeders	Municipal authorities Households NGO1 Mechanical service provider 2 NGO3 NGO4

Figure 15.8 First influence-interest matrix.

15.5.3 STEP 3: Reassessment of the key stakeholders according to the validated options

Once the selected options have been validated by all the stakeholders at the end of the feasibility study (Activity N), the process leaders should reassess the key stakeholders in order to select who will be closely involved in the preparation of the Action Plan/Detailed Project Development and define the roles, responsibilities and training needs (Activity O). At this stage, scenarios for the allocation of roles and responsibilities in the future FSM system are already roughly defined based on the detailed evaluation of selected options, and the selection of stakeholders is more at the individual than group level. In the stakeholder table, the 'Interests' and 'Opportunities/threats' columns can be replaced by a 'Roles and responsibilities' column, as shown in Table 15.5; the column 'Impacts' is no longer necessary.

Table 15.5 Stakeholder table adapted for the Action Plan development and implementation phases

Stakeholders	Roles and responsibilities	Strengths	Weaknesses	Relationships	Involvement needs
Stakeholder a					
Stakeholder b					
Stakeholder c					
...					

The roles and responsibilities of the key stakeholders can be categorised into four broad components, as illustrated in Case Study 15.3:

Construction, including the detailed design of the treatment plant.

Management, including 1) the detailed definition of the roles and responsibilities for implementation and O&M; 2) institutional arrangements and conventions between stakeholders; 3) securing financial and institutional mechanisms; 4) capacity building and required job creation.

Enduse, including endproduct marketing and sale channels.

Information: focus on the stakeholders that have to be regularly informed, be it for diplomatic reasons or awareness raising (e.g. households).

A stakeholder can be part of several of these components. These components, which may be further divided in sub-components, shape different groups; these groups, according to the involvement level needed, will become discussion groups (or 'focus groups') - related to the different aspects to be settled in the Action Plan (Section 17.4.3), target groups (in the case of information campaigns, for example) or groups to be invited to workshops (Chapter 16).

The outputs of Step 3 of the stakeholder analysis are:

- an adapted stakeholder table;
- an updated influence-interest matrix; and
- a list of stakeholders for each component.

This step has two main goals: firstly, to determine which stakeholder to involve in which aspect of the action planning and how; and secondly, to anticipate the involvement of stakeholders during the implementation phase, so that it is possible to give any necessary training early on enough in the process.

Case Study 15.3: Stakeholder analysis in a medium-sized West African city – part III
(Adapted from Reymond, 2008)

STEP 3 – Reassessment of key stakeholders according to the validated options
(Activity O in the FSM planning framework, Table 17.1)

The feasibility study showed that co-composting is not an option in this context, that cattle breeders are not interested in buying forage (a potential endproduct) and that a few sites identified at the beginning of the process are not appropriate for a FSTP. In other words, the two NGOs involved in solid waste management have lost influence and cattle breeders have lost interest in the project. As for the mechanical service provider 2, it has lost influence because of the inappropriateness of its site, but remains important, as it is still working with sludge. These changes are reflected in the grey blocks in Table 15.6, which results in an updated influence-interest matrix (Figure 15.8).

Table 15.6 Case study – reassessing stakeholders according to findings

	Criteria					
Stakeholder	C1 Activity FSM	C2 Political power	C3 Support threath	C4 Funding	C5 Ownership site	C6 Enduse
Municipal authorities		■	■	■		
Regional Directorates			■			
Traditional authorities			■		░	
Mechanical service provider 1	■					
Mechanical service provider 2	■				░	
NGO1	■		■	■	■	
NGO2			■	■		
NGO3			■			░
NGO4			■		░	░
Farmers						■
Cattle breeders						░
Households	■		■			

░ Options that are not appropriate in the given context

Planning

	Low influence	High influence
Low interest	Cattle breeders	Regional Directorates Traditional authorities NGO2 NGO3 NGO4
High interest	Mechanical service provider 1 Mechanical service provider 2 Farmers	Municipal authorities Households NGO1

Figure 15.9 The updated influence-interest matrix (step 3).

In preparation for action planning, the key stakeholders are categorised as shown in Figure 15.10. In this case, it is already clear that NGO1 will have an important role in the construction and management of the FSTP, in collaboration with the municipal authorities and the Regional Directorate for Public Works. As well as NGO1 and the municipal authorities, discussions concerning the management schemes will involve all the mechanical service providers (there are no manual service providers in this city). In terms of enduse, there is a strong interest on the farmers side, and discussions will involve representatives of the farmer associations, NGO1, the municipal authorities and the Regional Directorate for Health on the modalities for enduse of the sludge and treated effluent coming out of the new FSTP. Finally, the other influential stakeholders will be kept informed of the project developments, culminating in this phase with the official presentation and validation of the Action Plan (Activity Q in the FSM planning framework, Table 17.1).

Construction	Management	Valorisation	To be informed
Municipal authorities	Municipal authorities	Municipal authorities	Households
NGO 1	NGO 1	NGO 1	NGO 2
RD public works	Mechanical service provider 1	Farmers	NGO 3
	Mechanical service provider 2	RD health	NGO 4
	RD sanitation		Regional Directorates
			Traditional authorities

Figure 15.10 Categorisation of the key stakeholders into four groups for the detailed project development.

15.5.4 STEP 4: Reassessment according to the Action Plan

Once the Action Plan/Detailed Project Development has been validated (Activity Q), roles and responsibilities in the future FSM system are clearly defined and allocated. The reassessment of key stakeholders at this stage (Activity R) will help to identify the strengths, weaknesses and capacity-building needs before implementation. New key stakeholders may emerge, like contractors and future FSTP operators.

Section 17.3 describes the roles and responsibilities linked to the Action Plan and the implementation phase, while Sections 16.5 and 16.6 give information about the formalisation of the roles & responsibilities and training and capacity-building needs respectively. Chapter 12 focuses on the institutional frameworks, and gives further details about stakeholder involvement at this step.

In brief, important aspects for the stakeholders include:
Construction: recruitment of contractors for construction and O&M, monitoring of the construction and start-up of the system;
Management: organisation of the sector, transfer of roles & responsibilities and capacity-building;
Information: especially an information campaign on the future FSM system and its implications; and
Training and capacity-building.

The outputs of Step 4 of the stakeholder analysis are:

- an updated stakeholder table (Table 15.4); and
- an updated influence-interest matrix.

The main goal of this step is thus to finalise roles and responsibility allocation for the implementation phase and define the involvement needs, especially for information and training.

15.5.5 STEP 5: Reassessment before the inauguration of the faecal sludge management plant

This reassessment (Activity W) mainly aims at building on lessons learnt during the implementation stage, identifying any remaining needs in capacity-building and filling any gaps. It also ensures that the O&M plan is properly in place and to confirm roles and responsibilities for the monitoring of the system.
The output of Step 5 of the stakeholder analysis is an updated stakeholder table.

15.6 BIBLIOGRAPHY

Brugha, R., Varvasovsky, Z. (2000). Stakeholder Analysis: a review. Health Policy and Planning 15, p. 239-246.

Cacouris, J. (2012). Recognising and dealing with informal influences in water and sanitation services delivery. Topic Brief. G. Norman, WSUP.

Chowdhry, S., Koné, D. (2012). Business Analysis of Fecal Sludge Management: Emptying and Transportation Services in Africa and Asia. Bill and Melinda Gates Foundation.

Friedman, A., Miles, S. (2006). Stakeholders: Theory and Practice. Oxford, Oxford University Press.

IIED (2005). Stakeholder Power Analysis - Power Tools, p. 1-23.

Koanda, H. (2006). Vers un assainissement urbain durable en Afrique subsaharienne: approche innovante de planification de la gestion des boues de vidange. Lausanne, EPFL. Thesis n° 3530.

Luethi, C., Morel, A., Tilley, E., Ulrich, L. (2011). Community-Led Urban environmental sanitation planning: CLUES – Complete guidelines for decision-makers with 30 tools. Swiss Federal Institute of Aquatic Science and Technology (EAWAG), Dübendorf, Switzerland, WSSCC, Geneva, Switzerland and UN-HABITAT, Nairobi, Kenya.

NETSSAF (2008). NETSSAF Participatory Planning Approach, a guideline for sustainable sanitation planning.

ODA (1995). Guidance Note on How To Do Stakeholder Analysis of Aid Projects and Programmes. Social Development Department. London, Overseas Development Administration (now DfID).

Reed, M.S., Graves, A., Dandy, N., Posthumus, H., Hubacek, K., Morris, J., Prell, C., Quinn, C.H., Stringer, L.C. (2009). Who's in and why? A typology of stakeholder analysis methods for natural resource management. Journal of Environmental Management 90(5), p.1933-1949.

Rietbergen-MacCracken, J., Narayan-Parker, D. (1998). Participation and social assessment: tools and techniques. World Bank Publications.

Scott, P. (2011). Unbundling tenure issues for urban sanitation development, Loughborough University. PhD.

End of Chapter Study Questions

1. Explain why the stakeholder analysis process is vital in FSM.
2. Stakeholders who should be involved in FSM planning process can be classified in a number of categories. What are five of these categories?
3. What are the challenges faced by manual and mechanical service providers in the FS emptying business?

Chapter 16

Stakeholder Engagement

Philippe Reymond and Magalie Bassan

Learning Objectives

- Understand why it is important to engage stakeholders from the beginning of a project, and how this can ease implementation and enhance long-term sustainability.
- Understand how to use the information gathered during the stakeholder analysis to plan stakeholders' involvement.
- Understand the tools to inform, consult and collaborate with stakeholders, how to use these tools, and when to implement them.
- Understand how to distribute and formalise roles and responsibilities, and to identify training needs.

16.1 INTRODUCTION

Stakeholder 'engagement' or stakeholder 'involvement' is key for the successful implementation of faecal sludge management (FSM) projects. It is the art of including stakeholders in the planning process in order to take into account their needs, priorities and interests, to achieve consensus and to remove opposition; in other words, to make them 'participate'. Stakeholder engagement is largely about defining the participation level of people in the process and how to best answer their needs, through for example awareness-raising or training and capacity-building. In order to understand stakeholders, it is important to identify and characterise them (Chapter 15). The stakeholder analysis is a dynamic task that cuts across the entire planning process as the influence and interests of individual stakeholders change, leading to evolving stakeholder engagement strategies (Reed, 2008; Reymond, 2008). Involvement needs to be defined based on the context and characteristics of the key stakeholders. The dynamic nature of stakeholder involvement is put into context in Chapter 17, which contextualises stakeholder analysis and engagement activities within the whole planning framework (summarised in the Planning Framework Figure). At the end of the process, stakeholder engagement culminates with the distribution and formalisation of roles and responsibilities in the selected models of organisation and institutional framework (Chapter 12).

The planning approach proposed in this book can be described as a 'participatory approach', as will be described in Chapter 17. It engages stakeholders to 'participate' in the process, not to remain passive observers. The success of the participatory approach depends on the accountability and motivation of stakeholders, the recognition of the added value provided by this approach, the knowledge that the process leaders have of the local context, and the credibility and resources they have (Mosler, 2004; Koanda, 2006).

After discussing why it is important to engage stakeholders, this chapter provides an overview of the different participation levels that can be chosen and how they can be selected based on the stakeholder analysis. Possible involvement tools are then listed with insight into how to choose the most appropriate ones in a specific context. Along with the planning process, the chapter then describes engagement milestones, cross-cutting tasks, such as training and awareness raising, and how to distribute and formalise roles and responsibilities.

16.2 THE IMPORTANCE OF ENGAGING STAKEHOLDERS

Proper FSM brings benefits to everybody as it solves long-term urban problems. Authorities gain recognition by improving the population's welfare. Private collection and transport entrepreneurs gain formal disposal sites and better recognition, and the price of services may be reduced for the households. However, the benefits may not be clear to everybody from the beginning and some people may be reluctant to change some aspects of their daily routine or to make the necessary effort for the project. For these reasons, information and transparency are fundamental, and consultation, collaboration and empowerment are the keys to get all the stakeholders to work together and build a system that functions well.

Figure 16.1 Workshop with all the sanitation stakeholders in the town hall of Sokodé, Togo (photo: Philippe Reymond).

Planning

Planning for FSM is often characterised by stakeholders with conflicting interests and goals, for example minimising the distance to the discharge point for the private operators versus finding a treatment site outside the city for the authorities (see also Section 15.5.4). For successful implementation, involved parties need to learn about and understand FSM systems, which includes both infrastructure and people. Some FSM projects failed because the distance to selected disposal sites was too far for private entrepreneurs, because the latter were facing problems with the police, or because low-income areas were excluded (see also Case Study 17.2). Such failures would not have happened if the people involved had been consulted and their needs and constraints identified early in the process.

Participation increases the effectiveness of a project because by engaging a wide range of interested parties, the prospects for an appropriate project design and commitment to achieving objectives are more likely to be maximised. Participation empowers people (ODA, 1995). It also increases the sustainability of a project because it develops the skills, trust and confidence required for the people to run the system once it is in place. Awareness raising, communication and capacity development should also accompany the whole process and, as such, are considered *cross-cutting tasks* (Lüthi *et al.*, 2011).

In the end, people participate on a voluntary basis (some get compensation, but nobody is forced into a participatory process). People need to have an interest, and believe that their involvement is for their own benefit or serves a greater goal. It is important to note that poor stakeholder engagement can contribute to the development of opposition, but negative feedback from stakeholders occurring in the process should be regarded positively, as it demonstrates the social and political acceptability of the proposed actions.

Finally, if stakeholder involvement is important, it also has a cost. Adequate resources (budget, personnel and time) must be allocated to the project, while the leaders must be willing to share control (Mosler, 2004). Good participatory processes are difficult and take time, because often trust needs to be built between the participants. However, engaging stakeholders from the start often saves time later by uncovering and overcoming problems which would otherwise hinder implementation and/or operation of the system (ODA, 1995) and thus, has the potential to save money in later phases. Participatory approaches need to be regarded as an investment by the implementing agent.

16.3 PARTICIPATION LEVELS

Choosing how to engage a stakeholder means choosing the appropriate participation level. The level of participation depends on what needs to be achieved with the targeted stakeholders. For example, within a given context, households may be informed about the process, or consulted to understand their collection needs. Collection and transport operators may be consulted about their routes and to help define optimal disposal sites, or collaborate on regulation definition. Collaboration is most often sought from the beginning with municipal authorities, which which is a consequence from their interest in and influence on the project (see also Section 15.4.2).

Several aspects should be considered when developing the involvement strategy (Koanda, 2006):

- perception of involvement: indicates how involved stakeholders feel;
- willingness to contribute to the project;
- expected benefit from the project;
- level of obligation which the stakeholder feels towards their responsibilities in the project; and
- people influencing the willingness of the stakeholder and extent of the peer pressure.

These aspects can also be used as indicators to assess the efficiency of the participatory process.

16.3.1 From information to delegation

Four main participation levels can be distinguished (adapted from ODA, 1995) and are discussed here, in the order of increasing involvement:

Information: The objective is to enable the stakeholders to understand the situation, the different options and their implications. This is a *one-way flow* of communication. All the stakeholders concerned by FSM need to be well informed in order to understand their role and the objectives of the project. For some of the stakeholders, who are not involved in the decision-making process, the involvement is limited to receiving information, which can be done through awareness-raising campaigns or informative meetings (e.g. an initial launching workshop – see Section 15.4) and related field visits. In some cases, information also serves to persuade people to take part in the process, by showing the benefits and providing incentives.

Consultation: The objective is to obtain the stakeholders' feedback on the situation, options, scenarios and/or decisions. This is *two-way* communication. It allows interests, priorities, needs and concerns to be taken into account (e.g. through interviews carried out with the different stakeholders at the beginning of the planning process). However the stakeholders are not involved in decision-making.

Collaboration: The objective is to work as a partner with the stakeholder(s) on various aspects, including the development of scenarios and the identification of the preferred solution. The power for taking decisions is shared between the stakeholders.

Empowerment/delegation: The objective is to build the capacities of stakeholders so that they are able to make informed decisions, to take responsibility for final decision making, and to assume their role and responsibilities once the FSM system is implemented.

Each level includes the previous ones; for example, collaboration cannot be done without consultation, and there is no consultation without information.

Different forms of participation can be used at the same time with the same stakeholder, or during the different stages of the project cycle. For example, some stakeholders may be first informed about the project, and later consulted to get their point of view. Once a trust relationship has been established and the required capacity is there, collaboration can be started.

16.3.2 Determination of the participation levels based on the stakeholder analysis

Before being able to develop an involvement strategy, the stakeholders need to be analysed (see Chapter 15). The stakeholder analysis provides the basis for deciding which stakeholders should participate in the different stages of the process and at which level of participation; this also makes it possible in turn to define which involvement tool(s) should be used with each stakeholder. This step is important, as the success of a project may depend partly on the validity of the assumptions made about its various stakeholders, and the risks facing the project, such as conflicting interests (ODA, 1995).

The involvement strategy mainly stems from the interests, influence and involvement needs of stakeholders. The 'involvement needs' are part of the information to be collected during the stakeholder analysis as described in Chapter 15. The second step is to consider the stakeholders' influence on and interest in the project, which are the main drivers for the selection of faecal sludge (FS) key stakeholders, with selection criteria that are directly related to them. The degree of influence and interest determines the participation level; then, for each participation level, a certain number of involvement tools are available, to be used according to the context and the characteristics of the stakeholders.

The influence and interest of stakeholders vary during the planning process. For this reason, an iterative selection of key stakeholders is proposed in this book. It also means that the involvement strategy may be adapted along the process, with consequent modifications in the participation level and ways to involve the different stakeholders. The five steps proposed in Section 15.6 thus apply to the reassessment of the involvement strategy as well.

16.3.3 The stakeholder participation matrix

The stakeholder participation matrix, as proposed by ODA (1995b), provides a visual representation of the selected participation level for each stakeholder. Table 16.1 shows a theoretical example that can be seen as typical for a medium-sized city, without any national government representations. Such a matrix is developed step by step along with the process, according to the needs and the desired level of detail. At the end of the process, it is a good summary of who has been involved, how and when, and provides a good basis for comparison between different projects. The participation matrix is a dynamic tool and should be adapted regularly according to the results of the iterative approach of key stakeholder selection proposed in Chapter 15, Section 15.6. It should be seen as a way to sum up the available information and take decisions on involvement strategies, such as organisation of workshops and meetings. Case Study 16.2 features a stakeholder participation matrix built retrospectively to illustrate how stakeholders were involved in an existing project in Burkina Faso.

Table 16.1 Stakeholder participation matrix (theoretical example representing a medium-sized city)

		Participation levels			
		Information	**Consultation**	**Collaboration**	**Empowerment / delegation**
Planning	*Launch of the planning process*	All stakeholders		Municipality, utilities	
	Detailed assessment of current situation		Key stakeholders[1]	Municipality, utilities	
	Identification of service options		Key stakeholders[1]	Municipality, utilities	
	Development of an Action Plan	All stakeholders	Endusers	Municipality, utilities, FS operators, NGOs	Empower weak and non-organised groups
Implementation		Households, traditional authorities and opinion leaders	Endusers	Municipality, utilities, FS operators, NGOs	Empower and delegate to municipality, utilities, FS operators, NGOs
Monitoring & Evaluation		Key stakeholders	Households, FS operators, endusers	Municipality, utilities, selected NGOs	

1 The identification and selection of key-stakeholders is described in Sections 15.3 and 15.4.

16.4 INVOLVEMENT TOOLS

Once participation levels for each stakeholder have been defined, the involvement tools can be selected. For each participation level, there are a number of possible involvement tools, as shown in Table 16.2. There are many ways to involve people in a FSM process and there is no ready-made recipe for which tool to use and when. Decisions should be context-driven. The optimal selection of involvement tools varies from case to case, for example, involvement needs may differ according to the complexity and boundaries of the project (e.g. planning at the policy level in a country where no formal FSM organisation exists yet, or planning two faecal sludge treatment plants (FSTPs) in a city where FSM operators are well structured).The personality of the stakeholders is also very important.

16.4.1 List of involvement tools

A few tools that can fit well in a FSM process are presented below (adapted from Mosler, 2004). Each tool corresponds to one or several participation levels, as shown in Table 16.2.

Individual meetings, informal or semi-structured interviews (Section 14.2.2): Meetings with stakeholders are very important, as they allow information to be collected while at the same time build trust and personal relationships. They also provide an understanding of the needs, priorities and constraints. Meeting people individually may lead to more open discussion by avoiding peer pressure.

Focus groups: This tool consists of discussions in small groups, led by a moderator, whereby stakeholders express and discuss their opinions. Focus groups can contribute towards opinion-forming in the group and can be organised in order to elaborate documents.

Table 16.2 Stakeholder involvement techniques and participation levels

	Information	Consultation	Collaboration	Empowerment / delegation
Personal meetings	■	■	■	■
Focus groups		■	■	■
Workshops	■	■	■	■
Site visits	■	■		
Media campaigns	■			
Household surveys		■		
Advocacy / lobbying	■		■	■
Mediation		■	■	■
Logical framework		■	■	

Planning

Workshops: A workshop aims to gather selected stakeholders together in order to push the process forward. It can be an *information workshop*, such as the *initial launching workshop* (see Planning Framework in Chapter 17, Activity C), aiming to communicate the plans, activities and current stage of the process. It can also be a *consultative workshop*, aiming to collect stakeholders' opinions and concerns, build consensus and formulate solutions. In some cases, workshops or focus group meetings can be held to acknowledge and reinforce the importance of members in the process and strengthen associations. For example, the organisation of groups of stakeholders such as collection and transport operators into an association may greatly simplify the participatory process, increase the visibility of these stakeholders and lead to empowerment (Bassan *et al.*, 2011).

Site visits: (see Section 14.2.4) A site visit is a powerful tool to expose all stakeholders to reality. Very often, authorities and people working in the office do not fully realise the situation until they can actually see and visualise it. Once they understand, they are much more prone to action and change. Visits to informal (illegal) FS dumping sites may be particularly useful (Figure 16.2). Transect walks, where process leaders walk through the concerned neighbourhoods with relevant stakeholders, are also a recommended option.

Participatory mapping (often referred to as *community* or *social mapping*): Assisted by a facilitator, the stakeholders develop a map of the target area and represent the features related to FSM infrastructure and services. It helps to get an overview of the project, visualise the situation and get a common understanding (Figure 16.3).

Surveys: A representative sample of the population is questioned on a particular topic by means of a structured questionnaire. An example is household surveys used for the assessment of the current situation (e.g. Section 14.3.3).

Figure 16.2 Site visit to an informal disposal site, Sokodé, Togo (photo: Philippe Reymond).

Figure 16.3 Participatory mapping taking place in India (photo: Philippe Reymond).

Media campaigns: Carried out with posters, advertisements, on radios, TV, internet or cell phones, media campaigns aim to inform and sensitise the public. It is particularly useful in making the population understand the changes introduced by a new FSM system and promoting changes in habits (e.g. not to discharge solid waste into the latrines).

Advocacy/lobbying: The goal is to ensure that the interests of non-organised and/or socially disadvantaged and less articulate groups within the population are considered in the planning process. The groups receive advice, and their interests are represented in the appropriate committees and bodies, either through representatives or through the voice of the process leaders. It is a form of empowerment. Advocacy and lobbying are mainly about convincing and persuading stakeholders. They can be used, for example, to convince authorities or utilities of the benefits of an integrated approach.

Mediation: In conflict situations, mediation through neutral third parties is the attempt to reach mutually agreed-upon solutions. First, the key issues and areas of conflict are stated and clarified (interests, aversions, and blockages). Then there is an attempt to find mutually satisfactory ways to resolve the conflict (evaluate options, and check for fairness).

Logical frameworks: Logical frameworks can be elaborated for each of the strategic objectives of a project. This tool aims to facilitate the logical organisation of projects with well-defined objectives, and can be used to promote communication between the stakeholders and to focus their attention (Aune, 2000). This approach allows the identification of the means and activities needed to reach the defined project objectives, together with the risk indicators and outputs related to them.

16.4.2 Determining the most appropriate involvement tools

The selection of involvement tools should be done on a case-by-case basis as it depends on the goals, the personality and capacities of the local stakeholders. The best participatory approach is a combination of several techniques (Koanda, 2006), as illustrated in Case Study 16.1. Before selecting a tool, the process leaders should consider the practical aspects linked with the socio-economic conditions of the stakeholders and make sure that the tool is adapted to the target group. It is also important to clarify in advance the availability of the resources required for conducting an adequate stakeholder involvement program (time, budget, and know-how). Credibility of the process leaders, official legitimisation and transparency are indispensable for the process success (Mosler 2004). The process leaders need a good knowledge of the local context, in order to find appropriate tools and ensure a good facilitation and mediation. A minimum level of trust between stakeholders is also required (Koanda 2006).

The following questions should be considered by the process leaders (adapted from Mosler, 2004):

- Should the group always meet as a whole, or should there also be meetings of sub-groups?
- Is the frequency of meetings tolerable to all participants (or a time burden)?
- How binding should group decisions be for the individual members of the group?
- To what extent will hidden agendas and the previous experiences of stakeholders influence their voicing of opinions in the group?

At the same time, the process leaders have to make sure that the involvement tools fit the local context (adapted from Mosler, 2004):

- **Political framework:** Does the type of stakeholder involvement fit into the existing political system? Do the political leaders need to be involved or not?
- **Legal framework:** Does the type of stakeholder involvement conform to the laws?
- **Institutional framework:** Does the type of stakeholder involvement match the given institutional framework (i.e. role distribution among the stakeholders - see Chapter 12)? Are the right authorities involved? For example, the coordination of the stakeholders from the whole service chain needs to be well organised, and carried out by a competent authority.
- **Social framework:** Does the type of stakeholder involvement conform to social customs?

Finally, some personal aspects are critical for the success of the process (adapted from Mosler, 2004):

- If there are key stakeholders who are against the project or show distrust, extra care should be taken to inform, consult, and understand them and to discuss how they can benefit from the project.
- If important stakeholders have no interest in the project, their needs have to be identified in order to stimulate their interest.
- If the stakeholder analysis reveals strong conflicts between stakeholders' interests, or if interests of some stakeholders are not being represented, techniques for dealing with conflict should be implemented.
- Assistance can be provided for stakeholder groups or associations in order to better organise them, to ensure that leaders are internally accepted, and to improve their recognition level amongst other stakeholders (see Case Study 16.2).

The period corresponding to the launch of the planning process and the detailed assessment of the current situation is usually dominated by information and consultation with the key stakeholders, aiming at gathering information on needs and priorities (see Chapter 14 and Section 17.3). Full collaboration with some of the stakeholders may be difficult at this stage as the necessary know-how may not be available to make informed decisions. As a consequence of this, part of the preliminary and feasibility studies may be conducted by experts.

Context-specific involvement tools are used to make all the different stakeholder groups ready to make informed decisions and reach consensus where necessary. Information is a cross-cutting task, and workshops are the fruit of a continuous effort of awareness raising, capacity building, focus group discussions and empowerment, i.e. an appropriate mix of all participation levels. For example, collection and transport of FS is frequently not formally regulated, and is hence technically considered illegal, and so is not properly acknowledged by authorities. To ensure an efficient participatory process, meetings and visits can help authorities to understand the importance of the private operators and their constraints related to the transport and discharge at illegal sites. Another example is the empowerment of manual and mechanical operators. If they do not have a professional association or their association is not active, or the leader is not considered representative, assistance should be provided to strengthen the organisation, otherwise its members should be involved in the important participatory events until an efficient management has been ensured (Lüthi *et al.*, 2011).

Implementation marks the transfer (delegation) of roles and responsibilities to the respective stakeholders (e.g. the FSTP operators), with further empowerment where needed. All the stakeholders who do not have a defined role should be kept informed (e.g. the households and the different authorities), so that the system can start running with the support of the population and other key stakeholders.

Case Study 16.1 : Selection of participation levels and involvement tools for faecal sludge stakeholders in Ouahigouya, Burkina Faso

(Adapted from Koanda, 2006)

Koanda demonstrates a participatory planning process to set up a FSM system in Ouahigouya (Burkina Faso), in which the tool selection was adapted to the different stakeholder groups. Mechanical and manual operators, farmers and women associations were consulted in the form of focus groups and informal interviews. Information was provided through films, which established a link with reality. These tools were chosen to make these stakeholders feel at ease. In parallel, information was relayed and consultations were carried out with the municipal authorities and technical services in formal meetings with multimedia presentations. This type of tool was a better match for the formal environment of these stakeholders.

Collaboration was chosen as a participation level for the validation of options. In order to minimise the risk of marginalisation, the validation was first discussed in focus groups. Then, a validation workshop was organised in the municipal building, which gave the different stakeholder groups the opportunity to discuss the choices made in the focus groups and to build consensus. The main outcome of this two-day workshop was the participatory selection of one scenario and the first draft of an Action Plan. In addition, the understanding of the municipal authorities was reinforced so that they could assume their coordination role. In this way they were empowered, while building the basis for the future delegation of responsibilities.

This example shows that it is possible to bring together all FS key stakeholders, even in a very hierarchical society, with their different social and economic profiles, education level and negotiation capacity. The choice of participation levels for the different stakeholders resulted from the stakeholder analysis carried out at the end of the preliminary studies (STEP 2: *Characterisation and selection of the key stakeholders* – see Section 15.5.2). Mechanical and manual operators, farmers and womens' associations were considered to have a high interest in the project, whereas the municipal authorities had both high interest and influence. Figure 16.4 features the corresponding influence-interest matrix. Based on this and the characteristics of these stakeholders, the involvement strategy was developed (see also Figure 15.1 and 15.2).

	Low influence	High influence
Low interest		
High interest	Mechanical and manual operators Farmers Womens' associations	Municipal authorities Technical services

Figure 16.4 Case Study Ouahigouya - influence-interest matrix at STEP 2.

16.5 MILESTONES AND CROSS-CUTTING TASKS

Globally, the level of participation of key stakeholders should increase as the process develops. From information at the very beginning, it should move towards collaboration, so that when it comes to implementation and operation, smooth collaboration between key stakeholders is secured. The way in which participation levels evolve is context-specific. However, the process is marked out by milestones corresponding to the end of phases, where participation levels are formally re-thought and where important changes can be decided for the next phase.

In parallel, the planning process is marked by two participatory cross-cutting tasks, as illustrated by Lüthi *et al.* (2011): (i) awareness raising to a wide audience (i.e. not limited to key stakeholders) and (ii) capacity building, which first aims to enable the key stakeholders to take informed decisions, and then prepares the key stakeholders to endorse their role and responsibilities in the implementation and operation of the system.

16.5.1 Main milestones in the participatory process

A participatory planning framework is proposed in Chapter 17, which is characterised by different phases and milestone events. Three main milestones are identified for the involvement strategy:

1. **Initial launching workshop**, including a field visit with all the stakeholders. This consists mainly of an information workshop, aiming to communicate the plans, activities and current stage of the process. Afterwards, all the key stakeholders should have a common understanding.

2. **Validation workshop of selected options by all the stakeholders:** This event brings all the key stakeholders together to publically and officially seal the decisions taken up to this point. The technical options and management options are presented, discussed and validated.

3. **Validation workshop of the Action Plan:** This workshop seals the agreements reached on the validation of options and how to proceed further. The roles and responsibilities of the different stakeholders in the project are defined in a common understanding, which will facilitate the coordination of the various tasks.

Figure 16.5 Sign translates as, 'Managing one's wastewater and latrine without harming one's neighbours is a sign of good faith', Nzérékoré, Guinea-Conakry (photo: Philippe Reymond).

Further workshops can be added. For example, Lüthi *et al.* (2011) propose two workshops for the identification of technical service options. First, an *expert consultation workshop* to identify the feasible systems, and second, the selection of the most appropriate options by the key stakeholders in a *stakeholder consultation workshop*[1]. Case Study 16.2 illustrates a similar process.

16.5.2 Raising awareness

Awareness raising is a cross-cutting task throughout the whole process, enabling people to make informed choices and adopt good practices (Lüthi *et al.*, 2011). Awareness-raising activities may be needed at different levels, with different stakeholders. Very often, the advantages of a new FSM system may not be obvious to all. Some of them may be reluctant to support the project, or are not interested in participating, especially if a behavioural change is required. Some may be interested but lack the necessary information or skills to efficiently take part in the decision-making process. In this case, awareness raising can be developed through workshops or field visits which allow the different stakeholders to understand each other's potential and constraints. It is also imperative to educate households on how to use and maintain their onsite sanitation infrastructure, and make them understand why this is important for the whole FSM system, even if they are not involved in the decision-making process. Ultimately, it is crucial that households understand how improving the FSM system will contribute to protecting public health and raise their living standards. Training can be carried out along with awareness raising to increase the skills of the stakeholders involved in the participatory approach.

Awareness raising is critical to reach a common understanding of existing problems and to ensure that stakeholders agree on the goals (Figure 16.5). It can also be crucial when utilities or private operators who are already delivering services need to change their habits, for example, when collection and transport operators who have been discharging FS directly into the environment are required to transport FS to the FSTP after it has been built.

1 See CLUES guidelines p. 33 -37 for details on such workshops (Lüthi *et al.*, 2011).

Involvement tools like information workshops, field visits and community mapping can be very effective in raising awareness of the situation amongst stakeholders. In all cases, awareness-raising activities involve extensive communication, both at individual and collective levels. The stakeholders need to be informed of the initial situation, and the environmental and public health risks linked to the existing practices, as well as the project's aims, potential approaches and benefits in terms of the economics, environment and society. Objectives and benefits of the participatory approach need to be defined and presented to the stakeholders to increase their understanding and commitment (McConville, 2010).

16.5.3 Training and capacity building

Skills and capacities are important components of the enabling environment (Lüthi *et al.*, 2011). When it comes to implementation, the capacities of the key stakeholders at the technical, managerial, financial, commercial and social levels are crucial. Therefore, the capacities and skills of the stakeholders need to be assessed (Chapter 15) and if necessary reinforced to ensure the efficiency and long-term sustainability of the project.

Table 16.3 presents an example of potential training needs for the different roles in the service chain. The role distribution is further discussed in Section 16.7. Each role requires certain skills and knowledge. A training plan may be elaborated to define the type of knowledge required and the stakeholders concerned, and to suggest a training schedule.

Several tools and activities, such as workshops, practical exercises, participative document elaboration and field visits, can be used for training (Figure 16.6). It is also recommended that people are introduced to existing infrastructures and expertise so that they can see facilities and pilots (Lüthi *et al.*, 2011). The employees who will be in charge of the operations and maintenance (O&M) can benefit from training sessions in other cities, where there are existing FSM programmes.

Fig. 16.6 Training on sanitation systems, Ecuador (photo: Philippe Reymond).

Table 16.3 Training needs for the responsibilities involved at each level of the organisational scheme

Components of the supply chain	Responsibilities	Training needs
Collection & transport	Contact with the customer (service schedule and payment)	Training on marketing, commercial and financial management
	Collection and transport to the transfer station / treatment plant / disposal site	Training on risks, safety measures and good practice for sludge collection and conveyance;
	Quality control	Training on risk, safety measures and good practice for sludge collection and conveyance
Treatment	Collection of disposal fee	Training on financial management
	Reception and management of the trucks, operation and maintenance of the treatment plant	Training on operation principles, and on operation and maintenance procedures Training on treatment processes
	Monitoring of the treatment plant	Training on operation principles, and on operation and maintenance procedures Training on monitoring parameters for the FSTP, sampling procedures, and interpretation of the results
	External quality control	Training on operation principles, and on operation and maintenance procedures Training on the analysis of parameters for the FSTP, and on interpretation of the results
Disposal/enduse	Reception, treatment and conditioning of the endproducts	Training on operation principles, and on operation and maintenance procedures
	Customer management (sale schedule and payment) and sales	Training on operation principles, and on operation and maintenance procedures Training on marketing, commercial and financial management
	Quality control	Training on operation principles, and on operation and maintenance procedures Training on how to analyse parameters for the endproducts, and how to interpret the results
Coordination of the sector	Monitoring of system, enforcement of laws, regulations and contractual agreements, public relations, organisation of sector meetings	Training on group coordination, team leading and communication Support on document elaboration such as contracts, licences and partnership agreements Training on data collection, monitoring, and capitalisation

Financial mechanisms can be defined to answer capacity-strengthening needs. Several solutions can be explored including distribution of fees, subsidies, microcredits, community development funds and so on (Lüthi *et al.*, 2011). For example, a part of the discharge fees at the FSTP could be ear-marked for training of the operational staff. Also, a part of the registration fees for professional FS collection and transport associations could be reserved for capacity building. The budget for capacity building should be an integral part of the initial project budget, so that the infrastructure built can be efficiently operated on a long-term basis. Continuous training should then be planned, either by the organisation in charge of the coordination, or by each group of stakeholders.

SUMMARY: From stakeholder analysis to involvement strategy

Choosing the right involvement tools at the right moment and developing an involvement strategy is not an easy task. The following is a list of the necessary stages in the order they are carried out, as detailed in this chapter and in Chapter 15:

1. Look at the FSM planning framework (Chapter 17) to locate the main steps in the stakeholder analysis and involvement milestones in the whole picture of the integrated planning process.
2. Identify the stakeholders (Section 15.4).
3. Characterise the stakeholders (Section 15.5); draw up a stakeholder table (Section 15.5.1).
4. In particular, characterise the stakeholders' interest in and influence on the project (Section 15.5.2), as well as their involvement needs; the selection criteria for key stakeholders helps to define their interests and influence (Section 15.5.3). Common interests and involvement needs are described in the table featured in Section 15.5.4.
5. Establish an influence-interest matrix, as shown in Section 15.5.2.
6. Define the participation levels (Section 16.3) based on the influence-interest matrix and the specific characteristics of each stakeholder.
7. Select involvement tools for each stakeholder or group of stakeholders, according to the defined participation levels, involvement needs and specific characteristics (Section 16.4); Table 16.2 illustrates which involvement tool is appropriate for each participation level.
8. Adapt the strategy according to how the planning process develops, for example along the step-by-step approach proposed in Section 15.6.
9. Consider the cross-cutting tasks.

The case studies (Case Studies 15.1, 16.1 and 16.2) provide further illustrations on how this can be done.

16.6 DISTRIBUTING AND FORMALISING ROLES AND RESPONSIBILITIES

Once the technical options and organisational modes have been chosen, the roles and responsibilities need to be distributed and formalised. Defining how this should be done is one aspect of the development of the Action Plan (see Chapter 17, Section 17.4.3 and the Planning Framework). As mentioned in Chapter 11, special care needs to be taken not to generate responsibility overlaps between different stakeholders. A precise definition of the activities, conditions and sanctions is needed for each component of the supply chain.

All the stakeholders' points of view, constraints, and skills must be properly understood and represented during the process of distributing roles and responsibilities. This is facilitated by the stakeholder analysis, and a continuous, participative process that allows the process leaders to understand the capacities of each stakeholder. The main strengths of the different stakeholders in each component of the supply chain are further discussed in Chapter 12.

16.6.1 Formalisation documents

According to the particular situation and the stakeholders who are involved, formalisation documents can take different forms such as licences, contracts, partnership agreements, standards and laws. These different types of documents are described below.

Licences: In the context of FSM, licences can be issued by the authorities for services throughout the whole supply chain. A stakeholder can have a licence for one or more services, e.g. for collection, transport and treatment (see Chapter 12). In all cases, the official licence document should contain a list of requirements, the activities allowed and the validity of the licence. The conditions to obtain the licence can be defined either in the document itself, in standards or official decisions, or in specific terms of reference for the different operators. Licences can have a limited validity. A monitoring and enforcement system is required to ensure the conditions are respected, be it during the licence time, or when renewing it. A sanction system will discourage operators from running an activity without a dedicated licence.

Contracts: Contracts can be signed between stakeholders involved in the FSM supply chain for specific activities or services. In some countries, contracts are signed between national utilities and authorities to set their objectives, as well as financial and operating conditions. The validity and specific conditions for contracts are most often defined by national and regional regulations. Three types of contracts can be distinguished: 1) contracts linking a service provider to its customers (e.g. households, shops), that have to be defined according to the regulations; 2) contracts linking two operators undertaking different activities in the supply chain (e.g. between the treatment plant operator and user of treatment endproducts); 3) contracts between one operator and the authorities (e.g. for the delegated operation of public infrastructure by a private operator or an association).

Partnership agreements: Agreements can be signed between two stakeholders to provide a collaborative framework for the institutional or technical management of any component of the FS supply chain. For example, a partnership agreement can be signed between a private operator and a municipality to define their contribution in the enforcement of rules and the use of fees collected at a treatment plant. The partnership details are constrained by the legal framework. Specific cases of partnership agreement are public-private partnerships, where stakeholders from the public and private sector collaborate to provide services to the population. This allows a collaboration which makes the most of the strengths of the different stakeholders (see Chapter 12).

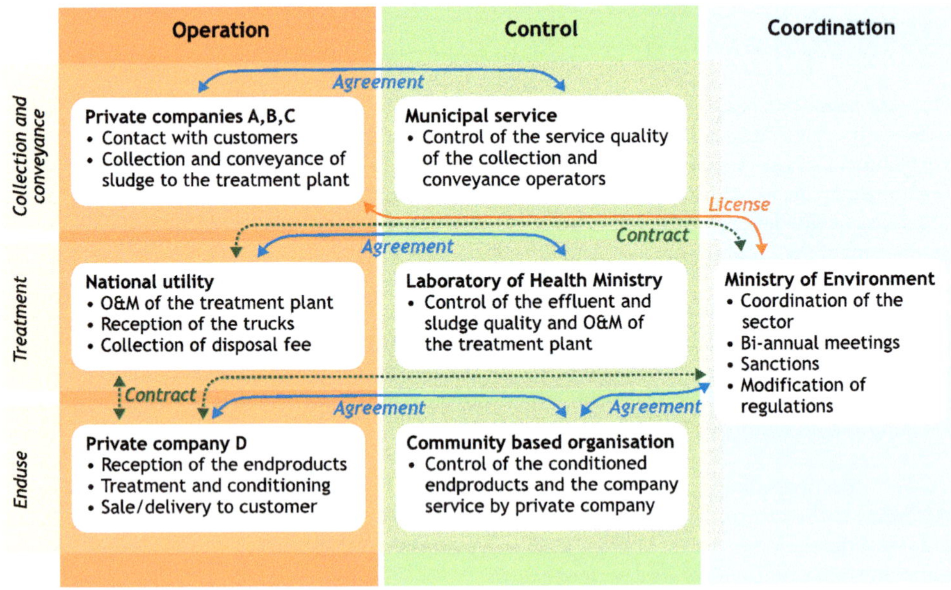

Figure 16.7 Theoretical example of a diagram of relationships with formal links between stakeholders.

Planning

Standards and laws: Standards and laws are defined by competent authorities to fix requirements, quality levels, obligations and sanctions for all components of the FSM supply chain. They may define the roles and responsibilities among stakeholders, are generally unlimited in time, can be modified through given legislative procedure, and further detailed by official decisions. They are discussed as a part of the enabling environment in Chapter 12 (Section 12.3).

16.6.2 Diagram of relationships

A diagram of relationships can be used as a tool to visualise and discuss potential formal links between stakeholders (see also Section 15.5).Positions can be added or removed from the diagram (e.g. when a stakeholder is in charge of treatment and by-product sale) (see Chapter 12). The contractual links between stakeholders can also be represented in a diagram of relationships. The diagram of relationships can be discussed during a workshop on the role distribution and institutional framework to ensure that each stakeholder has a full understanding of the organisational scheme. Several modes of organisation can be evaluated. Figure 16.7 shows a theoretical example of a diagram of relationships with different formalisation documents. In this example, partnership agreements link the operators with the control institutions, licences are delivered by the authorities to the private companies in charge of the collection, conveyance and enduse, and contracts link the authorities with the semi-private utility in charge of treatment.

Case Study 16.2: Participative definition of an organisational setup and technical strategy in Ouagadougou, Burkina Faso

(Adapted from Bassan and Strande, 2011)

Burkina Faso adopted a National Sanitation Strategy in 1996. Its principal innovation is the integration of FS management in the document. Based on this strategy and on Sanitation Plans for the main cities of the country, the National Utility for Water and Sanitation (ONEA) launched meetings to elaborate a strategy for FSM and site visits to identify potential sites for FSTPs (Bassan and Strande, 2011).

In Burkina Faso, the responsibility for the management of excreta and wastewater in the main cities is delegated to ONEA within its contract with the State, but the municipalities have the mandate to provide water and sanitation within their area. Sewered sanitation has been implemented in the centre and industrial areas of the two main cities (Ouagadougou and Bobo-Dioulasso). The rehabilitation of the simple latrines and the distribution of new ventilated improved pit (VIP) latrines is planned in other areas and mid-sized cities. Private informal entrepreneurs provide collection and transport services, given the lack of official organisation for this activity.

The elaboration of a FSM global organisation scheme for the capital city, Ouagadougou, was launched under the supervision of ONEA, the project leader. The stakeholder analysis and a succinct assessment of the initial situation were conducted, but no specific involvement strategy was developed for the choice of technical options or institutional framework. Between 2009 and June 2010, only the president of the collection and transport operators association was invited to participate in the site identification and to some of the important validation meetings. Members of the association were not informed of this process. As a consequence, key stakeholders were not consulted prior to the FSTP design, and the decisions were taken independently from the real needs and constraints of the collection and transport operators.

External process leaders were mandated from January 2010 to help develop the legal and contractual framework. After the elaboration of first drafts of the terms of reference and technical licences for the collection and transport entrepreneurs, a consultation and validation workshop was organised to discuss the current situation and the preliminary institutional framework. As no initial launching workshop had been organised at the beginning of the process to inform all the stakeholders about the aim and activities of the project, the different stakeholders did not have a common understanding of the situation. This event highlighted the need for longer discussions and for better involvement of the key stakeholders (collection and transport entrepreneurs and municipal services - district authorities, police, property management, legal authorities). Further information activities were needed before the elaboration of final versions of the documents setting the institutional framework. It was also acknowledged that the collection and transport association was not effective and that members did not recognise its committee.

Focus groups were led by the international advisors and ONEA in order to inform the stakeholders, reach a common understanding, and collaborate with them to refine the regulatory documents. First, the collection and transport operators were identified and contacted. The collection and transport association was then empowered through several focus groups, personal meetings, visits and workshops and a new committee was elected. In parallel, the members actively collaborated to define the institutional framework for the collection and transport activities. As a part of the empowerment activities, the participation of the association and some of its members in information workshops was also encouraged.

The mayors of the five districts of Ouagadougou were contacted, together with the municipal police, the property management, and the legal authorities. The project and the first draft developed with the collection and transport operators were presented in a first workshop. Several focus groups were then organised to modify the documents.

At the end of this process, a validation workshop with all local stakeholders was held, during which the documents elaborated within the focus groups were validated, the roles and responsibilities in the future FSM system pre-defined, and the transition phase to the implementation of the organisational scheme discussed. The initial project planning in this case study is typical of infrastructure projects where an international consulting company was mandated to develop the technical component only and the organisational setup was not addressed and no stakeholder participation was planned in the beginning. The involvement of the collection and transport operator, which makes the FSM project differ from wastewater projects where the transport is done by sewers, only became obvious late in the process.

This case study shows how important it is to involve all key stakeholders from the beginning of the process, and how time can be saved if this is done well. It also shows that the lack of involvement of all the stakeholders from the beginning results in complications when it comes to the distribution of roles and responsibilities. The overall process was time demanding, especially because the awareness raising, empowerment and involvement needs of the collection and transport operators had not been acknowledged early enough. The trust relationship with the collection and transport operators was hard to build as several decisions concerning their work had already been taken without them being consulted.

The main strengths of this project were the real involvement of the utility (ONEA), the willingness to define the organisational and regulatory setups before the implementation, and the recruitment of a long-term process leader for the stakeholder engagement, the definition of the role and responsibilities, and the elaboration of the institutional framework. Without the discussions and workshops conducted to elaborate these setups, it is possible that the FSTP would have been built without any consultation of the collection and transport operators, who would probably have rejected the obligation to discharge FS at the FSTPs.

Finally, the involvement tools applied, though late in the process, allowed an efficient awareness-raising process, the participation of some of the operators, and the empowerment of the association. After about one year of consultation workshops, the organisational setup documents set according to the local context were validated by all the stakeholders in a workshop (Bassan *et al.*, 2011).

The stakeholder participation matrix of this project, developed retrospectively for the sake of analysis, is presented below (Table 16.4). It highlights the separate development of the technical and organisational components. The planning process as proposed in this book (see Section 17.3 and the planning framework) is included for comparison in the last column. Figure 16.4 also presents the late but efficient activities to involve the key stakeholders, especially the municipal services and the collection & transport association, in the definition of the organisational scheme and the regulatory documents. This process started with information and consultation and ended with empowerment and collaboration.

Figure 16.8 Inspection of an inlet, Burkina Faso (photo: Magalie Bassan).

Planning

Table 16.4 Retrospective stakeholder participation matrix of the Ouagadougou case study with main involvement activities

Planning steps	Participation levels of Case Study 16.2				Activities	
	Information	Consultation	Collaboration	Empowerment/ delegation	Case Study 16.2	Ideal case *(see Planning Framework, Table 17.2)*
Launch of the planning process	-	Municipal services, President of C&T assoc.[2], MoUP[2]	ONEA (project leader), international technical consultant	-	Field visits for selection of treatment sites	START OF THE PARTICIPATORY PLANNING PROCESS FOR THE PLANNING OF THE WHOLE FSM SYSTEM **Launching workshop** Field visits
Assessment of the current situation	-	-	ONEA, international technical consultant	-	Brief technical assessment studies **Report of preliminary studies on treatment technologies**	Participative assessment of the initial situation Information/consultation with stakeholders **Report of preliminary studies**
Identification of service options	-	-	ONEA, international technical consultant	-	**Report of feasibility study on treatment technologies**	Involvement activities according to need **Report of feasibility study** **Validation workshop of options**
Development of an Action Plan	-	-	ONEA, international technical consultant	-	Detailed design for treatment options **Detailed project document on treatment technologies**	Participative action planning with key stakeholders; empowerment and awareness raising **Validation workshop of Action Plan** **Detailed project document**

Planning steps	Participation levels of Case Study 16.2				Activities	
	Information	Consultation	Collaboration	Empowerment/ delegation	Case Study 16.2	Ideal case *(see Planning Framework, Table 17.2)*
Assessment of the current situation	Municipal services, C&T operators, MoUP , Universities, NGOs, international tech. consultant	Municipal services, C&T[1] operators, ONEA	ONEA, international advisors	-	PARTICIPATIVE PLANNING PROCESS: PLANNING OF ORGANISATIONAL SETUP Brief institutional assessment studies **Preliminary organisational setup** **Information workshop**	
Identification of service options & development of an Action Plan	Municipal services, MoUP[2], Universities, NGOs	-	Municipal services, C&T operators, ONEA, international advisors	Municipal services, C&T operators, ONEA	Empowerment, awareness raising Focus groups to develop regulatory texts **Validation workshop on roles and regulatory texts**	

1 C&T: Collection and transport

2 MoUP: Ministery of Urban Planning

16.7 BIBLIOGRAPHY

Aune, J. B. (2000). Logical Framework Approach and PRA - mutually exclusive or complementary tools for project planning? Development in Practice 10(5), p.687-690.

Bassan, M., Mbéguéré, M., Zabsonré, F. (2011). Integrated faecal sludge management scheme for the cities of Burkina Faso. 2nd IWA Development Congress. Kuala Lumpur, Malaysia, IWA.

Bassan, M., Strande, L. (2011). Capacity strengthening in sanitation: benefit of a long-term collaboration with a utility and research institute. 35th WEDC International Conference. WEDC. Loughborough.

Bassan, M., Tchonda, T., Mbéguéré, M., Zabsonré, F. (2011). Processus d'élaboration d'un cadre institutionnel régulant l'activité de vidange mécanique de la ville de Ouagadougou, Burkina Faso. 16th African Water International Congress and Exhibition. Marrakech.

Koanda, H. (2006). Vers un assainissement urbain durable en Afrique subsaharienne: approche innovante de planification de la gestion des boues de vidange. Thesis n° 3530, EPFL.

Lüthi, C., Morel, A., Tilley, E., Ulrich, L. (2011). Community-Led Urban Environmental Sanitation Planning: CLUES. Complete Guidelines for Decision-Makers with 30 Tools. Dübendorf, Switzerland, Swiss Federal Institute of Aquatic Science and Technology (EAWAG).

McConville, J. (2010). Unpacking Sanitation Planning - Comparing Theory and Practice. PhD Thesis PhD, Chalmers University of Technology.

Mosler, H.-J. (2004). A framework for stakeholder analysis and stakeholder involvement. International Water Management Course. Rüschlikon, Zürich.

ODA. (1995a). Technical Note on Enhancing Stakeholder Participation in Aid Activities. n. D. Overseas Development Administration (ODA).

ODA. (1995b). Guidance Note on How To Do Stakeholder Analysis of Aid Projects and Programmes. Social Development Department. London, Overseas Development Administration (now DfID).

Reed, M.S. (2008). 'Stakeholder participation for environmental management: A literature review.' Biological Conservation 141(10), p. 2417-2431.

Reymond, P. (2008). Elaboration d'une méthodologie permettant de déterminer une option durable pour le traitement des boues de vidange dans une ville moyenne d'Afrique subsaharienne - Application à la ville de Sokodé, au Togo. MSc. Thesis, EPFL, EAWAG/SANDEC.

Chapter 17

Planning Integrated Faecal Sludge Management Systems

Philippe Reymond

Learning Objectives

- Understand the importance of an integrated approach for faecal sludge management.
- Learn how to plan a faecal sludge management project on a city level, including a logical framework of the necessary activities.
- Understand how to select context appropriate options and to determine the critical selection factors in each specific context.
- Be able to link the different aspects developed in the book (e.g. management, finance, stakeholders' interests, technology, local environment) and understand how they are connected and influence each other.

17.1 INTRODUCTION

The process leaders who are designated with the responsibility for planning and implementing a city-wide faecal sludge management (FSM) system (Case Study 17.1) often face a complicated situation, characterised by diverse levels of service and a patchwork of uncoordinated and independent stakeholders managing various activities. FSM planning aims to transform such a complex situation into a well-organised and coordinated management framework, which is usually initially expressed in the form of a city sanitation plan or citywide sanitation strategy (Case Study 17.3) and later translated into action plans and concrete implementation. This is no easy task as stakeholders have different and even conflicting interests, needs and constraints (Chapter 15). However, it is a crucial task, as urban sanitation planning is the key to sound investment (WSP, 2009) and clear action plans greatly assist in sourcing funding. If donor money is being sought, a detailed plan with a clear strategy will be necessary. The problem with urban sanitation is not only a lack of investment, but also the lack of a plan.

FSM planning is about understanding and matching stakeholders' interests, needs and constraints with an appropriate and accepted management scheme (Chapter 12) and financial mechanisms (Chapter 13). It is also about assessing capacity building and needs for empowerment (Chapters 15 and 16). Such understanding can only be acquired through a thorough assessment of the initial situation

(Chapter 14). Experience in FSM shows that every solution should be context-specific and integrated. Moreover, experience in Asia demonstrates that any number of approaches can be successful when implemented in conjunction with a comprehensive legal and regulatory framework, clear delineation and appropriate delegation of roles and responsibilities, and dedicated public funding (AECOM and SANDEC/EAWAG, 2010).

At the beginning of such a complex task, it is not easy to know where to begin, how to collect and structure the necessary information, or how to work with the key stakeholders in order to reach a sustainable system.

In this chapter, an integrated planning approach is proposed, in order to facilitate the work of a planner or engineer in a city and to bind together the different activities and ideas presented in this book in a logical and structured way. This integrated approach is described in Section 17.3 and reflected in two key elements: the FSM planning framework (Table 17.1) and the technology selection scheme (Figure 17.8). The Planning Framework highlights the essential tasks and activities and can be used as a template or an aid when getting started, whatever the city in question. As for the Technology Selection Scheme, it proposes an elimination-based approach based on the local context and the interest for enduse.

Figure 17.1 The city of Elmina, Ghana (photo: Philippe Reymond).

Table 17.1 Faecal sludge management planning framework

FSM planning from A to Z

Standard project phases		Activities	Outcomes	Chapters	Participatory stages: CLUES	Participatory stages: SAN21
Exploratory study	A	Preliminary assessment of the initial situation and first inventory of stakeholders	Overview of the situation; facilitators are identified	14 15	Process ignition	Establish a city sanitation task force
			Inception report			
Preliminary (pre-feasibility) studies	B	Identification and preliminary characterisation of the stakeholders and their relationships	All stakeholders are identified and characterised	15	Launch of the planning process	Understand the existing context
	C	Initial launching workshop, including field visit with all the stakeholders	Stakeholders are sensitised to sanitation reality and aware about the project's objectives	16.5	Launch of the planning process	Understand the existing context
	D	Assessment of: - Sanitation practice and needs, reuse interests - Institutional setup, government support - Legal and regulatory framework - Existing organisational modes - City structure and heterogeneity of sanitation practices - Existing financial flows - Climate	Sanitation practices are identified, as well as urban heterogeneity; Strengths, weaknesses, opportunities and threats are identified (SWOT analysis); The enabling environment is described	14	Detailed assessment of the current situation	Understand the existing context
	E	Selection of potential organisational modes	Orientation of the process towards realistic options	12	Detailed assessment of the current situation	Understand the existing context
	F	Identification of sites for treatment	Stakeholders have indicated existing and potential sites	14.4	Detailed assessment of the current situation	Understand the existing context
	G	Characterisation and selection of key stakeholders	Stakeholder who have interest in and/or influence on the process are identified	15.4 to 15.5	Detailed assessment of the current situation	Understand the existing context
			Preliminary studies report			
Feasibility study	H	Quantification and characterisation of sludge	Process leaders know what has to be treated	2	Identification of service options	Identify viable solutions
	I	Characterisation and selection of sites	Appropriate sites are selected	14.4	Identification of service options	Identify viable solutions
	J	Preselection of combinations of technologies, organisational modes and financial mechanisms	Scenarios are elaborated	5,11,12, 13,15,17	Identification of service options	Identify viable solutions
	K	Detailed evaluation of selected options, including: - Requirements of technology combinations, pros and cons, O&M - Organisational mode and institutional setup; roles & responsibilities; contractual arrangements - Capital and operation costs, financial mechanisms, estimated budget - Skills required to run each system - Environmental impact assessment	System scenarios are evaluated and optimised	4-17	Identification of service options	Identify viable solutions
	L	Preliminary presentation of the results to the key stakeholders	Stakeholders are consulted and agreement is secured	16	Identification of service options	Identify viable solutions
	M	Final selection of system options		17	Identification of service options	Identify viable solutions
	N	Workshop : Validation of chosen options by all the stakeholders	Proposals are validated by all stakeholders	16.5	Identification of service options	Identify viable solutions
	O	Reassessment of key stakeholders according to the validated options	Influence and interest of stakeholders are reassessed according to the previous decisions	15.5	Identification of service options	Identify viable solutions
			Feasibility study report			
Detailed project development	P	Detailed project development (Action Plan): - Detailed design of the treatment plant - Detailed definition of roles & responsibilities - O&M management plan with clear allocation of costs, responsibilities and training needs - Conventions between stakeholders, securing financial and institutional mechanisms - Strategy for control and enforcement - Definition of needs for capacity building and job creation - Definition of contracts and bidding processes - M&E strategy for the implementation phase - Timeline for implementation with distinct phases and an itemised implementation budget	The Action Plan is written; The whole system is described in detail	11 12 13 16 17	Development of an Action Plan	Elaborate Strategic Plan
	Q	Workshop : Presentation of the Action Plan	The Action Plan is validated by all stakeholders	16.5	Development of an Action Plan	Elaborate Strategic Plan
	R	Reassessment of key stakeholders according to Action Plan	Roles and responsibilities of stakeholders are redefined according to the Action Plan	15.5	Implementation of the Action Plan	Prepare for implementation
			Detailed Project Document			
Implementation	S	Recruitment of contractors for building and O&M		11	Implementation of the Action Plan	
	T	Organisation of the sector, transfer of roles & responsibilities	FS management is transferred to the corresponding stakeholders	11,12,13,16	Implementation of the Action Plan	
	U	Capacity building / information campaigns	Awareness is raised among users; Capacity is built where needed	16	Implementation of the Action Plan	
	V	Monitoring of construction	Building according to state-of-the-art is ensured	11	Implementation of the Action Plan	
	W	Reassessment of key stakeholders before inauguration of the FSTP	Capacity of stakeholders to deal with their new roles and responsibilities is assessed	15.5	Implementation of the Action Plan	
	X	Start-up of the system	The FSTP is brought to its state of equilibrium; stakeholders have acquired the necessary skills	11	Implementation of the Action Plan	
	Y	Official inauguration ceremony	The FSTP is officially transferred to the city authorities / private entrepreneurs		Implementation of the Action Plan	
M&E	Z	Monitoring of the running system (technical stability, satisfaction of stakeholders, cost recovery)	The system is monitored to ensure its sustainability	11		

Case Study 17.1 : Leadership in the planning process
(Adapted from Parkinson *et al.*, 2013)

A planning process needs process leaders. There is never one single person responsible for FSM, but rather a web of stakeholders. Experience shows that the creation of a **city sanitation task force** is an effective means to engage with different institutions from the public and private sectors, and with non-governmental organisations (NGOs).

In order to develop a citywide sanitation plan there must be one institution that provides leadership; this is essential to ensure that the planning process maintains a clear direction and subsequently achieves the objectives agreed by the key stakeholders. In most situations, the most appropriate leader of the sanitation planning process is the local authority. If the planning process is driven by external agencies and in too short a time scale, the plan will invariably lack local ownership and there will be no incentive to move forward with the implementation of the plan.

It is necessary to ensure that there is sufficient commitment and communication between the different stakeholders prior to embarking on the planning process. During the consultation process, there must be sufficient time and opportunity for all the stakeholders to be involved. There is also a need for the planning process to be properly facilitated in order to guide and support interaction and communication between stakeholders. It is therefore important to identify the right individuals and institutions that have these skills. This may be technical support or skills such as stakeholder coordination, conflict resolution and community organisation.

Figure 17.2 A representation of three key elements in the planning process: the city sanitation plan, local authority and a sanitation task force (adapted from Jonathan Parkinson).

17.2 NEED FOR AN INTEGRATED APPROACH

In the past, many water and sanitation projects have failed because of the lack of an integrated approach (Box 17.1). The development of physical infrastructure is only one component of a functioning FSM program, which also depends upon sustained public sector commitment and funding, effective policies, appropriate implementation and compliance enforcement (AECOM and SANDEC/EAWAG, 2010). Common reasons for failure are the implementation of infrastructure without consulting the main stakeholders or without planning adequate operation and maintenance (O&M) and financial schemes. Besides lack of institutionalisation of the system, lack of skills, insufficient organisational capacity and lack of cost-recovery mechanisms are also recognised as major factors in failure. Broadly speaking, it is possible to say that the enabling environment necessary for a functioning FSM system was either not there in the first place, or was not developed as an integral part of the project.

The lack of an enabling environment should not be considered as a reason for not engaging in a city because activities such as planning the O&M, defining roles and responsibilities, and structuring financial instruments for cost recovery can be structured into the project design. A bit more time and resources invested in the preliminary phase of the project can save a lot of time and money during and after implementation. In addition, a careful assessment of the initial situation (Chapter 14) and involvement of stakeholders (Chapter 16) will ensure a more appropriate selection of technical options and also provide insight into the presence (or absent) of the fundamental conditions necessary for an enabling environment (Section 17.2.1).

BOX 17.1 : Examples of project failure due to lack of an integrated approach

Successful citywide sanitation projects in low-income countries are few and far between. Failed projects are the norm rather than the exception. In most cases, the reason is to be found in the lack of an integrated planning approach. Here are a few examples:

- A faecal sludge treatment plant (FSTP) was constructed 15 km from the city centre. Given the local context and the situation of the faecal sludge (FS) collection and transport service providers, this distance was too great, and the FSTP never received any sludge. Involvement of the FS collection and transport service providers in the planning process would have avoided this failure.

- Lack of O&M of a well-functioning FSTP led to a total breakdown. This was due to the lack of a clear definition of roles and responsibilities in the FSM scheme and a strict O&M plan which resulted in the municipality destroying portions of drying beds, and prevented relatively minor repairs from taking place (Case Study 12.2). Instead of taking simple action, stakeholders allowed the situation to deteriorate.

- A co-composting facility closed because no financial analysis had been carried out beforehand. The project designers had not taken into account the significant O&M costs for such a facility or the market demand for compost, and the sale of compost alone could not cover the operating costs.

- Large donors did not coordinate among their projects, leading to patchwork action instead of a sound city-wide sanitation plan.

Experience can also be drawn from wastewater projects (ADB, 2006), especially regarding user needs and constraints and treatment plants' O&M requirements.

Figure 17.3 A non-functioning treatment plant in Yaoundé, Cameroon (photo: Linda Strande).

17.2.1 Understanding and working towards an enabling environment

The major barriers to progress in sanitation coverage lie within the institutions, policies and realities of low- and middle-income countries (Lüthi *et al.*, 2011a). The public sector is often weak in terms of skills, structures, planning capacity and bureaucratic procedures, and mechanisms are not always in place to recover investment, operation or management costs, leading to a degradation of service provision or even system failure. Developing a national capacity to bring about change is therefore crucial. This needs to include the building of capacity and skills, introducing changes to organisational culture, developing nation-wide policies and providing sufficient financial resources.

An enabling environment is critical for the success of any type of investment, whether this is for the improvement of a single public latrine or for a city-wide FSM system (AECOM and SANDEC/EAWAG, 2010; Lüthi *et al.*, 2011a; Lüthi *et al.*, 2011b). Without it, the resources committed to bring about change run the risk of not being effective. Understanding the conditions necessary in a particular context for the environment to be enabling is part of an integrated approach. Once those are understood, measures to fulfil them should be an integral part of the project for it to be sustainable.

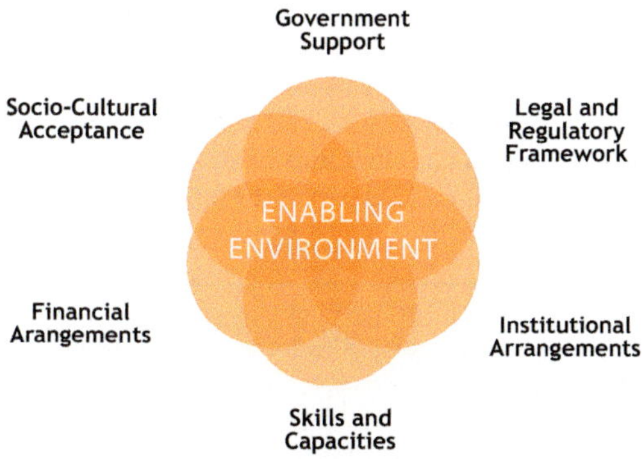

Figure 17.4 Components of an enabling environment (adapted from: Lüthi *et al.*, 2011a).

In order to understand the large variety of potential influences, the enabling conditions are classified into six categories, as shown in Figure 17.4 (Lüthi *et al.*, 2011a): 1) Government support, 2) Legal and regulatory framework, 3) Institutional arrangements, 4) Skills and capacities, 5) Financial arrangements, and 6) Socio-cultural acceptance.

Government support: Conflicting political priorities and therefore, a lack of explicit political support, is often the initial cause for project failure. Enabling government support includes not only relevant national policy frameworks and sector strategies, but also receptive local authorities and decision-makers.

Legal and regulatory framework: The technical norms and standards that influence the types and levels of service that are put in place are clearly important. Typical problems include regulatory inconsistencies, lack of regulations or unrealistic standards. A further issue in many countries is the poor enforcement of existing regulations. For the legal framework to contribute to the enabling environment, it must be transparent, realistic and enforced.

Institutional arrangements: Public institutions and private actors are integral to an enabling environment and getting the institutional environment right is a key ingredient for the sustainable delivery of sanitation services (see also Chapter 12). This encompasses the correct understanding of roles and responsibilities and capacities of each stakeholder, as wel as their influence and interest in improving service provision. A potential obstacle may be overlapping mandates between different institutions and ministries.

Skills and capacities: Developing the required skills and capacities at all levels is a key requirement and an issue that can take considerable time to develop. Identifying capacity gaps, particularly at district and municipal level, and then filling the gaps with tailored training courses, on-the-job training, etc. is a prerequisite (see also Chapter 16).

Financial arrangements: Implementing and maintaining environmental sanitation services is costly and requires an enabling financial environment. Financial contributions and investments are required from users, from government agencies and from the private sector (see also Chapter 13).

Socio-cultural acceptance : Achieving socio-cultural acceptance depends on matching each aspect of the proposed sanitation system as closely as possible to the users' preferences. Failure to ensure that the implemented solution is socio-culturally embedded is one of the most common reasons for past project failure (see also Chapter 15).

If they are absent, conditions for an enabling environment should be created before going any further into implementation and need to be addressed as part of the initial stage(s) of the planning process[1].

> **Case Study 17.2: Difficulties in creating enabling institutional arrangements**
> (Adapted from Lüthi *et al.*, 2009)
>
> Changing the status quo when dealing with institutional arrangements is not easy. Often the decision-making and selection of options is heavily influenced by vested interests and local politics.
>
> A challenge that informal or quasi-formal service providers often face when dealing with the leading sector institutions (i.e. commercialised or public utilities), which have a *de facto* monopoly, is to change the 'business as usual' mode of doing things. For example, in an East African town, the municipal utility carries the term 'sewerage' in its name and was mainly interested in expanding its sewerage network to all the planned areas of town, even though almost 90% of the city's citizens continued to rely on onsite systems such as septic tanks and simple pit latrines. Trapped in the rigid definition of its sewerage mandate, this municipal utility would not consider managing FS at the time. Changing the status quo would have meant starting a lengthy advocacy process with the utility and the overarching governmental agencies.
>
> A few years later, the government overcame this problem with new national legislation, entrusting national utilities, and not municipal utilities as before, with the mandate to manage FS.

17.2.2 The importance of a participatory approach

Imposing a blueprint system from the top down is invariably prone to failure, even if it has succeeded in other locations. Institutions and the mindsets of people working for these institutions may not be ready, skills may not be there, and a number of stakeholders that previously made a living out of activities related to sanitation are excluded. Blueprint systems usually turn out more expensive than systems designed specifically for the local context.

In order to find out the most appropriate and cost-effective system, all stakeholders should be involved (Chapters 15 and 16), and competencies should be sought. For example, the current emptying service providers, even if they do not have a legal status, will be the most knowledgeable people regarding the collection of FS. Sometimes it may be difficult for utilities or governmental agencies to give up or to delegate some of their power and responsibilities. However, it is the price to pay for a cost-effective service with broad coverage. It is important that FSM be government or utility driven, but inclusion of other stakeholders is usually necessary to fill in the gaps of state service provision. Private public partnerships (PPPs) are synergies that need to be investigated.

---

1 To learn more about the enabling environment, see also two reference publications, available at www.sandec.ch
- Lüthi *et al.* 2011a, Community-Led Urban Environmental Sanitation Planning: CLUES, p. 49-65.
- Lüthi *et al.*, 2011b, Sustainable Sanitation in Cities - a Framework for Action, p. 127-133.

Figure 17.5 A stakeholder workshop in Nepal (photo: Lukas Ulrich).

Sometimes sanitation development is either impeded or made impossible by institutional relationships resulting from stakeholders seeking personal prestige rather than public welfare. It is the role of external agencies (donors, consultants, and NGOs) to bring these stakeholders together and show that collaboration is a win-win process. The role of such agencies to advocate for integrated approaches and to push for positive change must be highlighted.

Many institutions and agencies may be reluctant to engage in a participatory process, arguing that it takes time and money. However, although participation has a cost, the benefits greatly outweigh the costs through increased potential for the system to succeed.

Case Study 17.3: City sanitation plans – the Indonesian case study
(Adapted from WSP, 2009 and WSP, 2010)

It is becoming clear that the problem of urban sanitation is not only a lack of investment; but also the lack of a plan. There is a growing awareness worldwide of the need for integrated 'city-wide sanitation strategies' or 'city sanitation plans' as a prerequisite for sound sanitation investment. India and Indonesia, for example, have taken important policy steps in this direction. Governments need to develop both incentives and obligations for municipalities to adopt comprehensive strategies, by linking sectoral funding to the adoption of a city sanitation plan.

The Indonesian Sanitation Sector Development Program (ISSDP) is an innovative response to the growing sanitation crisis. Instead of funding investments directly, it fosters an enabling environment for progress, with special focus on city-level planning, strengthening sector strategy and institutional arrangements, and advocacy and awareness-raising at all levels. Central to the process is collaboration between the various government organisations.

ISSDP has tried, through the planning process, to directly address the shortcomings of existing sanitation services in the cities, particularly poor inter-agency coordination, a history of *ad hoc*, supply-driven investments and a deficient information base for decision-making. Thus, it avoids 'blueprint' approaches to infrastructure development that treat the city as a blank sheet on which completely new services should be imposed. Instead, it starts from an analysis of what already exists, and then considers how this could be improved in incremental steps, if and when funds become available and municipal capacity grows. It seeks to enhance synergy among the key stakeholders, develop sanitation in all parts of the city and create opportunities and incentives for private sector initiatives. Finally, the plan is translated into recurring annual action plans that can be implemented in given budget years.

The process is led by a city-level **sanitation working group**, endorsed by city leaders and composed of members from the municipal agencies, the private sector, NGOs and community groups and it is supported by a full-time facilitator. The sanitation working group does not replicate the roles of existing sector agencies, but helps to coordinate stakeholders to accelerate sanitation planning. The main challenges lay in creating local ownership of the planning process (due to the expectation that donor-funded consultants will cater for everything) and in institutionalising the group. There is always a risk that the parent organisations (which control staff and budgets) will adopt a 'business as usual' stance, irrespective of the plans that have been developed.

The plan is helping to elevate the profile of sanitation issues in the eyes of urban stakeholders and to address the gap between huge up-front investment projects with poorly targeted investments and small-scale community-based programmes that are failing to make a significant impact.

Figure 17.6 A narrow street in Indonesia (photo: Maren Heuvels).

17.3 PROPOSAL OF A PLANNING APPROACH AND LOGICAL FRAMEWORK

For an integrated and participatory approach to be applied, it has to be embedded in the initial project design (e.g. project proposals, Terms of Reference). ADB (2006) proposes an example of a more integrated terms of reference for assisting in project preparation. Time and money have to be allocated for the activities aiming to involve stakeholders and a provision has to be made to hire relevant social consultants. It also implies careful planning of the activities in order to allow effective coordination between the different consultants. Coordination is often a major shortcoming of big-donor projects where the different components are financed by different agencies. For this reason, it is important to have a clear plan of activities and a timeline from the start, and which all stakeholders agree upon.

In order to support FSM project design, a basis for a FSM project logical framework is proposed, structuring the activities and outcomes of FSM planning chronologically (Table 17.2, Project phases and participatory process stages), with reference to the corresponding sections in the book. The framework includes the participatory activities, reflecting the stakeholder analysis approach proposed in Section 15.5 and the involvement milestones described in Section 16.5. The different activities are fitted into the 'traditional' project phases and in the stages of recognised participatory planning approaches such as CLUES and Sanitation 21 (Lüthi *et al.*, 2011a; Parkinson *et al.*, 2013) – the link with the two latter being developed in Case Study 17.5. 'Traditional' project phases (exploratory study, preliminary studies, feasibility studies, detailed project development, implementation and monitoring and evaluation) and stages in participatory planning approaches (launch of the planning process, detailed assessment of the current situation and user priorities, identification of service options, development of an Action Plan and implementation of the Action Plan) follow a different but complementary logic, with slightly different main steps. They are compared side-by-side in Table 17.2, which describes briefly the main components of the "FSM planning from A to Z" framework (Table 17.1), and the main outcomes of each phase or stage are highlighted.

In the following sections, the logical framework is explained alongside the 'traditional' project phases. Several activities may be carried out in parallel, or in a different order to that listed here, according to the local context.

Figure 17.7 Narrow street in Raipur, India (photo: Philippe Reymond).

Planning

Table 17.2 Project phases and participatory process stages

Project phases	Phase description	Main Outcomes – Agenda	Participatory planning stages: CLUES	Participatory planning stages: SAN21
Exploratory study	First contact with the field. Main objectives: identify FSM stakeholders, get an overview of the situation and identify facilitators.	**Inception report**	Process ignition	**STAGE 1** Establish a city sanitation task force
Preliminary (pre-feasibility) studies	The preliminary studies consist of a detailed assessment of the local context.	Initial launching workshop **Preliminary studies report**	Launch of the planning process Detailed assessment of the current situation	**STAGE 2** Understand the existing context
Feasibility study	The feasibility study consists of an in-depth analysis of the situation, leading to system scenarios. At the end of the feasibility study, the results are discussed with the different stakeholders and formally validated.	Workshop: Validation of selected options **Feasibility study report**	Identification of service options	**STAGE 3** Identify viable solutions
Detailed project development	This phase aims to define in practice the modalities of implementation of the validated scenario. It ends with a workshop that presents and validates these modalities.	Workshop: Presentation of the Action Plan **Detailed Project Document**	Development of an Action Plan	**STAGE 4** Elaborate Strategic Plan
Implementation	This is the implementation phase, ending up with the official delivery of a working system.	Official inauguration ceremony	Implementation of the Action Plan	**STAGE 5** Prepare for implementation
Monitoring & Evaluation	The system is monitored to ensure its sustainability.			

Case Study 17.4: Link with the global planning approaches CLUES and Sanitation 21

FSM planning is part of the larger framework of environmental sanitation planning, which includes city-wide wastewater, storm water and solid waste management. The prevalence of FS over wastewater varies from city to city. In West Africa for example, sanitation planning is clearly oriented towards FSM planning.

Several approaches have been developed to help planners in defining appropriate management strategies (McConville *et al.*, 2011). They differ in the spatial level they address (households, communities or city as shown in Figure 17.8) and the planning perspective (bottom-up vs. top-down). Such distinctions in level help the process leaders to simplify and organise their work by breaking down the process spatially and temporally. Each city unit should be dealt with separately in the first phase, while at the same time dealing with city authorities. For each component, corresponding methodologies may be used. Then the different pieces of the puzzle will come together to form a city-wide FSM system.

The planning model proposed in this book encompasses the entire system and tries to link the different global planning approaches such as Sanitation 21 and CLUES in the specific field of FSM. The Sanitation21 planning framework (Parkinson *et al.*, 2013) and the Citywide Sanitation Strategy (WSP 2010) are city-wide infrastructure planning approaches. The CLUES (Community-Led Urban Environmental Sanitation) guidelines (Lüthi *et al.*, 2011a) developed by EAWAG/SANDEC are complementary to these and address planning at the community level (Figure 17.8).

FSM planning links both levels, as the management needs to be organised city-wide, but in very close relationship with the users, e.g. the households, manual and mechanical operators or endusers. CLUES methodology may be used to assess user priorities in medium-sized cities and low-income areas. It should be adapted for cases where stakeholders are too numerous to be dealt with individually and have to be organised in associations, which is the case in most larger cities and districts within cities. In these circumstances Sanitation 21 can provide guidance on how to translate user priorities at the city-wide level while dealing with the municipal authorities.

In FSM, mechanical and manual emptying service providers and endusers stand at the interface between both levels. Building a management interface in this case means organising and giving a voice to those groups at the decision-making level. In parallel, the financial mechanisms bind the whole system together.

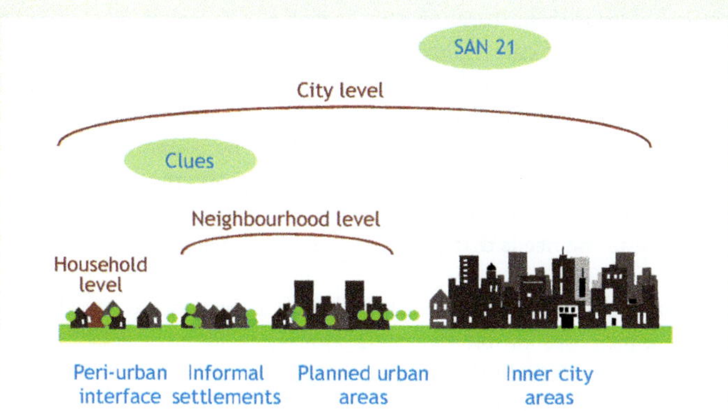

Figure 17.8 Planning contexts (figure: SANDEC).

Planning

17.3.1 Exploratory and preliminary studies

The exploratory study is usually short (e.g. two weeks) and should focus on establishing lines of communication with potential partners (including the authorities), a first inventory of the stakeholders and a preliminary assessment of the current situation (Chapter 14). Process leaders should make field trips and transect walks (Section 16.4), and visit disposal sites. It is important to try and obtain the opinion of as many stakeholders as possible. It is also important to identify facilitators for the next steps of the project, i.e. people who are familiar with the situation and have easy access to the different stakeholders.

Preliminary studies should start with the identification and preliminary characterisation of the stakeholders (Sections 15.3, 15.4 and 15.5.1) and the relationships between them. It is recommended that they are all invited to a launching workshop (Section16.5), where everyone can develop a common understanding of the situation in the intervention area, as well as agree on the process of how to solve the problems (Gutterer *et al.*, 2009; Lüthi *et al.*, 2011a). This common understanding is very important and can never be assumed. Experience shows that many stakeholders are usually not aware of the situation, especially among local authorities. For this reason, it is recommended that a field visit for all stakeholders is included as part of the workshop in order to raise awareness.

Government and utility representatives should be involved in the launching workshop to avoid potential conflicts with existing policies, regulations and municipal by-laws. It will also help clarify the available support and skills at municipal or district levels (Lüthi *et al.*, 2011a). This workshop should also aim at (Gutterer *et al.*, 2009):

- creating awareness amongst decision-makers on legal requirements, required resources and institutional backing;
- developing a supportive environment and getting different stakeholders and authorities to offer their competencies; and
- launching a process for the provision of financial and human resources at different government levels.

Once all stakeholders are sensitised to the sanitation reality and aware of the project objectives, it will be much easier to collect information for the preliminary and feasibility studies. It is often not easy to access data, especially where 'information is power', and it is important to make key contacts and build trust through transparancy.

Chapter 14 describes how to carry out a detailed assessment of the current situation. Outputs should include a refined stakeholder analysis, baseline data, as well as a thorough assessment of the enabling environment (Section 17.2.1) and current levels of service provision (Lüthi *et al.*, 2011a). It is also important to select possible organisational modes (Chapter 12) and potential treatment sites (Section 14.4), as the range of options available may significantly influence the technology selection. This is illustrated in the Technology Selection Scheme (Figure 17.10). The assessment should be based upon a participatory process (using semi-structured interviews, focus group meetings or household surveys, Section 16.4), because each stakeholder has specific experience and knowledge of the situation. They also have particular interests and needs that must be taken into consideration (Section 15.4).

At the end of this phase, the process leaders should have a clear idea of feasible options and key stakeholders. This book proposes an iterative methodology to identify, select and characterise the relevant stakeholders (Section 15.5). Such a systematic approach is important in order to identify any threat or opportunity and assess the influence and interest of stakeholders in the project, especially those linked with potential management and institutional arrangements. This approach can avoid many problems in later stages of the project.

17.3.2 Feasibility study

The main result of the feasibility study is the identification of viable system options. This phase starts with the quantification and characterisation of sludge (Chapter 2), as a prerequisite for the selection and design of technical options (Section17.4). Quantities and characteristics are specific to each city. Such a study requires time and should be planned accordingly. A characterisation and selection of sites (Section14.4) should also be made as this may influence the viability of different technical options.

Based on this and the information collected during the preliminary studies, technical, organisational and financial options can be pre-selected and studied in-depth, and discussed in a participatory manner. It is critical to recognise that FSM planning is about the combination of services, not about single technologies (Section17.4.1). Each aspect influences and is influenced by the others. Each scenario should be evaluated in detail as follows:

- requirements of technology combinations, pros and cons, O&M (Chapters 4 to 11);
- management and institutional setup; roles and responsibilities; contractual arrangements (Chapter 12 and Section 16.6);
- capital and operation costs, financial mechanisms, estimated budget (Chapter 13);
- capacity building and training needs (Section 16.5.3); and
- environmental impact assessment (often legally required).

Each scenario should be examined in conjunction with the checklist of the enabling environment (Section17.2.1). If an aspect of the scenario does not fit, it should either be adapted or activities should be carried out to create the missing conditions. In particular, support from the authorities should be secured, training should be planned to provide all the skills required and the system should be financially sustainable. The strengths, limitations and implications of the preselected systems should be assessed.

It is important to involve the key stakeholders in the evaluation (Chapter 16), as they will have to take over the roles and responsibilities of the system. It is also important that stakeholders are properly informed. The agreement reached by the stakeholders should be based on an understanding of the management and financial implications of the selected systems.

At the end of the phase, it is recommended to organise a **validation workshop** with all the key stakeholders, in order to **publicly** and **officially** present and validate the decisions taken. Any disagreement from important and/or influential stakeholders should be cleared **before** the public workshop.

The feasibility study report should state clearly who the main stakeholders for the next phase will be. For this reason, it is recommended at this point to reassess the interest and influence of the key stakeholders according to the validated options. A categorisation of the stakeholders is proposed in Section 15.4.3 in order to structure the involvement needs for the action planning.

17.3.3 Detailed project development – Action Planning

Based on the options validated in the previous phase, a Detailed Project Document, or Action Plan, can be developed. This document should include the following items:

- detailed design of the treatment plant;
- detailed definition of roles and responsibilities in the new system and terms of references;
- O&M management plan with a clear allocation of costs, responsibilities and training needs;
- agreements between stakeholders, securing financial and institutional mechanisms;
- strategy for control and enforcement: including the frequency of control, means needed and sanctions;

- definition of needs for capacity building and job creation;
- definition of contracts and bidding processes;
- monitoring and evaluation strategy for the implementation phase; and
- timeline for implementation with distinct phases and an itemised implementation budget.

O&M is often a cause of failure in development projects and thus the O&M management plan is particularly important (Chapter 11). It should include (Lüthi *et al.*, 2011): 1) O&M tasks, including routine inspection and maintenance, periodic maintenance, and urgent maintenance; 2) administrative tasks, including book-keeping, collecting fees, annual budgeting, paying employees, dealing with complaints; 3) reporting procedures; 4) responsibilities of all parties concerned; and 5) training activities for responsible persons. Key stakeholders should be reassessed according to the definition of roles and responsibilities (Section 15.5.4).

The Detailed Project Document/Action Plan should be presented, discussed and validated in a workshop with all the key stakeholders. Several workshops may be needed until a consensus is reached.

17.3.4 Implementation

This phase is mainly about translating the Action Plan into work packages that will ultimately become contracts for implementing the FSM system (Chapter 11). Several arrangements are applicable for the implementation of the plans, the most common being through private sector contractors based on competitive tendering and bidding procedures.

In parallel to this process, stakeholders should be organised according to the Action Plan. If needed, the legal and regulatory framework should be adapted. According to the identified needs, capacity building should be provided for a smooth transfer of roles and responsibilities (e.g. Section 16.5.3). The public should also be properly informed about the new FSM system and the improvements being carried out in their municipality. This will increase awareness and ownership by the public as well as by the authorities.Before the inauguration of the FSTP, the strengths, weaknesses and training needs of key stakeholders should be reassessed (Section15.5.5). At this point there is still time to organise further training and adapt the capacity-building strategy.

After the finalisation of the construction works, the whole system (i.e. infrastructure and stakeholders) requires a start-up phase for acclimatisation (Chapter 11). For the FSTP, the start-up period will last until the system reaches its state of equilibrium and expected performance. For example, with planted drying beds, the acclimation of plants is a delicate operation that should not be neglected (Chapter 8). As for the stakeholders, they will need some time to get used to their new roles and responsibilities, and some adjustments will certainly be needed in the first few months. Support from the project team is essential at the beginning of the operational phase.

Finally, an inauguration ceremony can be organised. Such an event can generate public interest and increase awareness and can also have a positive influence on institutional decision-makers (Lüthi *et al.*, 2011a).

17.3.5 Monitoring and evaluation

Any FSM system should be monitored and evaluated (Chapter 11 and Figure 17.9). Many development projects have failed because there was no follow-up after commissioning of the FSTP. The stability of the FSTP treatment units, the satisfaction of stakeholders, the functioning of the organisational scheme, the cost recovery level and the sustainability of financial mechanisms should be monitored. Adjustments will probably still have to be made after commissioning.

Figure 17.9 Sampling in the field with a portable laboratory, Egypt (photo: Philippe Reymond).

Monitoring throughout the whole year is necessary, especially for the FSTP. Climate (heat, rain) can positively or negatively affect the treatment performance. The quantity and characteristics of sludge differ from one season to the other (especially between dry and rainy seasons - Chapter 2). This can also influence the performance of the FSTP, as well as the whole supply chain, as the demand for emptying services varies as well.

In addition, dissemination of lessons learnt is important to the development of FSM.

17.4 SELECTING CONTEXT-APPROPRIATE TECHNICAL OPTIONS

Setting up a FSM system is not only about the selection of single technological options, but more importantly, about finding a sustainable combination of services that guarantees the appropriate collection, conveyance, treatment and disposal or enduse of FS, in a way that ensures household satisfaction, broad coverage and cost recovery. In this book, an elimination-based approach is proposed, based on selection criteria and related critical parameters. This approach is context-specific and focusses on the enduse of endproducts.

17.4.1 Combination of services

A sound selection of combination of services can only succeed after a thorough assessment of the initial situation (Chapter 14), feasible modes of organisation (Chapter 12), potential financial arrangements (Chapter 13), existing sites (Section14.4) and stakeholder analysis and involvement (Chapters 15-16).

Table 17.3 Criteria for selection of treatment options

Treatment performance	Local context	O&M requirements	Costs
• Effluent and sludge quality according to national standards	• Characteristics of sludge (dewaterability, concentration, degree of digestion, spreadability) • Quantity and frequency of sludge discharged at the FSTP • Climate • Land availability and cost • Interest in enduse (fertiliser, forage, biogas, compost, fuel)	• Skills needed for operation, maintenance and monitoring available locally • Spare parts available locally	• Investment costs covered (land, infrastructure, human resources, capacity building) • O&M costs covered • Affordability for households

The choice of a combination of services is influenced among others by the type of onsite sanitation systems (pit latrines, septic tanks, etc., Tilley *et al.*, 2014), the sludge quantity and characteristics (Chapter 2), rain patterns (quantity, distribution over time), the existing FS private sector and the institutional setting.

The assessment of available capacities and gaps is crucial. Ultimately, the success of a FSM plan largely depends on:

- the **capacity** of the stakeholders to **enforce the financial mechanisms** that have been planned, allowing cost recovery for the FSTP; and
- the **capacity** of the stakeholders to **operate and maintain the FSTP**.

Only an integrated approach can guarantee that, in the end, these capacities are present.

17.4.2 Criteria for selection of treatment options

A FSM system should be efficient and flexible, i.e. able to function normally and adapt to the frequency of sludge delivery and sludge quantities and characteristics, cope with climatic variations, produce endproducts that are safe for use, be able to guarantee that the investment and O&M costs are acceptable and that there are skilled employees for operation (adapted from Klingel *et al.*, 2002). Options for enduse and resource recovery (Chapter 10) should be promoted where there is an observed demand. This way, uncontrolled discharge of endproducts into the environment is avoided and nutrient enduse is maximised. Designing treatment technologies with the intended enduse in mind also helps to ensure that the technologies are not over- or under-designed to achieve the appropriate level of treatment.

Eleven criteria for the selection of a combination of technologies are proposed, divided into four categories: treatment performance, local context, O&M requirements and costs (Table 17.3). They should serve as guidelines. If one of these is not matched or not taken into account in the planning, a revision should be considered, as it could impact on the sustainability of the project.

17.4.3 Elimination-based approach

The integrated approach presented in this book emphasises the importance of the assessment of the initial situation, the financial, organisational and O&M realities, as well as the characteristics of the available treatment sites for selecting context-appropriate technology. The proposed Technology Selection Scheme (Figure 17.10 and Case study 17.6) takes the existing practices, user priorities and the enabling environment as the basis for selection. Once this assessment has been completed, the elimination-based approach, based on qualitative technical decision factors, can be applied. After this, the process leaders must check if the selected options match the available skills, financial and organisational realities, as well as the potential treatment sites. If they do not match, the selection should be modified until a suitable combination is reached. It is therefore an iterative process, and ties various chapters of this book together by linking the different aspects of technology selection.

The selection process may take time, as all activities are run in parallel and defining a management scheme and an O&M plan with the FS stakeholders, or finding appropriate treatment sites are difficult tasks. However, as discussed in various case studies, these tasks must be finalised before taking the final decision on technical options as setting up a management scheme once the infrastructure has already been built, will result in failure of the FSTP.

Demand for endproducts is also highlighted in the scheme. Enduse can be, but is not necessarily, significant for the financial balance of the system (Chapter 13), but it increases safe disposal and the motivation of FSTP operators to deliver quality services, and offsets disposal costs. However, demand for enduse is one thing, satisfying it on the ground is another (Murray *et al.*, 2010). It is often forgotten that marketing an endproduct and making it available for the endusers also has a cost and that the endusers themselves, even if interested in the endproduct, may not be willing to pay for it, or may not have the capacity. Hence, distribution and logistics of treatment products to users is an important consideration.

Box 17.2: How to use the Treatment Technology Selection Scheme (Figure 17.10)

The elimination-based approach is represented as a Technology Selection Scheme (Figure 17.10). The technical options are featured in blue boxes, and the potential endproducts in green boxes; the options in light blue and endproducts in light green represent promising perspectives that have not yet fully been validated for FS treatment. For the purpose of clarity, only key technical decision factors are represented and streams other than sludge, such as the liquid fractionfrom each option, organic waste or energy, are not represented. Decision factors are qualitative, not quantitative, as well-defined thresholds do not yet exist. More detailed information on technologies is found in Chapter 5.

The technologies are categorised according to their function, within the treatment process:

1 Solid-liquid separation
2 Stabilisation
3 Dewatering/drying
4 Pathogen reduction

If a technology has two functions, it lies at the interface between the corresponding sub-sections of the scheme. Technologies that can be combined are linked with arrows, representing sludge transfers.

The decision factors are studied during the detailed assessment of the initial situation (Chapter 14). The main factor is the quantity and quality of sludge (Chapter 2). Dewaterability of sludge is crucial, whether for intrinsic (concentration, degree of digestion) or external parameters (rain patterns) (Chapter 3 and Chapters 5-9). It determines if a solid-liquid separation step is needed or not. If yes, digesters, settling-thickening basins or onsite anaerobic baffled reactors (ABR) should be considered.

Sludge 'spreadability' and amount of precipitation are two decision parameters when choosing between planted and unplanted drying beds. Where there are periods of intense rainfall, planted drying beds are preferred, as the sludge residence time is much higher. However, where sludge is too thick to be easily distributed on planted drying beds, unplanted drying beds are preferred.

Dry sludge, humified sludge, biomass, biogas and compost are all possible endproducts (Chapter 10). Biogas can be produced if fresh sludge or sludge that is not fully stabilised is available and if the temperature is high enough. If organic waste is sorted and available, it may be possible to add it to a biogas digester or to co-compost.

Selecting options may be an iterative process, until the combination of technologies matches the requirements of the local context.

17.4.4 Sanitation system proposal

Communicating the results of the technology selection in a clear and systematic manner is key for the discussion of the proposals with the project stakeholders. The Compendium of Sanitation Systems and Technologies (Tilley *et al.*, 2014) provides a clear and easily readable way to show the value chain, from the type of latrines used to the selected treatment and enduse/disposal options. The value chain is divided into five sections: 1) user interface; 2) collection and storage/treatment; 3) conveyance; 4) (semi-) centralised treatment; and 5) use and/or disposal. Each existing or planned option is represented in its respective functional group and is linked to the following step by arrows and input/output product(s).

Box 17.3 provides an example of how a system proposal can be presented.

Box 17.3: Example diagram of a sanitation system proposal

The feasibility study for the implementation of a new FSM plan is being conducted in a Sub-Saharan city. The assessment of the initial situation (Chapter 14) showed that the population mainly relies on single pit and VIP latrines, with dry or pour-flush toilets according to the cultural background. Some well-off families as well as administrative and commercial buildings have septic tanks. Greywater is disposed of separately in open stormwater drains or soak pits, as is septic tank effluent. FS from the pits is mainly collected by mechanical providers, but a few neighbourhoods remain inaccessible to motorised vehicles and in these areas pits are emptied manually.

The Sanitation Task Force, in charge of elaborating a City Sanitation Plan, is proposing the construction of two FSTPs, one in the north and one in the south of the city. FS is to be conveyed by the existing private service providers. Transfer stations are to be built at the interface between the main roads and neighbourhoods served by manual service providers. Treatment in both FSTPs would be based on a combination of a settling/thickening basin followed by unplanted drying beds. The effluent is to be treated in a series of waste stabilisation ponds (WSP). Sludge collected from the drying beds is to be stored for at least six months before being sold to farmers. The effluent of the WSP would be discharged in a nearby stream or, during the dry season, used for irrigation. In the last pond, plant operators may possibly introduce aquaculture.

Planning for a meeting with the local authorities, the Sanitation Task Force has prepared a diagram of their sanitation system proposal in order to facilitate the presentation and discussion of the results. This is shown in Figure 17.11.

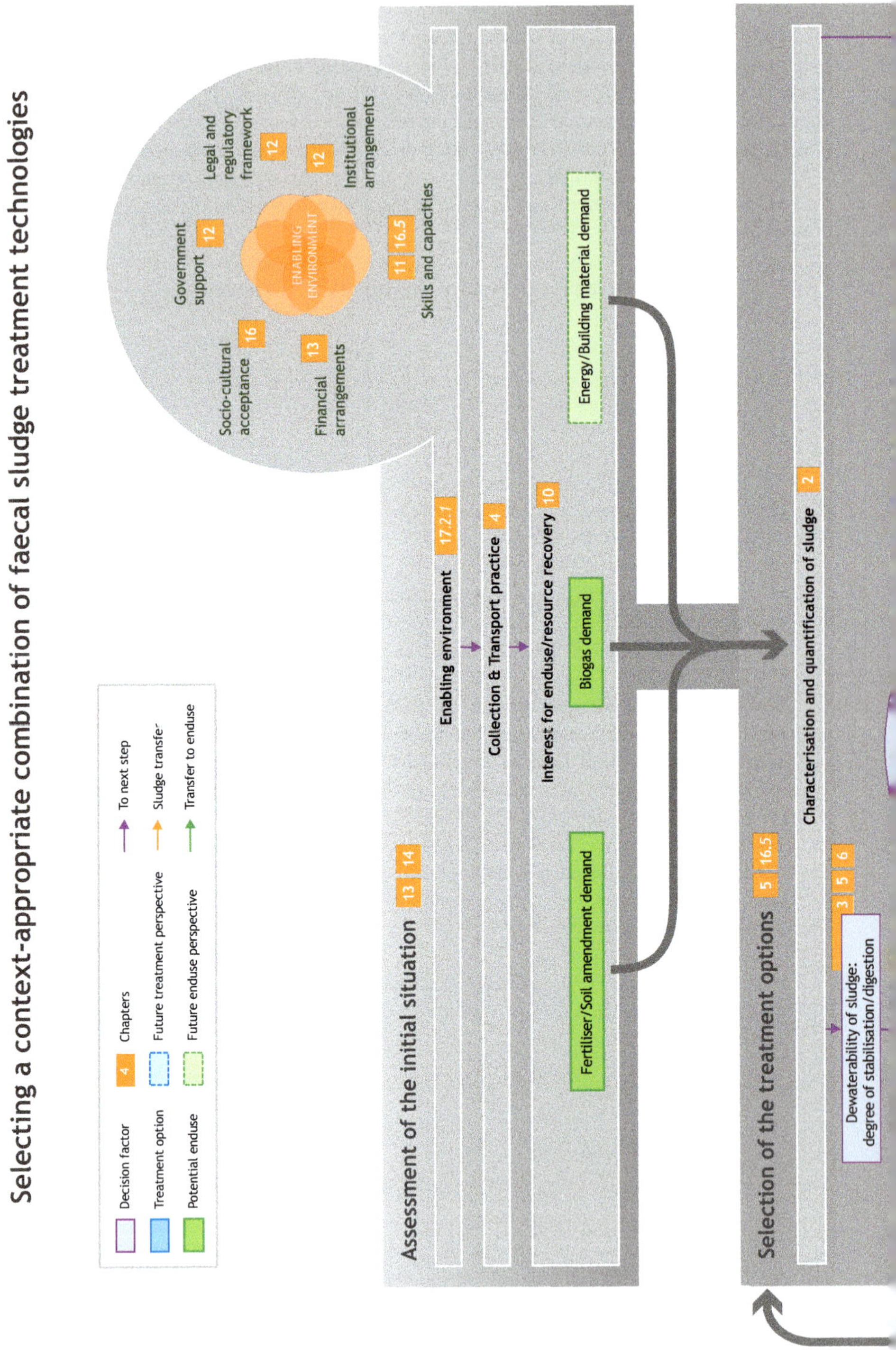

Figure 17.10 Technology Selection Scheme: How to select a context-appropriate combination of faecal sludge treatment technologies.

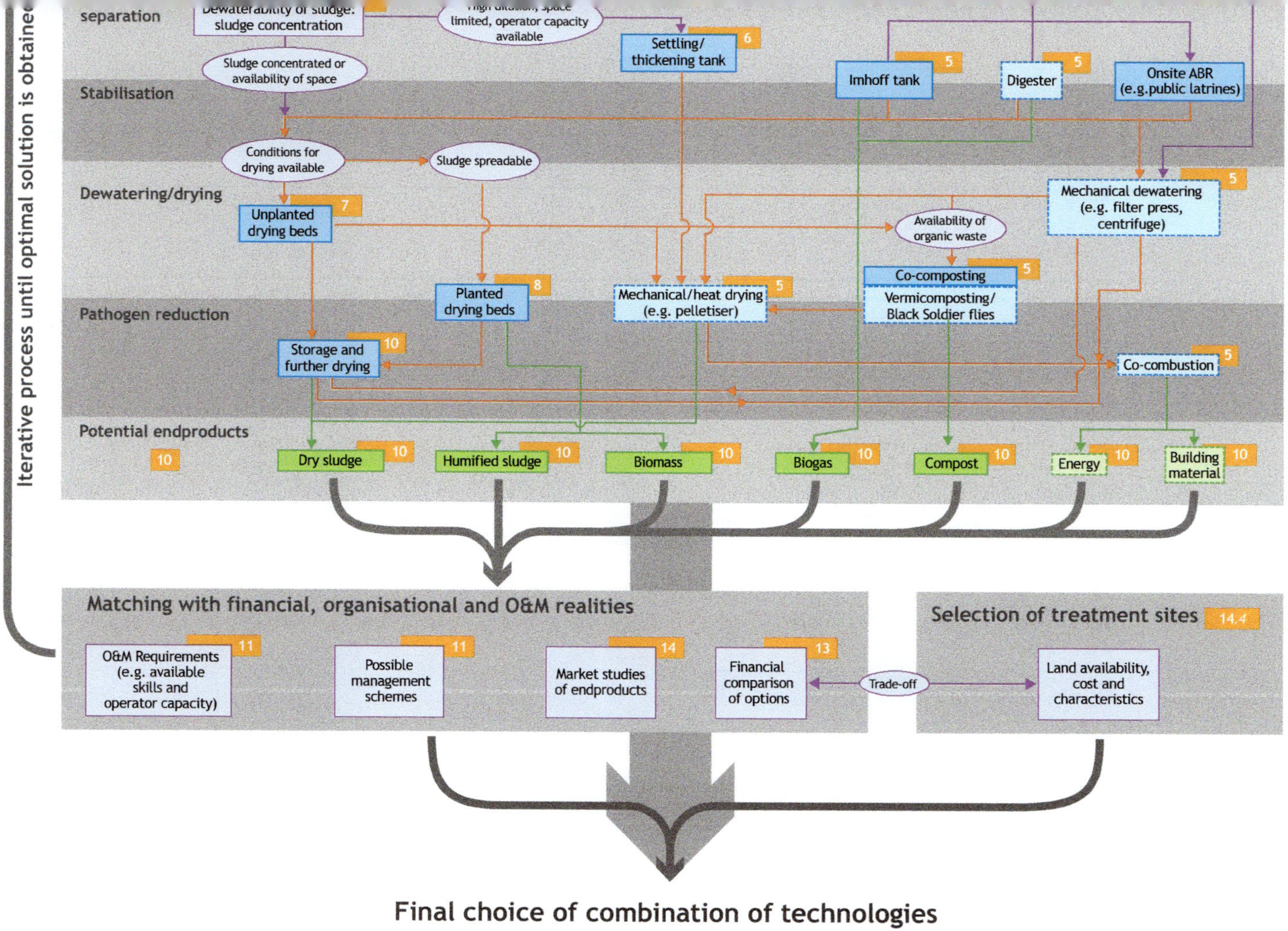
Iterative process until optimal solution is obtained
separation
Dewaterability of sludge: sludge concentration
limited, operator capacity available
Settling/ thickening tank
6
Imhoff tank
5
Digester
5
Onsite ABR (e.g.public latrines)
Sludge concentrated or availability of space
Stabilisation
Conditions for drying available
Sludge spreadable
Dewatering/drying
Unplanted drying beds
7
Mechanical dewatering (e.g. filter press, centrifuge)
5
Availability of organic waste
Planted drying beds
8
Mechanical/heat drying (e.g. pelletiser)
5
Co-composting
Vermicomposting/ Black Soldier flies
5
Pathogen reduction
Storage and further drying
10
Co-combustion
5
Potential endproducts
10
Dry sludge
10
Humified sludge
10
Biomass
10
Biogas
10
Compost
10
Energy
10
Building material
10
Matching with financial, organisational and O&M realities
O&M Requirements (e.g. available skills and operator capacity)
11
Possible management schemes
11
Market studies of endproducts
14
Financial comparison of options
13
Trade-off
Selection of treatment sites
14.4
Land availability, cost and characteristics
Final choice of combination of technologies

Figure 17.11 Example diagram of a faecal sludge sanitation system proposal (adapted from Tilley *et al.*, 2014).

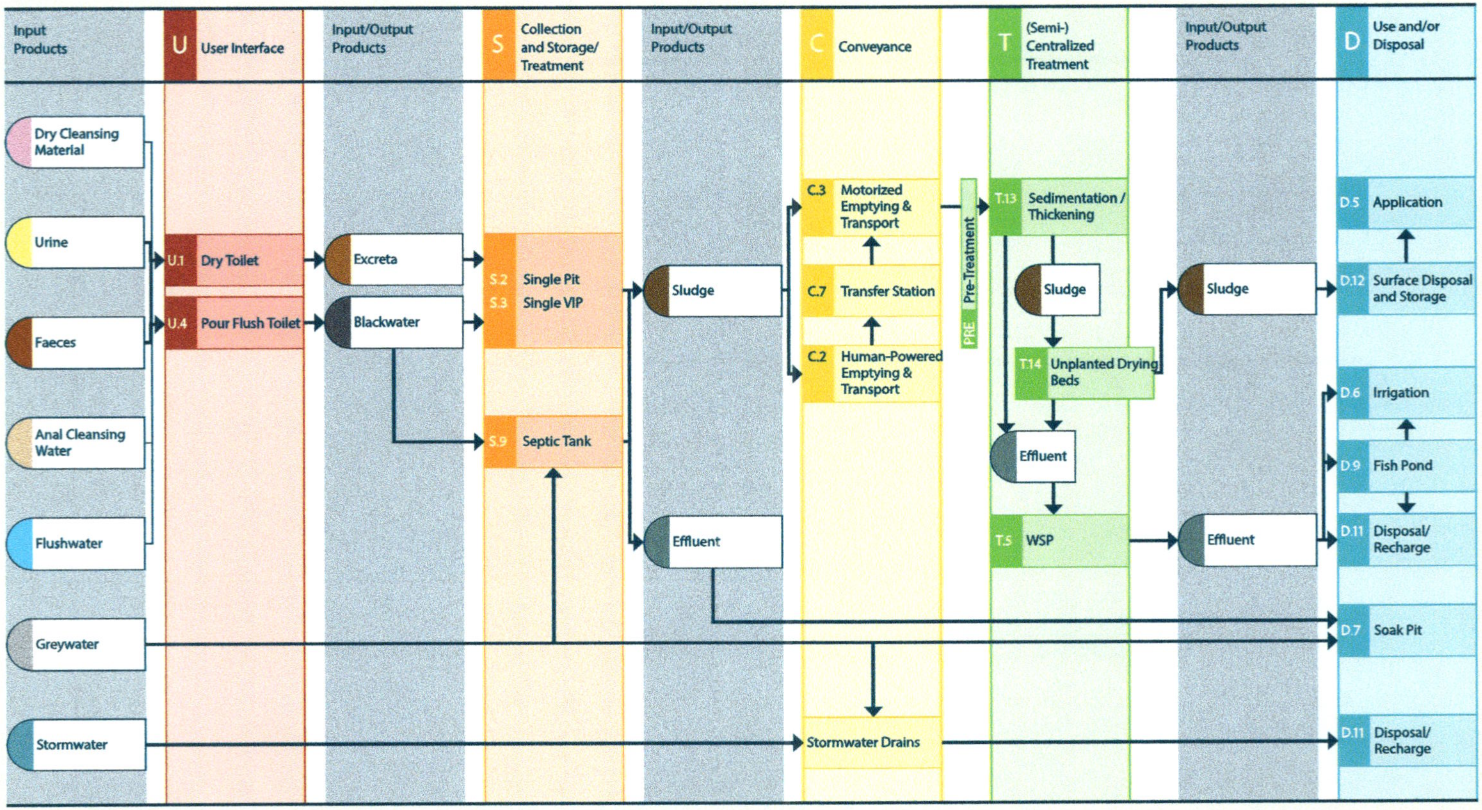

17.5 BIBLIOGRAPHY

ADB (2006). Model Terms of Reference - Planning Urban Sanitation and Wastewater Management Improvements. Manila.

AECOM and SANDEC/EAWAG (2010). A Rapid Assessment of Septage Management in Asia: Policies and Practices in India, Indonesia,Malaysia, the Philippines, Sri Lanka, Thailand, and Vietnam. USAID. Bangkok, Thailand.

Drechsel, P., Scott, C., Raschid-Sally, L., Redwood, M., Bahri, A., (2009). Earthscan, IDRC, IWMI.

Gutterer, B., Sasse, S., Thilo, P., Reckerzügel, T. (2009). Decentralised Wastewater Treatment Systems (DEWATS) and Sanitation in Developing Countries - A Practical Guide. BORDA, WEDC.

Klingel F., Montangero, A., Koné, D., Strauss, M. (2002). Faecal Sludge Management in Developing Countries - A Planning Manual / Gestion des boues de vidange dans les pays en voie de développement - Un manuel de planification. Dübendorf, EAWAG/SANDEC.

Lüthi, C., Morel A, Kohler, P., Tilley, E. (2009). People's Choice First. A 4-Country Comparative Validation of the HCES Planning Approach for Environmental Sanitation. NCCR: 129.

Lüthi, C., Morel, A., Tilley, E., Ulrich, L. (2011a). Community-Led Urban Environmental Sanitation, Complete Guidelines for Decision-Makers with 30 Tools. Dübendorf, Switzerland, Swiss Federal Institute of Aquatic Science and Technology (EAWAG).

Lüthi, C., Panesar, A., Schütze, T., Norström, A., McConville, J., Parkinson, J., Saywell, D., Ingle, R. (2011b). Sustainable Sanitation in Cities: A Framework for Action. Rijswijk, Papiroz Publishing House.

McConville, J., Norström, A., Lüthi, C., Panesar, A., Schütze, T., Parkinson, J., Saywell, D., Ingle, R. (2011). Planning for Sustainable Sanitation. In: Sustainable Sanitation for Cities – A framework for action. Rijswijk, Papiroz Publishing House.

Murray, A., Buckley, C. (2010). Designing Reuse-Oriented Sanitation Infrastructure: The Design for Service Planning Approach. Wastewater Irrigation and Health.

Parkinson, J., Lüthi, C. (2013). Sanitation 21 – a planning framework for improving city-wide sanitation services. London, UK.

Tilley, E., Ulrich, L., Lüthi, C., Reymond, P., Zurbrügg, C. (2014). Compendium of Sanitation Systems and Technologies. Dübendorf: Swiss Federal Institute of Aquatic Science & Technology (EAWAG) 2nd revised edition.

WSP (2009). Urban Sanitation in Indonesia: Planning for Progress. WSP Field Notes.

WSP (2010). Marching together with a citywide sanitation strategy. ISSDP, Indonesia.

Chapter 18

The Way Forward

Linda Strande

18.1 INTRODUCTION

The systems level approach to faecal sludge management (FSM) developed in this book should be seen as a building block for the future design and operation of functional and sustainable FSM systems. FSM is a new and rapidly growing field, and improvements and gains in knowledge are rapidly occurring. These advances will continue to build upon each other and improve solutions and approaches for FSM. Each section of this book has drawn important conclusions and has proposed steps to take in the fields of technology, management and planning to develop sustainable FSM systems. Some highlights include:

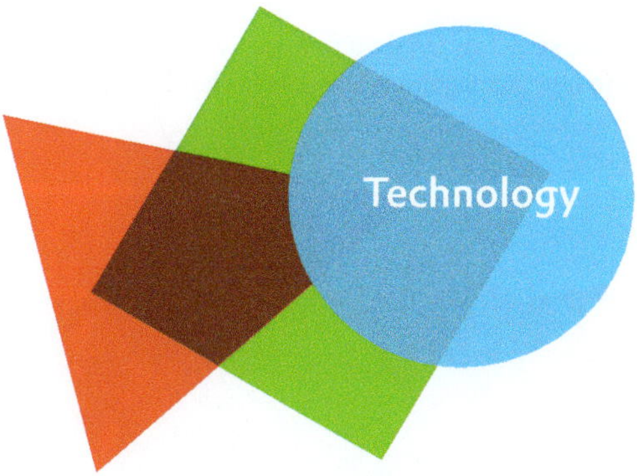

- Designing for the final enduse or disposal option of treatment products.
  This approach will ensure that effluents and endproducts achieve adequate and appropriate levels of treatment; that systems are not over-designed, wasting financial resources; and that systems are not under-designed, risking public and environmental health.
- Designing for the actual quantity and characteristics of faecal sludge.
  This approach will ensure that technologies are effectively designed and that faecal sludge (FS) can be treated on a citywide scale; however, methods for better FS quantification and characterisation still need to be developed.
- Creating onsite storage technologies and transfer stations, and emptying methodologies.
  This is a critical link in the FS service chain. Having safe, efficient and affordable collection and transport of FS will help to ensure that FS is delivered to (centralised or de-centralised) treatment plants and not discharged untreated into the environment.
- Developing an understanding of treatment mechanisms.
  This will be the basis for developing new FS treatment technologies, and adapting existing ones from wastewater and sludge treatment practice.

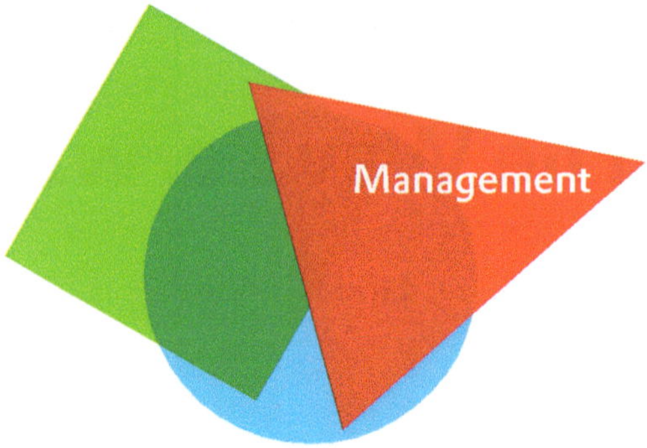

- **Incorporating management concerns from the beginning of the project planning.**
  Linking factors such as management to decisions on technology options and incorporating ongoing operations, maintenance and monitoring procedures into technology design and planning are key to ensuring a long-term sustainable operation.
- **Setting up legal and regulatory frameworks for faecal sludge management and introducing funded incentive and enforcement mechanisms.**
  This is necessary to ensure that regulation and enforcement of public health and environmental standards occur.
- **Considering different models of financial transfers.**
  This will help to formalise the sector and make it financially sustainable, and could include incentives as a method of transition to new management models in the short-term.

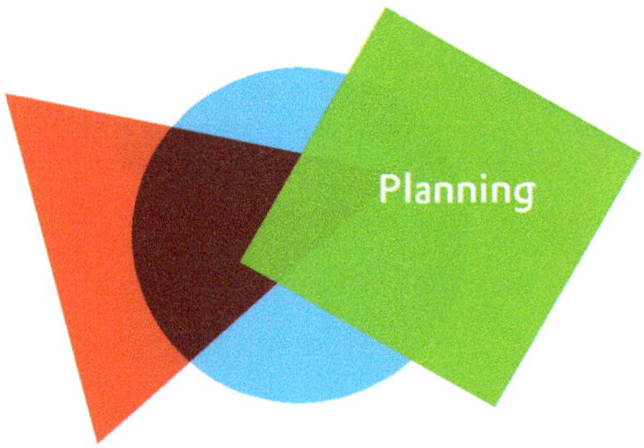

- **Assessing and understanding the initial situation in a specific contexts.**
  Sanitation practices are very heterogeneous, not only between countries and among cities, but also within the cities themselves. Different situations require different solutions. A thorough assessment ensures that solutions are tailored to meet the actual needs, builds on what is existing and takes into account the context-specific strengths and constraints.
- **Integrating stakeholders into faecal sludge management and understanding their interests and influence.**
  This is key for FSM project design: analysing and engaging stakeholders should be carried out throughout the entire project as it is a continuous and iterative process. This will help to build consensus, identify needs, define capacity building requirements, and empower traditionally

neglected groups. Above all, it will allow the stakeholders to make informed decisions, understand the implications of their choices and be ready to fulfil their roles and responsibilities in the FSM system.

- **Fitting the participatory process within traditional project cycles.**
  Any extra costs resulting from additional meetings incurred during the participatory process are quickly offset by savings during implementation and operation from factors and complications that were identified and alleviated during the process, and success is enhanced by more effective management schemes, better institutional setups and integration of the private sector (FSM Planning Scheme "From A to Z", Table 17.1).
- **Applying an integrated planning approach at the city level.**
  This is imperative for understanding critical factors for selecting context appropriate options. The entire enabling environment must be considered. In particular, management and financial schemes must be defined and validated prior to making final decisions on technical options.

The strength of the approach in this book is considering all three fields technology, management and planning together in deriving sustainable FSM solutions. The FSM Planning Scheme 'From A to Z' as well as the technology selection scheme (Figure 17.10) illustrate this approach and help to navigate through the book; they should be considered as a check-list and as a visual tool to structure planning processes, to include all necessary components and to communicate with non-expert stakeholders.

The successful implementation of each of the above steps requires knowledge of all three fields. Deriving sustainable FSM infrastructures requires tackling large, complicated issues that are interrelated. It is necessary to understand how these fields fit together, and to understand the connections and influences of each field upon the others. Six critical bottlenecks are identified here that are all at the crossroads of technology, management and planning, and which all need to be addressed to successfully move the field forward:

1. Acknowledging the importance of FSM
2. Setting up frameworks and responsibilities
3. Increasing knowledge dissemination and capacity development
4. Creating sustainable business models and fee structures
5. Implementing integrated planning methodologies
6. Developing appropriate technologies

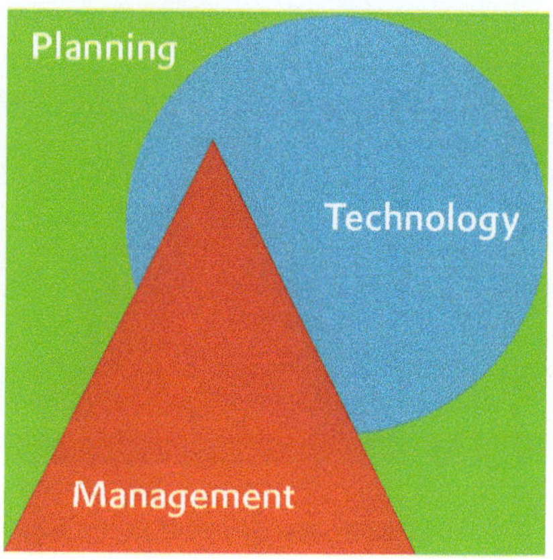

18.1.1 Acknowledging the importance of FSM

For development of sustainable FSM systems, a significant step requires the acknowledgement of its importance by stakeholders in all fields of technology, management and planning. This includes governments taking responsibility for providing FSM, donor agencies providing funding for feasible and appropriate FSM solutions (Figure 18.1), and large intergovernmental organisations promoting FSM together with the goal of ending open defecation. As FSM is acknowledged as a real need and legitimate solution, it will naturally result in significantly greater amounts of attention and resources being focused on FSM. An example of acknowledgement is provided by the Philippine Government, which in 2012 was the first national government in SE Asia to approve a FSM plan (National Sewerage and Septage Management Program (NSSMP)) (Robbins *et al.*, 2012). By installing this program, the government not only accepted and acknowledged the importance of FSM, but also that FSM and hybrid forms of combined centralised wastewater treatment and FSM are considered viable solutions.

Highlighting economic costs related to lack of sanitation services, in addition to public health aspects, is another way to promote the value of investments in FSM. The lack of access to sanitation has a global impact of 260 billion USD annually (Hutton, 2013). The Water and Sanitation Program (WSP) of the World Bank has identified through its Economics of Sanitation Initiative (ESI) (www.wsp.org/content/economic-impacts-sanitation) that sanitation also has an economic impact on sectors that are unrelated but important for the economy. For example, in India tourism-related losses due to insufficient sanitation services amounted to 266 million USD per year (Hutton *et al.*, 2008).

The Millennium Development Goals (MDGs) have been very successful in raising international attention on the need for sanitation. The inclusion of sludge management as part of the post-2015 Development Agenda with the Sustainable Development Goals (SDGs) would build upon this momentum to increase awareness of the importance of 'environmental sanitation' and the importance of considering all water systems together; i.e. wastewater, drinking water, irrigation and drainage, together with solid waste management (EAWAG, 2005).

Figure 18.1 Drying beds for faecal sludge treatment under construction at Lubigi faecal sludge treatment plant in Kampala as part of the Lake Victoria Protection Stage I Project funded by KfW, EU and Government of Uganda/NWSC (photo: Lars Schoebitz).

18.1.2 Setting up frameworks and responsibilities

Having one entity of a city government solely responsible for sanitation, regardless of technology deployed, increases a sense of responsibility that can be lost in more fragmented management models where different agencies manage parts of the service chain. This also facilitates efficiency in citywide planning. Streamlining eliminates any responsibility overlap between stakeholders, and also avoid gaps in responsibilities (Bassan *et al.*, 2013a). A successful example of defining roles and responsibilities is provided by Indonesia in collaboration with WSP through the Sanitation Sector Development Program (ISSDP). Before implementation of this program, Indonesia had one of the lowest wastewater and FS treatment coverage rates in SE Asia, but now the government has a strong commitment to sanitation with a national strategy. The National Planning Development Agency (Bappenas) plays the lead role in decision making, with local governments implementing urban sanitation within their jurisdictions (WSP, 2011).

Institutional frameworks are necessary to set requirements and ensure compliance. A balance needs to be found between standards that are too stringent and hence prevent any action from being taken because they cannot be met, and providing adequate and appropriate protection of public and environmental health. One possibility is implementing step-wise improvements that are more cost-effective and can continue to be built upon in the future (Parkinson *et al.*, 2013). Metrics are then needed to evaluate the 'effectiveness' of solutions beyond the household level, at the overall outcome level. To this purpose the WSP is currently developing their *Diagnostics and Guidelines for FS Management in Poor Urban Areas*, which are diagnostic and decision-making tools for the development of improved citywide FSM in urban areas (Blackett, 2013).

Resource recovery from FS treatment products can increase management performance by treatment facility operators as they attempt to maximise revenue streams from sale of by-products. However, resource recovery always comes with a certain level of risk regarding safety of the products and enduse. To address this, the World Health Organization (WHO) is currently developing Sanitation Safety Plans (SSPs), to aid the responsible government entity in minimising health risks associated with resource recovery by facilitating the implementation of the 'Guidelines for the Safe use of Wastewater, Excreta and Greywater in Agriculture and Aquaculture' (Medlicott 2013, WHO 2006). Another project by the International Water Association (IWA) includes development of a Participatory Rapid Sanitation System Risk Assessment (PRSSRA) methodology for rapid risk assessment through stakeholder engagement to prioritise interventions that reduce risks. Finally, some countries are establishing guidelines and certification programs to help shape and formalise the resource recovery sector.

18.1.3 Increasing knowledge dissemination and capacity development

As FSM is a relatively new field, much of the existing knowledge remains with practitioners in the field without a written record, and there is a lack of affordable and accessible reference materials. Developing methods that increase local expertise is imperative as many shortages within the FSM service chain are the result of lack of institutional capacity, management deficiencies, insufficient staff and inadequate technical capacity, and all aspects within the service chain are likely to require support to develop human resources capacity (Parkinson *et al.*, 2013). To address this, there is a need for easy to digest material to enable non-technical people to access information (Parkinson *et al.*, 2013). Hopefully new knowledge-sharing tools can help to bridge the gap in distribution of current research results, for example SuSanA (Sustainable Sanitation Alliance – www.susana.org), which since 2007 has provided through an open international network of members a working platform for sustainable sanitation and a forum for policy dialogue. Additional online resources are presented in Chapter 1. Another highly effective strategy is increased south-south interactions among city officials and practitioners for learning and sharing of experiences. A good example is the professional FS collection and transport associations in Kampala, Uganda and Dakar, Senegal. Based on their success, the directors of these associations are routinely asked to present and share knowledge at conferences and meetings throughout sub-Saharan Africa. Another successful example is the MILE (the Municipal Institute of Learning) in Durban, South Africa which was set up to transfer knowledge and experiences from Durban to other municipalities throughout Africa. MILE offers training courses and field visits on a regular basis with funding from the United Nations Institute for Training and Research (UNITAR) and the eThekwini municipality in Durban. eThekwini Water and Sanitation (EWS) also partners with municipalities throughout Africa to share knowledge and bring about improvements in service provision. The senior management of EWS also interacts and shares experiences with the management of other water and sanitation organisations in low- and middle-income countries with funding provided by the World Bank and the WSP.

The value of capacity building and more applied research in the field of FSM is nowadays widely recognised and the number of ongoing research projects is rapidly increasing (Figure 18.2). For example, since establishment of the Water, Sanitation and Hygiene (WSH) program by the Bill & Melinda Gates Foundation (BMGF), a large number of projects have been funded on FSM especially focusing on the urban poor. One of their projects is the SaniUP project ('Stimulating local innovation on sanitation for the urban poor in sub-Saharan Africa and South-East Asia') which has two principal objectives: (i) to stimulate local innovation on sanitation for the urban poor through research, and (ii) to strengthen the sanitation sector in developing countries through education and training. First outputs of the project included development of a three-week course on FSM (www.unesco-ihe.org) in the curricula of the UNESCO-IHE Sanitary Engineering Programme, editing and publishing this FSM book (with co-funding from the Swiss Agency for Development and Cooperation (SDC), and a full online FSM course that will be available in 2015 (www.unesco-ihe.org/online-course-faecal-sludge-management).

Figure18.2 PhD fellows performing faecal sludge characterisation at the Sanitary Engineering laboratory of UNESCO-IHE under the framework of the project financed by the Bill & Melinda Gates Foundation (photo: UNESCO-IHE).

18.1.4 Creating sustainable business models and fee structures

Overall, depending on local circumstances, FSM can be much less expensive than centralised sewer-based solutions (Dodane *et al.*, 2012). However, there still need to be adequate financial flows throughout the entire service chain or the system will not work. Frequently fee structures are not equitable with the poorest households having to pay often twice for sanitation services, through wastewater treatment tariffs being included in drinking water provision, and when paying to have onsite sanitation facilities emptied. Different business models other than the traditional municipality-driven model for sanitation services need to be considered to reduce the financial burden at the household level.

Although one entity should be responsible for the overall responsibility and framework for FSM, this entity does not have to be responsible for conducting every activity in the FSM service chain. From a business model perspective, different customers and value propositions are possible. Customers for services include the household level user who desires FS removed and taken away, but ultimately is not concerned with its final fate as long as it is removed, municipalities or public entities that are responsible for the protection of public health, and endusers of treatment products who gain value from resource recovery. One model of business development that is effective in the informal sector is 'coopetition', a combination of cooperation and competition, where small scale businesses spring up to fill a need, and even though they are competing against each other they mutually benefit through their association (cooperation). An example is collection and transport of FS in Bangalore where competition amongst companies benefits the household level by keeping prices for emptying services lower. But at the same time, the collection and transport association and the subsequent demand for technology has also resulted in improved supply chains for truck parts and local shops that have the capacity to build and repair vacuum trucks, greatly reducing costs to the businesses. In addition, the providers deliver FS to farmers who appreciate the value of it and are competing with each other to obtain cheap manure, which ultimately increases revenue to their business (Gebauer *et al.*, 2013).

Another possibility is public private partnerships (PPP), which also create new opportunities and challenges in urban planning for municipalities when managing potential conflicts between private and public interests. Strategies include tariffs being set that encourage producers to sell waste-to-energy derived electricity to the grid, guaranteeing a price and market to make financing available for capital investments and technology development. Municipalities could also make multi-year agreements with private sector partners to 'guarantee waste feedstock supply' to ensure the financial feasibility of large scale production/treatment facilities. Public entities could cross-subsidise collection and transport companies to facilitate their revenue generation when emptying and transporting FS, while also setting and enforcing maximum emptying fees at the household level. A reasonably successful PPP is functioning in Kampala, Uganda between the NWSC (National Water and Sewerage Corporation), KCCA (Kampala Capital City Authority), NEMA (National Environment Management Authority) and the PEA (Private Emptier Association). The PEA, registered in 1999, is responsible for providing the critical link for all FS collection and transport in Kampala (although an official PPP agreement has not yet been signed).

Examples of current research in this area include Waste Enterprisers based in Kenya, that is using resource recovery to reinvent the economics of FS treatment and disposal. Rather than thinking of reuse as an add-on to an otherwise costly treatment plant, the company is building 'factories' that will use FS as a raw material and convert it to solid fuel for sale to industries. By streamlining processing costs and designing its system to maximise energy recovery, Waste Enterprisers has created a profitable business model that aims to turn FSM into the by-product of producing renewable energy. They are currently building their first commercial-scale plant in Kenya (www.waste-enterprisers.com). The national sanitation utility (ONAS) in Dakar, Senegal is piloting a call centre, where all household level users call for FS collection and transport services. The call centre then puts out a notification to the collection and transport companies who bid for the job with the lowest bid winning, competitively

Figure 18.3 Implementation of the FAQ method (Faecal Sludge Quantification and Characterisation) in Kamapla, Uganda (photo: Lars Schoebitz).

reducing costs to the household level user. In the future the pilot study plans to implement GPS tracking and SMS notifications. The RRR (Resource, Recovery and Reuse) project is evaluating the feasibility of implementing large scale waste-based business models with resource recovery of water, nutrients and energy. Feasibility studies are currently being evaluated in Lima, Peru; Hanoi, Vietnam; Bangalore, India; and Kampala, Uganda (www.sandec.ch/RRR). Another example is Sanergy, an NGO in an informal settlement in Nairobi that has 260 toilet installations. They are applying a business model that involves manufacturing and selling of toilets to the local community, collection of fees from toilet users, daily emptying and cleaning of individual toilet facilities, transport of urine and faeces to a centralised treatment location, and centralised urine and FS treatment. Sanergy is researching best options for resource recovery, including biogas and compost.

18.1.5 Implementingintegrated planning methodologies

The implementation of integrated planning approaches for citywide FSM systems are imperative to successfully address the urban sanitation challenge. However, they can be quite difficult to implement due to the heterogeneity of urban areas in low- and middle-income countries, characterised by rapid growth rates, and very diverse landscapes in terms of income level, sanitation technologies and formal and informal settlements, in addition to weak enabling environments (Hawkins *et al.*, 2013). Planning methodologies need to continue to be developed that create (Parkinson *et al.*, 2013):

- a vision of the need for sanitation improvements which is shared between different stakeholders within the city;
- a definition of clear and realistic priorities for improvement across the entire city;
- a comprehensive sanitation development plan for the entire city that corresponds to the users' demands and the different physical and socio-economic conditions within the city; and
- an enabling environment with regard to governance, finances, capacity enhancement, technology and inclusiveness.

Understanding annual accumulations and characteristics of FS on a citywide scale is a requirement for the design of adequate and appropriate treatment technologies; however, there are no existing reliable methods to achieve this. Characterising and quantifying FS is difficult due to the wide range of existing technologies (e.g. VIP latrines, unlined pit latrines and septic tanks) in use at the household level, in addition to public toilets, commercial entities, restaurants and schools. In addition, there is typically no reliable information available on the number or types of existing technologies. FS characteristics and production are highly variable, and not well understood. Sampling and analysing at a citywide scale is very time and resource intensive. To address this, methods such as FAQ (Faecal Sludge Quantification and Characterisation) are being developed, to provide a logical and affordable approach for quantification and characterisation at the city level. FAQ is based on the assumption that demographic data can be a predictor of FS characteristics (e.g. income level, legal status of housing, population density, and age of building), and that it is also influenced by physical factors (e.g. water table, soil type and elevation). Income, for example, could be a predictor because it impacts diet and quality of construction. This data can be then be analysed spatially with GiS to develop a representative sampling plan based on available resources. FAQ is now being field tested in Kampala, Uganda and Hanoi, Vietnam (Figure 18.3; www.sandec.ch).

Another example of planning is with emergency sanitation. The eSOS® (emergency Sanitation Operation System) is a BMGF funded activity being conducted by UNESCO-IHE (Brdjanovic *et al.*, 2013). eSOS® addresses the entire emergency sanitation chain in situations where external aid is required to meet sanitation demands (Figure 18.4).

Figure 18.4 Example of setting for eSOS® application (photo: Peter Greste, Al Jazeera, smart eSOS® toilet illustration: FLEX/the INNOVATIONLAB).

The core of any emergency management effort is integration, sharing, communication and collaboration.Information and Communication Technologies (ICT) are uniquely qualified to address these core issues and improve them at each step in the service chain. In the future, eSOS® will also be modifiable for (i) sanitation management under challenging conditions usually prevailing in urban-poor areas, such as informal settlements, (ii) sanitation provision to visitors of major open-air events such as concerts, fairs, etc., and (iii) solid waste management. The primary goal of eSOS® is to provide efficient and effective sanitation service during and after emergencies through minimising risk to the public health of the most vulnerable members of society. The secondary goal is to reduce investment, operation and maintenance costs of emergency sanitation facilities and service as a prerequisite for the sustainability of solutions, especially in the post emergency period.

Another important planning tool for the implementation of FS treatment on a decentralised or semi-centralised level are methodologies to evaluate appropriate levels of centralisation and decentralisation. Higher levels of decentralisation are more affordable when considering costs associated with transporting FS, and reducing distribution associated with resource recovery. However, the increased management demands and capital costs can result in less decentralised options being more cost effective. The correlation between scale and cost is not linear, and typically a breakeven point can be found (Gaulke, 2006). All of these factors are dependent on the local context and specificities of each city. Another way to address this need is through improved technologies that can remove/immobilise pathogens onsite, making collection and transport safer, and disposal or resource recovery less complex. This is one of the major goals of the BMGF *Reinvent the Toilet Challenge* (RTTC) (see below).

18.1.6 Developing appropriate technologies

There is a great need for the development of appropriate FSM technologies, even though solutions for entire FSM systems will not rely on technology alone, and must be considered within the local context. New technologies are in general based on pioneering developments in research, and historically research agendas have been driven by countries where centralised sewer-based sanitation solutions are the accepted norm. This points to a need for solution-oriented FSM research to be conducted in countries where it is directly relevant. In addition, for new knowledge to get taken up and influence policy, it requires local researchers working together with the urban governments that are responsible for FSM (Bassan and Strande, 2011). Due to the urgent need for technical solutions, research and implementations need to continue to be conducted in parallel, getting to scale as rapidly as possible. For example, transferring experience from planted and unplanted drying beds for dewatering of wastewater sludge to implementation of full-scale FS treatment, with optimisation of the technology transfer continuing following implementation (Dodane *et al.*, 2011). Technologies also need to be selected not only based on the specific characteristics of FS, but also on factors such as the local market demand for resource recovery of treatment products, or the potential for co-treatment (Diener *et al.*, 2014). Provided here are some examples of current research in the following areas:

- characterisation of faecal sludge;
- collection and transport;
- semi-centralised treatment technologies;
- onsite treatment technologies; and
- resource recovery.

18.2 CHARACTERISATION OF FAECAL SLUDGE

As presented in Chapter 2, FS is highly variable and characteristics of FS are not well understood. To design optimal treatment technologies, this variability and factors that influence it need to be understood (Bassan *et al.*, 2013b). The PURR project (www.sandec.ch) is being conducted to understand factors of onsite technologies and methods of collection and transport that influence FS characteristics. Initial stages of this project include a characterisation study and development of synthetic FS recipes that can be used to evaluate factors that impact biological degradation at the laboratory scale. Other researchers have also developed synthetic sludge recipes to evaluate physical properties that influence mechanical emptying (Radford and Fenner, 2013). Another reason for the current variability for results of FS characterisation is the lack of standardised methods. Methods have been adapted from wastewater and soil analyses, but the accuracy of methods for FS needs to be evaluated, and then standard methods taken up by the sector to ensure comparability of research results. The Pollution Research Group (PRG) at the University of KwaZulu Natal (UKZN) has conducted extensive research in this area and put together a collection of standard operating procedures (SOPs) for the analysis of the chemical (e.g. pH, potassium, ammonia) and mechanical (e.g. thermal conductivity, calorimetric analysis) properties of FS. This type of fundamental laboratory research is necessary to develop a detailed understanding of FS characteristics, and to provide mechanisms for comparable and standardised research to be conducted worldwide.

Figure 18.5 Drying bed research: mixing device for unplanted faecal sludge drying beds at Bugolobi wastewater treatment plant in Kampala, Uganda; evaluating potential plant species for planted drying beds in Dakar, Senegal; and planted drying bed pilot for treatment of drying bed leachate in Yaoundé, Cameroon (photo: Linda Strande).

18.3 COLLECTION AND TRANSPORT

Currently, the best available technology for sludge removal is vacuum trucks, but they are typically expensive and cannot reach households located on narrow streets and alleys. The BMGF-funded Omni-ingestor project aims to develop equipment that is more dexterous, evacuate FS more quickly, can remove dense FS efficiently (> 40% solids) and are able to dewater FS onsite. Water is heavy and therefore expensive to transport; dewatering FS and treating the effluent onsite would allow for the treated water to be directly reclaimed or safely disposed of in drains. This would greatly reduce transport costs and allow for more emptying operations performed between trips to the FSTP, as well as reducing time spent in traffic. Various prototypes are currently being developed by the private sector.

18.4 SEMI-CENTRALISED TREATMENT TECHNOLOGIES

The PURR project is evaluating the potential for co-management of FS together with wastewater sludge in Vietnam. The potential for biogas production from co-digestion of wastewater sludge and FS is being evaluated, together with the feasibility of co-digestion with other high strength waste streams. The DAR (De Déchet à Ressources) project in Dakar Senegal is evaluating drying bed technologies through optimisation of planted and unplanted drying beds (Figure 18.5). Drying beds require relatively low capital and operational costs, but are space intensive. Increasing efficiency could reduce the required space, increasing their applicability in space-limited urban areas. Research is currently being conducted on alternative media (e.g. crushed glass), mixing regimes, and greenhouses to increase drying rates. Research for planted drying beds is being conducted to identify previously unused plant species that could increase treatment performance and increase the potential for resource recovery through production and sale of fodder plants (www.sandec.ch). A steam engine-based community-scale waste processing technology is currently being developed by Janicki Industries. The concept is that a 150 kW combined heat and power plant will utilise FS as the fuel source for electricity generation. The heat generated from combustion within a fluidised sand bed will produce high-pressure steam that is expanded in a reciprocating piston steam engine connected to a generator, producing electricity. The exhaust from this engine (process heat) will also be harnessed to dry the incoming FS. The concept for this treatment plant comes from the careful re-design of basic power plant components, making them economical in mass production for small-scale plants.

18.5 ONSITE TREATMENT TECHNOLOGIES

Achieving reliable levels of treatment with onsite sanitation technologies presents a very challenging problem due to factors such as the lack of technical management, demands for reliable energy and high costs. The RTTC currently has multiple research projects addressing this challenge. The first round of technologies were presented at the RTTC fair in Seattle in 2012 and the second in Delhi in March 2014.

Some examples of technologies include hydrothermal carbonisation, microwave technology, supercritical oxidation, pyrolysis, and electrochemical processes. The Research Triangle Institute (RTI) is developing an integrated toilet technology that will separate solid and liquid waste, dry and burn solid waste using a combination of mechanical, solar, and thermal energy (primarily driven by down-draft gasification), disinfect liquid waste, and convert the resulting combustion energy into stored electricity (www.rti.org). The California Institute of Technology (Caltech) is developing a comprehensive, human waste treatment and toilet system that has at its core a photovoltaic-powered (PV), self-standing electrochemical chemical reactor that generates hydrogen for energy and nitrogen for fertiliser as by-products of treatment. The treatment process is a multistep oxidation of the organic waste and the bacteria present in the mixture. The fully integrated treatment system will include:

in-situ waste disinfection, residual solid waste processing, by-product extraction, generation of hydrogen as a by-product of waste treatment, a solar energy battery storage system, solar arrays, and a microfiltration component for final polishing of the water before reuse and recycling. Loughborough University is developing a system that is comprised of a draining balance tank; filters; high temperature pressure reactor; and evaporator-sodium chloride separation. The system operates in three stages: solids-liquid separation, followed by auto-thermal treatment of the solids to provide heat for water and salt separation. The main part of the solids treatment and the liquid evaporator will be constructed within the same unit as plug-together modules.

Figure 18.6 FaME (Faecal Management Enterprises) project pilot scale kiln for co-combustion of faecal sludge in brick production in Kampala, Uganda (photo: Pitman Ian Tushemezibwe).

18.6 RESOURCE RECOVERY

Research in this area includes the FaME (Faecal Management Enterprises) project, which is attempting to identify large-scale markets for resource recovery to provide a significant and reliable cash flow for enduses (Figure 18.6). The project is identifying innovative methods of resource recovery, and is also focusing on scaling up the use of dried sludge as a fuel in combustion. Results of the FaME research project are providing evidence of the promising technical and financial potential of FS products and filling knowledge gaps for the full-scale implementation of its use as an industrial fuel sludge based on calorific value (Murray Muspratt *et al.*, 2014), market demand of end products (Diener *et al.*, 2014), viable financial flows for collection and transport, and optimisation of drying bed technologies (www.sandec.ch).

18.7 FINAL REMARKS

Creativity is essential in every aspect of technology, management and planning to continue to advance solutions that are globally transferable and applicable for the currently 2.7 billion people worldwide served by onsite sanitation technologies and the billions more that will need to be served in the decades to come. Keeping an open mind will be key to developing innovative and optimal solutions, learning from the past, but also not limiting future possibilities through biases of what has or has not worked in the past in other situations. As highlighted by this chapter, there is currently lots of innovative research being conducted at scales of laboratory, pilot, and implementation level. There is a wealth of information rapidly becoming available, some that is scalable for implementation, and much more that is still in the development pipeline. Recent efforts put into research and capacity development will no doubt result in innovations concerning all aspects of the FSM chain and will create a new generation of scientists and engineers as a driver of change towards integrated FSM. Undoubtedly, this is a very exciting and promising time for the advancements in FSM research and education and their application in practice. The FSM field will continue to advance, and hopefully the next edition of this book will contain much more information on success stories on design and implementation of comprehensive FSM systems based on the newly acquired experiences.

18.8 BIBLIOGRAPHY

Bassan, M., Strande, L. (2011). Capacity strengthening in sanitation : benefits of a long-term collaboration with a utility and research institute. Refereed paper presented at 35th WEDC International Conference, Loughborough, UK.

Bassan, M., Mbéguéré, M., Tchonda, T., Zabsonré, F., Strande, L. (2013a) Integrated faecal sludge management scheme for the cities of Burkina Faso. Journal of Water, Sanitation and Hygiene for Development 3(2), p.216–221.

Bassan, M., Tchonda, T., Yiougo, L., Zoellig, H., Mahamane, I., Mbéguéré, M, Strande, L. (2013b). Characterization of faecal sludge during dry and rainy seasons in Ouagadougou, Burkina Faso. Refereed paper presented at 36th WEDC International Conference, Nakuru, Kenya.

Blackett, I. (2013). FS Management in 12 Cities Review Findings and Next Steps. Presentation at Stockholm World Water Week.

Brdjanovic, D., Zakaria F., Mawioo P.M., Thye Y.P., Garcia H.A., Hooijmans C.M., Setiadi T. (2013). eSOS® – Innovative Emergency Sanitation Concept. In Proceedings: 3rd IWA Development Congress and Exhibition, 14-17 October 2013, Nairobi, Kenya.

Diener, S., Semiyaga, S., Niwagaba, C., Muspratt, A., Gning, J.B., Mbéguéré, M., Ennin, J.E., Zurbrugg, C., Strande, L. (2014). A value proposition: resource recovery from faecal sludge – can it be the driver for improved sanitation? Resources Conservation & Recycling (in press).

Dodane, P.H., Mbéguéré, M., Kengne, I.M., Strande, L. (2011). Planted Drying Beds for Faecal Sludge Treatment: Lessons Learned Through Scaling Up in Dakar, Senegal SANDEC News 12, p.14-15.

Dodane, P.H., Mbéguéré, M., Ousmane, S., Strande, L. (2012). Capital and Operating Costs of Full-Scale Faecal Sludge Management and Wastewater Treatment Systems in Dakar, Senegal. Environmental Science & Technology 46, p.3705-3711.

EAWAG, 2005. Household-Centred Environmental Sanitation: Implementing the Bellagio Principles in Urban Environmental Sanitation. Published by EAWAG: Swiss Federal Institute of Aquatic Science and Technology.

Gebauer, H., Larsen, T., Lüthi, C., Messmer, U., Schöbitz, L., Strande, L. (2013). Business Model Innovations for Transformative Services: Doing Well Through Doing Good?, Presentation at QUIS 13 Conference, June 10-13, Karlstad University, Sweden.

Gaulke, L.S., (2006). Johkasou: On-site Wastewater Treatment and Reuses in Japan. Proceedings of the Institute of Civil Engineers - Water Management, 159(2), p.103-109.

Hawkins, P., Blackett, I., Heymans, C. (2013). Poor-inclusive Urban Sanitation: An Overview. Published by The World Bank.

Hutton, G., Rodriguez UE, Napitupulu, L., Thang, P., Kov, P. (2008). Economic impacts of sanitation in Southeast Asia. World Bank, WSP. 144 pages. Also in Summary form for policy makers (23 pages).

Hutton, G., (2013). Global costs and benefits of reaching universal coverage of sanitation and drinking-water supply. Journal of Water and Health 11(1), p. 1-12.

Medlicott, K., (2013). Sanitation Safety Planning (SSP) – Step-by-step Guide for Safe Use and Disposal of Wastewater. Presentation at the IWA Development Congress, Nairobi, Kenya.

Murray Muspratt, A., Nakato, T., Niwagaba, C., Dione, H., Kang, J., Stupin, L., Regulinski, J., Mbéguéré, M., and Strande, L. (in press). Fuel potential of faecal sludge: calorific value results from Uganda, Ghana and Senegal. Journal of Water Sanitation and Hygiene in Developing Countries.

Parkinson, J., Lüthi, C., Walther, D. (2013). Sanitation 21: A Planning Framework for Improving City-wide Sanitation Services. Published by IWA.

Radford, J., Fenner, R. (2013). Characterisation and fluidisation of synthetic pit latrine sludge. Journal of Water, Sanitation and Hygiene for Development 3(3), p. 375–382.

Robbins, D., Strande, L., Doczi, J. (2012). Opportunities in FS Management for Cities in Developing Countries: Experiences from the Philippines. Water 21 (14.6), p.22-25.

World Health Organization (WHO) (2006). Guidelines for the safe use of wastewater, excreta and greywater; Volume 4: Excreta and greywater use in agriculture. Published by the World Health Organization.

Water and Sanitation Program (WSP), 2011. Lessons in urban sanitation development : Indonesia sanitation sector development program 2006-2010. Water and Sanitation Program: field note. Washington, DC: World Bank.